Nanoscience and Nanotechnology for Smart Prevention, Diagnostics and Therapeutics

Scrivener Publishing
100 Cummings Center, Suite 541J
Beverly, MA 01915-6106

Publishers at Scrivener
Martin Scrivener (martin@scrivenerpublishing.com)
Phillip Carmical (pcarmical@scrivenerpublishing.com)

Nanoscience and Nanotechnology for Smart Prevention, Diagnostics and Therapeutics

Fundamentals to Applications

Edited by

Sathish-Kumar Kamaraj

Instituto Politécnico Nacional (IPN)-Centro de Investigación en Ciencia Aplicada y Tecnología Avanzada (CICATA-Altamira), Mexico

Arun Thirumurugan

University of ATACAMA, Vallenar, Chile

Muthuchamy Maruthupandy

Department of Health Sciences, Dong-A University, Busan, South Korea

Mercedes Guadalupe López Pérez

Department of Biotechnology and Biochemistry, Centro de Investigación y de Estudios Avanzados del IPN - CINVESTAV, Guanajuato, Mexico

and

Shanmuga Sundar Dhanabalan

School of Engineering, RMIT University, Australia

Scrivener Publishing

WILEY

This edition first published 2024 by John Wiley & Sons, Inc., 111 River Street, Hoboken, NJ 07030, USA and Scrivener Publishing LLC, 100 Cummings Center, Suite 541J, Beverly, MA 01915, USA

For more information about Scrivener publications please visit www.scrivenerpublishing.com.

Wiley Global Headquarters
111 River Street, Hoboken, NJ 07030, USA

For details of our global editorial offices, customer services, and more information about Wiley products visit us at www.wiley.com.

Library of Congress Cataloging-in-Publication Data

ISBN 978-1-394-17457-7

Cover image: Pixabay.Com
Cover design by Russell Richardson

Set in size of 11pt and Minion Pro by Manila Typesetting Company, Makati, Philippines

Printed in the USA

10 9 8 7 6 5 4 3 2 1

Contents

Preface

Within the realms of biomedical research and technology, the implementation of nanoscience and nanotechnology is acquiring an ever-increasing amount of significance. The scope of this multidisciplinary scientific field includes studies of the nanoscale behavior of chemistry, physics, materials science and engineering, biology, and medicine. Fascinatingly, nanoscale dimensions with a high surface-to-volume ratio, simple surface modification, improved physicochemical stability, specialized optical properties, and targeted and controlled release capabilities can result in lower toxicity and higher efficacy, making them more appropriate for efficient smart prevention, diagnosis, and treatment. Further, nano-dimensional materials interact efficiently with biological functional molecules (such as proteins, nucleic acids, carbohydrates, lipids, and complexes) to answer numerous day-to-day difficulties that are related to a bio-medical theme. This is a significant advancement in the field. In addition, the impact that the nanomaterial has on optical, photothermal, electrical, electronic, and magnetic parameters paves the way for the development of smart and intelligent technologies. These technologies are based on external stimuli to combat preventive, diagnostic, and therapeutic issues in a commendable manner. This can be achieved by combining the nanomaterial's optical, photothermal, electrical, electronic, and magnetic parameters.

This book focuses on the fundamental features of various nanomaterials that are related to the development of biomedical technologies. These fundamental qualities are broken up into three parts: prevention, diagnostics, and therapeutics. When it comes to infectious diseases, prevention is of utmost importance. Highly advanced nanomaterials including silver, titanium, graphene-based filters, and copper nanoparticles are used to fight infectious illnesses. Once the symptoms have been recognized in the patients, through the use of effective and straightforward nanodiagnostic techniques, the diseases can be accurately localized in either a qualitative or quantitative manner. Nanodiagnostics tools currently dominate the field of biomedical diagnostics because of their high degree of accuracy, low requirement for samples and reagents, user-friendliness, portability, and capacity to perform point-of-care (POC) applications. Nanomaterials are widely used in imaging due to many factors, including: their signal generation and amplification abilities; the ongoing development of reliant new imaging techniques, such as photoacoustic imaging and Raman imaging; their targeting potential, due to the possibility of functionalizing their surface with cancer-targeting moieties; their multimodality, since some nanomaterials can generate signals for more than one imaging technique); and their affordability.

The most notable area of focus in contemporary therapeutics is the investigation into numerous breakthroughs that have been made possible by applying nanotechnology in the treatment of site-specific cancers. In a similar vein, intelligent nanodrug delivery systems make it possible to improve medicine delivery at a particular point of care while simultaneously underrating the adverse effects that are associated with drugs and drug carriers. In the initial generation of nanocarriers, liposomes and straightforward polymers played important roles. Phospholipids are the building blocks of liposomes. Phospholipids have a hydrophobic tail and a polar head, and they can self-assemble into spheres with a diameter varying from tens to hundreds of nanometers. Liposomes are used to transport lipids throughout the body. Stealth liposomes are liposomes that have been functionalized with polyethylene glycol (PEG) to increase their half-life in circulation. Drugs that are either hydrophilic or hydrophobic can be delivered via liposomes. In most cases, the nanocarriers that are utilized in the process of drug delivery are functionalized by a PEG that has a molecular weight (MW) ranging anywhere from 1 to 40 kDa. This coating is useful for nanocarriers because it decreases the nonspecific interactions that nanocarriers have with serum proteins. These interactions flag nanocarriers for internalization by cells that are part of the reticuloendothelial system (RES). They become less immunogenic as a result, and cells inside the RES absorb them less specifically. As a consequence, their phagocytosis is reduced, which in turn leads to a prolongation of the nanocarriers' duration in circulation. PEG functionalization of nanocarriers can reduce the toxicity of the nanocarriers, as well as prevent them from clumping together. Examples of current nanocarriers include viruses, nanoparticles consisting of gold, magnetic nanoparticles, quantum dots, titanium dioxide, zinc oxide, and silica, graphene, carbon nanotubes and fullerenes, and hybrids of materials (such as lipid-coated or polymer-coated nanoparticles).

This book presents the fundamentals of nanomaterials and discusses the direct applications of nanomaterials to the biomedical sector. In addition, it explores the potential therapeutic applications of nanotheranostics in the far future. We are deeply grateful to everyone who helped with this book and greatly appreciate the dedicated support and valuable assistance rendered by Martin Scrivener and the Scrivener Publishing team during its publication.

Sathish-Kumar Kamaraj
Arun Thirumurugan
Muthuchamy Maruthupandy
Mercedes Guadalupe López Pérez
Shanmuga Sundar Dhanabalan

Acknowledgements

First, we want to thank God for blessing us with good health and the ability to edit this book. Our deepest gratitude goes to the series editor and advisory board for believing in our work and accepting our book for publication. Thanks to everyone who helped make this book a reality—the authors, the reviewers, and everyone in between. We are grateful to the many publishers and authors who permitted us to use their work, especially the figures and tables.

Sathish-Kumar Kamaraj would like to thank the director of the Centro de Investigación en Ciencia Aplicada y Tecnología Avanzada, Unidad Altamira (CICATA Altamira) and the director general of the Instituto Politécnico Nacional (IPN) for their ongoing support and facilities that enable the research activities. I appreciate the work of Secretara de Investigación y Posgrado (SIP)-IPN on project SIP:20231443. Additional extensions have been granted to the Secretary of Public Education (SEP-Mexico) and the National Council of Humanities, Sciences, and Technologies (CONAHCyT-Mexico). He expresses his gratitude to Mrs. Mounika Kamaraj and Bbg Aarudhra for providing family assistance.

Arun Thirumurugan would like to express his gratitude to Dr. Justin Joseyphus (NIT-T, India), Prof. P.V Satyam (IOP, India), Dr. Ali Akbari-Fakhrabadi (FCFM, University of Chile, Chile), Prof. RV Mangala Raja (University of Adolfo Ibanez, Santiago, Chile) for their kindness and guidance. Arun Thirumurugan would like to thank Dr. R. Udaya Bhaskar and Mauricio J. Morel (University of ATACAMA, Chile), Carolina Venegas, Yerko Reyes, and Juan Campos, Sede Vallenar, University of ATACAMA, Chile for their support. Arun Thirumurugan acknowledges ANID for the financial support through SA 77210070.

Muthuchamy Maruthupandy would like to thank Dr. Anand (DMCS, Madurai Kamarai University, India), Prof. Ji-Ming Song (SCCE, Anhui University, P.R. China), Dr. Aravind Kumar Rengan (IIT-H, India), Dr. Franck Quero (FCFM, University of Chile, Chile), Prof. Wan-Seob Cho (LAMS, Dong-A University, South Korea) for their guidance and support.

MM acknowledges the Brine Korea 21 (BK-21) and Basic Science Research Program through the National Research Foundation of Korea (NRF), funded by the Ministry of Science, ICT & Future Planning (NRF-2022R1I1A1A01069472) for financial support.

Mercedes Guadalupe López Pérez would like to thank each and every one of the students and postdoctoral fellows who have passed through my laboratory during the last three decades, without them, my scientific and academic life would not be the same. She also wants to thank her institution (CINVESTAV) and, in a very special way, the many Agave and Tequila companies, without their financial support, her entire career would be less rewarding. Finally, she would like to thank this group of co-authors, who have made possible the culmination of this project. Thank you for including her on this journey.

Shanmuga Sundar Dhanabalan would like to express his sincere thanks to Prof. Sivanantha Raja Avaninathan (Alagappa Chettiar Government College of Engineering and Technology, Karaikudi, Tamil Nadu, India), Prof. Marcos Flores-Carrasco (FCFM, University of Chile, Chile), Prof. Sharath Sriram, and Prof. Madhu Bhaskaran (Functional Materials and Microsystems, RMIT University, Australia) for their continuous support, guidance, and encouragement. He would like to extend his gratitude to Mrs Preethi Chidambaram and his family for support.

1

Bio–Nano Interface Technology for Biomedical Applications

Ana Luisa Gómez-Gómez[1], Deyanira del Rosario Moguel-Concha[1], José Eduardo Borges-Martínez[1], Alma Leticia Martínez-Ayala[2] and Gloria Dávila-Ortiz[1]*

[1]*Departamento de Ingeniería Bioquímica, Escuela Nacional de Ciencias Biológicas, Instituto Politécnico Nacional (IPN), Av. Wilfrido Massieu S/N, Unidad Profesional Adolfo López Mateos, Zacatenco, Delegación Gustavo A. Madero, Ciudad de México, México*

[2]*Centro de Desarrollo de Productos Bióticos, Instituto Politécnico Nacional, Carretera Yautepec-Jojutla, Col. San Isidro, CEPROBI, Yautepec, Morelos, Mexico*

Abstract

Nanoencapsulation protects biologically active compounds from deterioration due to system conditions such as oxidation, temperature, and pH changes, among other interactions that occur at the interface. Thus, biomedical applications of nanometric structures require the evaluation of a complex delivery system, in which the interactions that exist at the biomolecule–nanostructure interface, as well as their physicochemical properties, will determine the scope of the delivery system. A wide range of nanostructured materials exists. However, nanoencapsulation of bioactive compounds is a novelty. One advantage of nanoencapsulation is that nanostructures can be coated with biomolecules such as lipids, proteins, and polysaccharides, resulting in reduced surface energy and providing biological benefits to the organisms they interact with. It is essential to mention several challenges to implementing nanomaterials in biomedicine, among which toxicity and decreased efficacy stand out. These disadvantages occur mainly due to a lack of understanding of the interactions between nanomaterials and their biological environment. Currently, the use of nanomaterials is based primarily on the functionality of biomolecules. While nanomaterials are often designed to take advantage of the functionality of biomolecules, it is important to consider the potential

**Corresponding author*: gdavilao@yahoo.com

Sathish-Kumar Kamaraj, Arun Thirumurugan, Muthuchamy Maruthupandy, Mercedes Guadalupe López Pérez and Shanmuga Sundar Dhanabalan (eds.) Nanoscience and Nanotechnology for Smart Prevention, Diagnostics and Therapeutics: Fundamentals to Applications, (1–22) © 2024 Scrivener Publishing LLC

impact of biomolecule–nanomaterial binding. Failure to account for such binding could lead to changes in the structure of the biomolecule, resulting in an altered or lost biological function of the compound. Additionally, binding could cause negative interactions between the bionanomaterial and the biological environment. The purpose of this chapter is to show the benefits of nanomaterials in conjunction with biomolecules in providing biological activity to help address various applications in the field of biomedicine, which will help to provide better and timely control of human health; as well as to identify the physicochemical properties of nanomaterials, which allow us to figure out what kind of interactions are involved at the bio–nano interface, due to their influence on the pharmacokinetic system stability associated with some parameters, such as the payload, release, and delivery efficiency. In addition, the effect of the physicochemical properties of nanomaterials and other factors that influence the structure, composition, and function of nanomaterial–bioactive compound complexes will be addressed, leading to a better understanding of the role of bioactive compound–nanomaterial interactions in controlling or predicting the biological fate of nanomaterials.

***Keywords*:** Nanoencapsulation, bio–nano interface, bioactive compounds, fluidized bed

1.1 Physicochemical Properties of Nanoencapsulated Systems

Some physical and chemical aspects of nanoparticle carriers and encapsulated drug molecules have a substantial impact on the basic attributes of nano-sized drug products, such as drug circulation, drug release from site-specific dosage forms, and absorption into bodily membranes. Particle size has a significant impact on the stability of nanoemulsion complex, it has been reported that decreasing particle diameter increases the bioavailability of encapsulated compounds [1, 2]. Therefore, these are chief elements that correlate robustly concerning the stability of the encapsulated system. As a result, the particle characteristics are required to have an appropriate delivery system [3]. Smaller particle sizes provide a greater mass transfer area, leading to an improved drug diffusion rate. Conversely, the rate of drug dissipation within bigger particles is lower, as they offer a reduced mass transfer surface area. Smaller particles, on the other hand, tend to agglomerate when kept and moved [1]. The size of the particles can range from 10 nm (nanoemulsion) to 1 mm (hydrogel droplets). Colloidal reliable encapsulated particles are typically spherical. In contrast, distinct forms, among them cylinders, deformed spheres, or irregular shapes, have been noticed, influencing changes in the properties of particles and the interest compound delivery process [3]. The polydispersity of an enclosed system indirectly shows its aggregation status. Higher polydispersity implies the existence of aggregates, and this can cause destabilization and breakage in emulsion-based encapsulating systems. When the polydispersity of an encapsulating system is lower than 0.2, it is said to be monodisperse; nevertheless, polydispersity under the value of 0.5 is also regarded for pharmaceutical applications [3]. Droplet size is a relevant characteristic of emulsions because it promotes emulsion stability as droplet size and polydispersity decrease. The simplest and most common technique for measuring particle size and polydispersity is dynamic light scattering (DLS) [3]. It measures the intensity fluctuation of dispersed light. This fluctuation is the result of the interference of scattered light by individual particles just because of Brownian motion. The key advantages of DLS are its fast analytical speed, lack of calibration requirements, and excellent sensitivity to submicrometer particles [1]. Generally, the scattering angle is set to 90 degrees in most DLS techniques. For a monodisperse sample, the particle size should not change upon increasing the light scattering angle. Due to the extent of scattering at different angles being affected by particle size, the intensity-averaged mean particle size varies for polydisperse samples [3]. Now, the external charge of captured materials corresponds strongly with their dispersion stability. The zeta potential is frequently employed to analyze the surface charges of encapsulated materials, demonstrating the dominance of electrostatic forces indirectly [3]. In this connection, the use of the zeta potential makes it possible to visualize in encapsulations the influence of the charge of active molecules on the surface properties of the packing material. These enable the investigation of the stability of encapsulated materials besides the study of the electrostatic forces that occur between the active molecules and the encapsulating material. Colloidal stability is usually analyzed from the zeta potential of a nanoparticle. These measurements are performed with a zeta potential analyzer or zeta meter and allow prediction of the storage stability of various colloidal dispersions. To ensure stability and avoid particle aggregation, absolute zeta potential levels must be high, either positive or negative. The zeta potential measurements can be used to estimate the degree of surface hydrophobicity. The zeta (ζ) potential may provide additional details about the material enclosed in nanocapsules or coated on their surface [1]. A study on encapsulation used spray drying as a technique to obtain vitamin E-loaded nanocapsules using modified starches such as octenyl succinic anhydride (OSA), with two purposes as emulsifier and barrier material. The ζ-potential, size distributions, mean particle size, and polydispersity index (PDI) of the initial and reconstituted nanoemulsions were measured via DLS [1]. The mean particle diameters ranged from 208 to 235 nm. Although the mean hydrodynamic diameters were quite different, the PDI and ζ-potential values were similar. Furthermore, the reconstituted nanoemulsions retained their original trim monomodal distribution with a modest increase in mean particle sizes, according to the authors. It is noteworthy that the reconstituted emulsions kept their polydispersity values (PDI < 0.250) and particle sizes (< 250 nm) indicating that the spray-drying technique had no effect on the nanoemulsions' properties [2]. As a result, the authors reported that OSA-modified starches with low molecular weight are efficacious in producing steady vitamin E nanocapsules for usage in pharmaceutical and food applications [2]. In another study, the authors evaluated the liberation of bioactive compounds with antioxidant and antihypertensive ability from packed extracts of Gulupa and Cholupa peel and seeds in an *in vitro* gastrointestinal model to simulate the process of digestion [4]. The encapsulates were constructed using wall material rice starch enzymatically modified. Characterization of the encapsulates revealed a range of electrical potential values between −6.34 and −6.66 mV. In addition, the DLS method determines the dispersion stability and disclosure PDI measurements from 1.33 to 1.51. The authors mentioned that the increment in the surface charge is due to the phenolic compounds on the particles. They observed that the wrapped extracts had an electronegative charge ζ-potential. As a result, the microcapsules showed high stability, so the encapsulates have an enormous amount of opportunities in the fields of food and medicine [4]. Advanced microscopic techniques such as scanning electron microscopy (SEM), transmission electron microscopy (TEM), and atomic force microscopy (AFM) are used to determine the general properties of nanoparticles such as size, shape, and surface charge. Physical stability and in vivo distribution of nanoparticles are important qualities that are heavily influenced by average particle diameter, size distribution, and surface charge. The geometric shape of polymeric nanoparticles is determined using electron microscopy techniques, which is an essential consideration that influences their toxicity. The physical stability, dispersion dispersibility, and *in vivo* performance of nanoparticle drug delivery systems are all determined by their surface charge capability [1, 5]. SEM gives information about a sample's exterior shape, chemical composition, and crystalline structure through direct visualization by the incident effect of the beam electron. Despite the benefits of morphology and size analyses, SEM provides only a limited amount of information on the particle size distribution and pores [1]. Most electrically insulating samples have a tiny layer of conductive substance on them, such as gold, carbon, other metals, or alloys. Usually, the most common conductive coating used for elemental analysis is carbon, although metallic overlays are more useful for high-resolution electronic imaging applications. With respect TEM operation principle, it is distinct from SEM, even though the type of results obtained are similar. The procedure for TEM samples is elaborate and takes considerable time. Surface features of the sample are obtained when an electron beam is incising on a thin slide of the sample, and the electrons interacting are transmitted through it [1, 5]. Finally, AFM offers some advantages compared to electron microscopy (TEM and SEM); it is based on the calculation force between the material surface and the probing tip. The approach is characterized by simple and easy sample preparation, quick picture capture, and appropriate finest resolution [1]. The analyzed particles need to be immobilized as a first step before viewing, in the case of larger particles is easy to get it due to their sedimentation rate. However, in small particles, the immobilization process faces troubles given to their Brownian motion. In this vein, sample dehydration can solved the troubleshooting, but at the same time, it may originate in cluster formation, lipid crystallization, shrinkage, and other unwanted changes. The main profit of AFM is its proficiency to image non-conductive samples without any specific covering treatment, which allows envisaging of careful biological and polymeric nano- and microstructures [1]. In one of the studies mentioned above [4], along with their characterization of encapsulates of Gulupa and Cholupa husk and seed extracts using modified rice starch as wall material, the authors employed SEM as part of the morphological characterization by obtaining images applied to an accelerating voltage of 12.5 kV and different magnification powers. They reported that the micrographs showed small sizes of 2 to 7 μm with a polyhedral arrangement and irregular shapes [4]. Moreover, it showed a broad range of particle aggregation, swelling, and solid heterogeneity appearance. In addition, some lack of sharpness was observed, with some granules showing barely rough surfaces. These characteristics could be attributed to the frequency of shrinkage due to water diffusion taking more, allowing the structures to shrink and deform to some extent. Therefore, these surface characteristics may be beneficial for improving the rehydration of powders [4].

Zhu *et al.*, evaluated nanoparticles (NPs) produced with xanthan gum and lysozyme, which contained two different peptides: TSeMMM (STP) and SeMDPGQQ (SHP), both containing selenium. TEM was used to notice the nanostructure of each sample. The images of the NPs were acquired at 15k magnifying power and 100 kV accelerating voltage. The NPs were observed to have a spherical morphology with a relatively constant state. NPs sizes were 153 nm somewhat greater than that of NPs-STP (145 nm) and NPs-SHP (148 nm). However, an important feature the authors comment on between TEM and DLS is that the particle size (hydrodynamic diameter in DLS) of NPs, NPs-STP, and NPs-SHP measured by TEM was lower diameter values than that measured by DLS analysis. This is due to sample conditioning in the course of TEM analysis, inasmuch as NPs, NPs-STP, and NPs-SHP are in lyophilized powder form measured, while by DLS average particle sizes measurements of NPs, NPs-STP, and NPs-SHP were in solution [6]. Meanwhile, in another study, Luo *et al.*, used ultrasound treatment to encapsulate within zein and gum Arabic (GA) the peptide TSeMMM (T) with selenium (zein@T/GA), which has immunomodulatory functions, that they obtained from selenium-enriched rice protein hydrolysates [7]. They studied the structural and morphology characteristics of three nanoparticle formulations: zein, zein, and gum Arabian (zein/GA); and the last was zein with gum Arabian and the peptide TSeMMM (zein@T/GA). SEM and AFM were used to determine micromorphological features. To this purpose, zein, zein/GA, and zein@T/GA nanoparticles were examined at 1 mg/mL concentration. The SEM results showed a smooth surface in zein nanoparticles with a majority of spherical shapes and a size distribution particle of 119 nm. Nevertheless, by SEM analysis were noticed aggregation and adherent of NPs. Thus, it was not possible to determine the size distribution. Therefore, information was complemented with the AFM technique. Determining, through 3D and 2D morphology images analysis the average nanoparticle size for zein/GA (90.9 nm) and zein@T/GA (43.7 nm). Moreover, the AFM results suggested that zein@T/GA NPs were smaller and showed less aggregation because the ultrasonic treatment generated changes in the zein structure. Which in turn resulted in the reduction of the nanoparticle size and improved stability of NPs in the system by the gum Arabian capacity to bind to zein via electrostatic interactions to generate more stable NPs. In this sense, the emphasis is on the stable and homogeneous system obtained via ultrasonic process to increase the interactions between the NPs components. The forces that influenced the zein@T/GA NPs formation were dipole-dipole attraction (hydrogen bond), electrostatic repulsion, and hydrophobic interactions. On the other hand, the encapsulation efficiency was improved, due to the micromorphological features given by the ultrasonic method [7].

Various methods have been used to research the crystalline arrangement and surface physicochemical properties of the encapsulation systems. Some of these techniques are X-ray diffraction (XRD), differential scanning calorimetry (DSC), and Raman spectroscopy. Regarding the structural parameters, XRD is helpful to describe the structural order degree in nanomaterials. Moreover, at the same time, allow classifying the material in the function of its crystallinity lattice properties. The structural complexity influences the material behaviour, for example, mechanical and thermal features. In this sense, DSC allows evaluation changes by the effect of temperature and thermal degradation in encapsulated. The encapsulation of molecules bioactive is a successful method to protect and improve the bioavailability of these molecules. However, it is possible the bioactive molecule wrapped does not show the crystallinity and the melting peaks patron, whereas, the mixture of the transporter material and the bioactive molecule, without encapsulation, demonstrate the features in combination and individual encapsulated components. The bioactive molecule encapsulated modifies the carrier lattice crystallinity diminishing the melting point. A complete diffusion of the bioactive compound in the encapsulating material lattice is estimated by a linear correlation betwixt melting temperature and the amount of trapped bioactive molecules. By contrast, and following this association, when the melting temperature of the encapsulating material lattice does not change, it is possible that the bioactive compound is not encapsulated. Therefore, with DSC analysis it is feasible to acquire valuable data to describe the matrix behaviour between the encapsulating material and trapped bioactive molecules, moreover to determine encapsulate thermal features [3]. Liu *et al.*, created ternary NPs derived from zein and egg white peptides (Z-EWDP-CS) by antisolvent precipitation method, to improve curcumin (Cur) stability and biological activity.

So they measured the structure crystallinity lattice parameters through XRD. The crystallographic patterns measured the Cur, NPs+Cur (nanoparticles with curcumin), and NPs-wCur (NPs without curcumin). Showing six diffraction peaks (8.9°, 12.2°, 14.5°, 17.3°, 25.5°, and 27.3°) for Cur, demonstrating high organized crystalline state. For NPs+Cur and NPs-wCur each XRD patron shows two broad diffraction peaks at 8.6°and 19.3° for the first and the second 9.1°and 19.0°. To summarize this result indicate that zein and EWDP have an amorphous patron. In contrast, the XRD diffraction patron of Cur-Z-EWDP-CS NPs shows a crystalline state compared with the other formulations. The combination of DSC, FTIR, and

XRD analysis confirmed the change in the crystalline lattice organization of Cur when encapsulated. They concluded that the encapsulation method was effective, enhancing the stability, thermal properties, and antioxidant activity of Z-EWDP NPs [8].

Emulsions cannot be overlooked as great encapsulation and release technologies; they have benefits such as controlled delivery and thermodynamic stability of the encapsulated nutrients. Single, multiple, and nanoemulsions have been used to encapsulate several types of molecules. Modifications in the composition and fabrication technologies for emulsion-based encapsulation result in a diverse variety of emulsions with novel features. In the food area, two types of emulsions are used, primary water-in-oil or oil-in-water emulsions and double emulsions water-oil-water [9]. The enclosed system's final use is determined by its rheological behavior. Functional foods contain encapsulated 'functional' foods that can improve the quality, flavour, stability, texture, or some other food techno-functional properties. Encapsulated materials are micrometer or submicrometer in size, they significate impact on the rheology properties of the final product, which is essential for several applications [3]. An example is Buyukkestelli and El's characterization of double emulsions (W1/O/W2) to encapsulate ferric chloride inside a water phase (W1). The formulated emulsion with a 40:60 proportion presented the maximum viscosity value and encapsulation effectiveness percent, while the creaminess index and droplet size were the lowest. They reported that the supposed viscosity of all the double emulsions increased proportionally according to the increment shear rate values. On the other hand, the double emulsions showed shear thinning behaviour as a function of n-value, which provides details about the flow pattern. A fluid behaves as a pseudoplastic (shear thinning) if the n-value is less than 1. The primary emulsion consisted anexterior aqueous phase (W1/O:W2) ratio of 40:60 showed a higher fraction of the primary emulsion and smaller particle size than did the other emulsions examined. As can be seen, this study allows the utilization of encapsulation via double emulsion in the iron fortification of food items [9].

1.2 Nanoencapsulation of Bioactive Compounds by Fluidized Bed Drying

Bioactive compounds (BCs) are present in small amounts in foods, mainly in fruits, vegetables, and grains. These compounds provide health benefits beyond the basic nutritional value. BCs are molecules that may have therapeutic potential by influencing energy consumption while reducing the proinflammatory state, oxidative stress, and metabolic disorders. Epidemiological studies indicate that high consumption of foods rich in BCs with antioxidant activity, such as phenolic compounds, bioactive peptides, vitamins, and various phytochemicals, has a positive effect on human health and could decrease the risk of numerous diseases such as cancer, heart disease, diabetes, and various chronic degenerative diseases [10].

Nevertheless, the bioactivity of these compounds is very short-lived, due to their rapid oxidation or insufficient residence time in the stomach, limited permeability, and intestinal solubility. The BCs are unstable under the usual conditions of food manufacturing and preservation (exposure to light, temperature, humidity, and oxygen levels) or gastrointestinal digestion (pH, temperature, enzymatic activity, absorption, and degradation of nutrients), causing partial or complete loss, which limits their use in human food. It is also substantial to mention that the efficacy of BCs to ward off illness hinge on the preservation of the active compounds. As a result, oral BCs administration requires extra protection in order to maintain bioavailability. Therefore, the administration of CBs needs extra protection to maintain their structural integrity and increase their bioavailability and biological targeting, making encapsulation. Particularly the nanoencapsulation process is a technological advance to employ [11, 18].

Encapsulation commonly is described broadly as the method of encapsulating a specific molecule (BCs) into a carrier (covering material). A method uses a physical barrier to shield BCs from the environment during storage or from other components when added to a food matrix. Another advantage is the capability to command the release rate of the BCs encapsulated, which increases the efficacy of the BCs by not having to over-dispense them. The compound encapsulated is referred to as the central, core, or active phase. The encapsulating material may be known as coating, capsule, carrier material, matrix, etc., and is made up of several biopolymers including carbohydrates like sugars, complex carbohydrates like gums and polysaccharides, proteins, synthetic polymers, etc. [12, 18].

BCs such as phenolic compounds and bioactive peptides have been of increasing interest due to a better understanding of their role at the pathophysiological level, and advances in biotechnology and biochemical synthesis. However, their use has been hampered by their poor biological availability, derived from their low stability against proteolytic breakdown, limited ability to pass across barriers, and a brief biologic half-life in bloodstream systems. Most BCs are still delivered parenterally due to their insufficient absorption in the gastrointestinal tract. Due to formulation versatility, sustained release qualities, cellular size, and biological compatibility with cells and tissues, NPs appear to be a potential for BC delivery [12, 13].

Nanoencapsulation provides particles of < 100 nm. According to the size, the particles encapsulated are classed as nanocapsules (< 0.2 μm), microcapsules (0.2–5000 μm), and macro capsules (> 5000 μm). Among the main advantages of the application particles below 100 nm its their homogeneity, which translates into higher encapsulation efficiency of the BCs. Moreover, their physicochemical properties are improved. Thus, nanoencapsulation is an attractive option to reformulate bioactive food components, which can have a significant impact on product processability, stability, and bioavailability [14]. The use of nanoparticulate BCs has been used in the treatment and vaccination of various diseases because of many benefits when compared to traditional dosage routes. The importance of size has also been highlighted and the advantages of NPs over microparticles have been revealed. It has been observed that more NPs pass through the epithelium than microparticles. NPs have received greater attention due to their potential for therapy and good stability in biological fluids. Their small size makes them suitable for parenteral delivery and they might be beneficial as injectable prolonged-release systems for organ or site-specific delivery [15, 16]. Tables 1.1 and 1.2 show some examples of compounds that are amenable to the nanoencapsulation process as well as some applications that have been carried out with them [14].

The technologies that have been used for the nanoencapsulation of BCs are classified into chemical and mechanical methods. Coacervation, ionic gelation, and emulsion belong to chemical methods. The main mechanical methods are spray drying, fluidized bed drying, and extrusion; a method

Table 1.1 Common food BCs (core chemicals) that have been nanoencapsulated [14].

Bioactive compound	Encapsulation application(s)	Implementation
Phenolic compounds	Protection. Enhancement of antioxidant activity. Inclusion of extra favourable features. Targeted delivery	Delivery systems for water-insoluble compounds (Quercetin) Oral delivery of tea catechins. Fortification and release system of folic acid

(Continued)

Table 1.1 Common food BCs (core chemicals) that have been nanoencapsulated [14]. (*Continued*)

Bioactive compound	Encapsulation application(s)	Implementation
Carotenoids	Stabilization improvement. High-efficiency release controlled. Expanded utilization in the industry.	Nanoencapsulation of saffron hydrophilic apocarotenoids (lycopene, crocins), and food dyes
Essential fatty acids	Improve stability and solubility properties. Volatility decreased. Favourable impact on the sensory quality of the final product.	Antimicrobial efficacy Masking unwanted flavours Good sensory attributes. Fortification with DHA, linolenic acid
Vitamins	Protection against oxidation. Shelf life increased.	Pharmaceutical and beverage sector. Milk fortification. Food supplement. Antioxidant additives Thiamine, riboflavin, vitamin D3, vitamin E, vitamin B9, vitamin B2.
Peptides and proteins	Enhancement of antibacterial activity, and antioxidant activity. Absorption and bioavailability improved.	Oral delivery of bioactive peptides Nisin encapsulation against Gram-negative bacteria
Probiotics and prebiotics	Promotion of gastrointestinal health. Control of cholesterol. Increment microorganism's viability.	Encapsulation of *Lactobacillus acidophilus* and *Lactobacillus casei* Incorporation in foods (cheese), and drink product (fermented milk, yoghurt)
Others	Improve physicochemical properties. Improve organoleptic characteristics. Controlled release.	Minerals, colorants, micronutrients, and flavouring compounds.

Table 1.2 Commonly used wall materials for encapsulation and representative characteristics to bear in mind [14].

Carbohydrates	Proteins	Fats and waxes	Polymers
Chitosan	Whey protein	Hydrogenated vegetable oils	Polyglycolides
Pectin	Sodium caseinate	Bee wax	Polylactides
Maltodextrin	Soy proteins	Lecithin	Polyethylene glycol
Alginates	Gelatin	Medium chain triglycerides	Polyorthoesters
Cyclodextrins	Caseins		Polyanhydrides
Gum Arabic	Zein		Polyacrylamide
Cellulose derivatives			Polycaprolactone
Modified starches			Polyacrylic acid
Carrageenan			Polyvinyl alcohol
Properties that should be considered:	Solubility, edibility, film-forming ability, emulsification, viscosity, degree of crystallinity, compatibility with the core compounds, cost, taste, and flavour		

that is commonly used in the pharmaceutical industry is fluidized bed drying [17, 18].

1.2.1 Fluidized via Bed Drying

Fluidized via bed drying is an encapsulation process based on the covered compound and additional shell material in powder particles; this process is performed continuously or in batches. First, BCs in conjunction with a binder solution are hung in a hot environment in a fluidization drying chamber with a temperature-humidity controlled, humidity, and drying air velocity; as the particles (support) enter the chamber, they begin to be in a state known as fluidization. Next, the binding agent mixed with the BC to be nanoencapsulated is atomized onto the support material, and a layer is formed. Each particle enters the spraying zone, it is gradually covered each time the atomization process is performed through a nozzle that may be located at the top, bottom, or side of the fluidization chamber. Subsequently, an evaporation process occurs, and the process ends when the support material binds to the binding agent and BCs through complexes known as bridges [18–20].

One of the advantages of the fluidized bed drying method compared to spray drying is that it can be achieved at lower temperatures because of better mass and energy transfer, and uniform temperature distribution. It has also been observed that a product with better compaction and sphericity is obtained. One of the fundamental parts of the fluidized bed to achieve a good coating and to obtain a controlled release of the BCs is the integrity and solubility of the shell material; among the materials used are gum Arabic, whey protein, and maltodextrins [18, 21].

There are precedents in which a fluidized bed has also been utilized to provide an extra layer of bioactive compounds for payload release in the intestine [22]. One of the types of compounds that have been nanoencapsulated with a fluidized bed are phenolic compounds. Oehme *et al.*, assessed the viability of encapsulated anthocyanins upon reaching the colon [23]; in this work, nanoencapsulated compounds were developed by ionotropic gelation and subsequently dried in a fluidized bed to reduce the water content and the size of the capsules.

On the other hand, Soh and Lee proposed using a fluidized bed with supercritical fluid to produce micro- or nanoencapsulated BCs by tailoring the properties of the BCs, the wall material, and the support compound [24].

The fluidized bed drying and coating methods have been recognized as an important means to be able to obtain BCs used in biomedicine (e.g., Wurster's fluidized bed coating [25]). Moreover, using supercritical CO_2 as the fluidization medium for the core particles, as well as the drying agent to remove moisture or solvent, respectively, from the core particles coated with the binder solution and BCs, is of great interest, as it allows the coating to be performed under oxygen-free conditions and at low temperature. One such study was performed by Niu *et al.*, who developed a Wurster-type coater employing near-supercritical CO_2 to remove moisture and solvent from the coated particles [26]; studies on particle fluidization with supercritical CO_2 have shown that conventional correlations for fluidization can be applied to supercritical fluids for the prediction of minimum fluidization and terminal velocities. Supercritical fluidized bed coating utilizes solid core particle fluidization using supercritical CO_2 as the fluidization medium and, at the same time, a solution with the coating material will be sprayed onto the fluidized particles. Supercritical CO_2 plays the role of drying medium (for aqueous solutions) or antisolvent (for organic solutions) to dry the fluidized particles in the bed [25, 26]. Coating of BCs in ingredients such as curcumin by supercritical fluid coating processes has been demonstrated. This technique has promising development to produce controlled multilayer coatings on BCs to achieve formulations with desired sustained release or surface properties [27].

1.3 Protein and Peptide Nanoencapsulation in Biomedical Applications

From a biomedical perspective, proteins and peptides have been suggested as drugs due to their biological activity, which is associated with their high selectivity and strength, a broad range of targets, and longer duration of their effect due to slower clearance from the body and cutting down the innate toxicity [28]. In this regard, the challenge for their use as drugs is their physicochemical properties, which influence poor membrane permeability, poor metabolic stability, and oral bioavailability [29]. One of the leading properties that affect the rate and increase protein and peptide absorption is permeability, which is directly related to molecule bioavailability. The gastrointestinal tract environment, such as pH and protease enzymatic action, can decrease the permeability and influence the absorption of drugs administered orally.

In the last years, oral administration has been improved to enhance BC absorption, and strategies for the development of nanoscale encapsulation provide better adherence to the mucus layer covering intestinal epithelial cells; hence, the retention time in the gastrointestinal tract increases. Specifically, these nanocarrier processes have considered the physiological environment to improve biomolecule bioavailability as the main focus in drug delivery systems. Nanoencapsulation allows protection of BCs, avoids degradation, improves bioavailability, and provides a stable system to release the compound. In connection with bioavailability, there are two terms, bioaccessibility and bioactivity. The first term can be outlined as the part of assimilate component discharge from the food composition which comes out available for absorption in the intestine, while the next term refers it as material available after the transformation by the gastrointestinal metabolism of food for assimilation, transport by the target tissue, absorption into intestinal epithelium cells, metabolism, and physiological response. *In vitro* assays have been refined to evaluate the percentage of bioaccessibility; these include *in vitro* digestion and dialysis methodologies for replicating gastrointestinal settings [30].

The particle adhesion process to the cell surface phospholipid bilayer considers adhesive forces and surface tension as a part of an interfacial tension phenomenon that determines the nanocapsule interface in contact and membrane wrapping. And for this, the adhesiveness and envelope at the adherence spot necessitate binding interactions as much as specific and nonspecific to get better of the resistive forces that impede particle uptake [31]. Hydrophobic and electrostatic forces, as well as hydrogen bond interactions, all play a role in peptide adsorption. Therefore, when existing an increment in the hydrogen bonds between lipid molecules in the cell membrane with loading molecules implies a spontaneous diffusion thus allowing a molecule penetrates more easily through the membrane. In order to pass across membranes, it ought to consider the features of the absorbed molecules that need to go through the hydrophilic head groups, which are charged and have a considerable viscosity; so if the extensive complex of hydrogen bonds is made, the absorbed molecule permeability cut down [32].

The permeability of a molecule across a membrane depends on many criteria, such as electrical charge, size, lipophilicity, and hydrogen bonding capacity [33]. This parameter is governed by the diffusive transport of the load-carrying along the crosslinked polymer shell, which includes the diffusion coefficient of the payload capacity in water, the average mesh size of the complex, the radius of the load-bearing, and the protein chain radius [34, 35]. Consequently, an important factor is the wall material, which should be highly biocompatible and have low immunogenicity, considering the therapeutic target. The wall material is a barrier that protects the BCs, is present in the diffusive process in the mechanism of release, and related to efficiency and targeting ability. Therefore, encapsulation efficiency (EE) is a measure of this process that means the highest amount of BC load inside or on the surface of the nanostructure; the percentage of EE is defined as: $EE\% = (Wt/Wi) \times 100$), where Wt is the total amount of the incorporated material and Wi is the total quantity of incorporated material added initially during the preparation [36]. Some factors that determine the EE are the partition coefficient and size of the payload molecule, the degree of crosslinking, the method spent to perform the encapsulation process, and the size and density of the nanocapsule shell. All these parameters are key to controlling the release of the encapsulated payload [37].

To summarize, nanoencapsulation increases the hydrogen bonds between the biomolecule and biological membrane, thus enhancing absorption across the membrane in concert with such surface charge, shape, and size. Research contributions in drug delivery nanosystems have demonstrated the effect of nanoencapsulation on protein and peptide bioactivity. Some of this work's conclusions indicate the positive enhancement of *in vitro* activities with biomedical purposes (Table 1.3). The nanoencapsulation process reaches greater efficiency values for peptide fractions with a minor degree of hydrolysis without regard to the enzyme used [38]. On the other hand, the properties of Ran blockade peptides with anticancer and antimetastatic activity were improved through a drug delivery nanosystem, and the results showed the anticancer action of the peptide-uploaded NPs, making an advantageous peptide delivery to the tumour and 42% inhibition of the tumour growth [39, 40]. In a diabetic model study, the nanoencapsulation establishment beat several barriers to the oral delivery of insulin, such as mucus layer penetration, drug stability in physiological conditions, and retention in the gastrointestinal tract, which was desirable for clarifying the insulin effectiveness through the release rate on drug absorption [40]. Nevertheless, the wall material has an important role in enhancing BC release; biopolymers such as chitosan exist in a protonated form in an acidic environment (gastrointestinal digestion) which allows for an upgrade of absorption. This means chitosan, via its positively charged amino groups, has the ability to interact with cellular membranes and cause structural modifications that open the tight junctions and allow the penetration of proteins and peptides across the membranes [41]. All of this has contributed to established synthesis methods and optimal operating conditions for producing highly stable bioactive nanocapsules.

On the other hand, the bio–nano interface (BNI) between nanoscale material and biological systems is a space where a series of dynamic interactions occur. BNI formation includes membrane and receptor–ligand binding interactions, which are mainly regulated by forces such as hydrodynamic, electrodynamic, electrostatic, and steric interactions in this space. In addition, the interaction between peptides and proteins and nanoscale material produces conformational changes that affect their secondary structures and consequently modify some physicochemical properties [42]. Therefore, it is important to understand the environment or conditions that regulate these BNI, for example the pH and ionic strength of the biological fluids, and their effect and influence on the absorption barrier and mechanism of the gastrointestinal tract and the mechanism

Table 1.3 Nanoencapsulation of proteins and peptides.

BCs	Wall material	EE*	LE**	Activity evaluation	Ref.
White shrimp peptides	Phosphatidylcholine	58.1–72.2%	-	Antioxidant activity, antihypertensive activity	[38]
Synthetic peptides	PEG PEG-PLGA	50–60%	-	*In vitro* antitumor assay in MDA-MB-231 and A549 cells	[39]
Insulin	n-Butylcyanoacrylate	100%	20–60%	*Ex vivo* retention effect, *in vivo* hypoglycaemic effect	[40]
Insulin	Nanoparticles of zein coated with a poly (anhydride)-thiamine conjugate (GT).	-	80 μg/mg	A diabetic rat model was used for *in vivo* nanoparticle evaluation.	[43]
Coenzyme-Q10 (CoQ10)	β-Lactoglobulin (CoQ10-BLG) Lactoferrin (CoQ10-LF)	5–10% w/w	62.79 ± 4.13% 86.77 ± 11.15%	*In vitro* release and permeability in Caco-2 cells. To assess CoQ10 permeability athwart the small intestinal epithelium. *In vitro* anti-ROS assay in J774A.1 cells	[44]
Octaarginine	Chitosan alginate	87.57 ± 3.65%	14.26 ± 0.59%	*In vitro* biocompatibility and bioactivity	[45]
Apigenin	Whey protein isolate	98.15%	196.21 mg/g	*In vitro* activity against colorectal cancer cells	[46]

*Encapsulation efficiency
**Loading efficiency

to improve the circulation rate of proteins and peptides and their biological activity. Hence, the synergy between the physicochemical properties of nanostructures and BCs provides strong evidence of nanoencapsulation advantages and their pharmacokinetic release behaviour in terms of scope in biomedical applications.

References

1. Yaneva, Z. and Georgieva, N., Physicochemical and morphological characterization of pharmaceutical nanocarriers and mathematical modeling of drug encapsulation/release mass transfer processes, in: *Nanoscale Fabrication, Optimization, Scale-Up and Biological Aspects of Pharmaceutical Nanotechnology*, A.M. Grumezeescu (Ed.), pp. 173–218, William Andrew Publishing, United States, 2018.
2. Hategekimana, J., Kingsley, G., Jianguo, M., Fang, Z., Encapsulation of vitamin E: Effect of physicochemical properties of wall material on retention and stability. *Carbohydr. Polym.*, 124, 172, 2015.
3. Gumfekar, S.P., Physicochemical characterization techniques in the encapsulation of active molecules, in: *Encapsulation of Active Molecules and Their Delivery System*, S.H. Sonawane, B.A. Bhanvase, M. Sivakumar (Eds.), pp. 9–22, Elsevier, Amsterdam, 2020.
4. Taborda, J.A.V., Arango, W.M., Arteaga, J.J.M., Almonacid, C.M.G., Encapsulation of bioactive compounds from byproducts of two species of passionflowers: Evaluation of the physicochemical properties and controlled release in a gastrointestinal model. *Heliyon*, 7, e07627, 2021.
5. Pal, S.L., Jana, U., Manna, P.K., Mohanta, G.P., Manavalan, R., Nanoparticle: An overview of preparation and characterization. *J. Appl. Pharm. Sci.*, 01, 228, 2011.
6. Zhu, Y., Sun, X., Luo, X., Ding, J., Fan, F., Li, P., Fang, Y., Encapsulation of selenium-containing peptides in xanthan gum-lysozyme nanoparticles as a powerful gastrointestinal delivery system. *Int. Food Res. J.*, 156, 111351, 2022.
7. Luo, X., Fan, F., Sun, X., Li, P., Xu, T., Ding, J., Fang, Y., Effect of ultrasonic treatment on the stability and release of selenium-containing peptide TSeMMM-encapsulated nanoparticles *in vitro* and *in vivo. Ultrason. Sonochem.*, 83, 105923, 2022.
8. Liu, J., Li, Y., Zhang, H., Liu, S., Yang, M., Cui, M., Du, Z., Fabrication, characterization and functional attributes of zein-egg white derived peptides (EWDP)-chitosan ternary nanoparticles for encapsulation of curcumin: Role of EWDP. *Food Chem.*, 372, 131266, 2022.
9. Buyukkestelli, H.I. and El, S.N., Development and characterization of double emulsion to encapsulate iron. *J. Food Eng.*, 263, 446, 2019.

10. Marcillo-Parra, V., Tupuna-Yerovi, D.S., González, Z., Ruales, J., Encapsulation of bioactive compounds from fruit and vegetable by-products for food application–A review. *Trends Food Sci. Technol.*, 116, 11, 2021.
11. Bazana, M.T., Codevilla, C.F., de Menezes, C.R., Nanoencapsulation of bioactive compounds: Challenges and perspectives. *Curr. Opin. Food Sci.*, 26, 47, 2019.
12. Reis, C.P., Neufeld, R.J., Ribeiro, A.J., Veiga, F., Nanoencapsulation II. Biomedical applications and current status of peptide and protein nanoparticulate delivery systems. *Nanomedicine*, 2, 53, 2006.
13. Rezaei, A., Fathi, M., Jafari, S.M., Nanoencapsulation of hydrophobic and low-soluble food bioactive compounds within different nanocarriers. *Food Hydrocoll.*, 88, 146, 2019.
14. Pateiro, M., Gómez, B., Munekata, P.E., Barba, F.J., Putnik, P., Kovačević, D.B., Lorenzo, J.M., Nanoencapsulation of promising bioactive compounds to improve their absorption, stability, functionality and the appearance of the final food products. *Molecules*, 26, 1547, 2021.
15. Li, H., Li, X., Ji, J., Mixed-charge bionanointerfaces: Opposite charges work in harmony to meet the challenges in biomedical applications. *Wiley Interdiscip. Rev. Nanomed. Nanobiotechnol.*, 12, e1600, 2020.
16. Gagner, J.E., Shrivastava, S., Qian, X., Dordick, J.S., Siegel, R.W., Engineering nanomaterials for biomedical applications requires understanding the nano-bio interface: A perspective. *J. Phys. Chem. Lett.*, 3, 3149, 2012.
17. Desai, K.G.H. and Jin Park, H., Recent developments in microencapsulation of food ingredients. *Dry. Technol.*, 23, 1361, 2005.
18. Pimentel-Moral, S., Verardo, V., Robert, P., Segura-Carretero, A., Martínez-Férez, A., Nanoencapsulation strategies applied to maximize target delivery of intact polyphenols, in: *Nanotechnology in the Agri-Food Industry*, A.M. Grumezescu (Ed.), pp. 559–595, Academic Press, United States, 2016.
19. Foroughi-Dahr, M., Mostoufi, N., Sotudeh-Gharebagh, R., Chaouki, J., Particle coating in fluidized beds, in: *Reference Module in Chemistry, Molecular Sciences and Chemical Engineering*, École Polytechnique de Montréal, Montréal, QC, Canada, 2017.
20. Wu, C., Yang, H., He, X., Hu, C., Yang, L., Li, H., Principle, development, application design and prospect of fluidized bed heat exchange technology: Comprehensive review. *Renew. Sustain. Energy Rev.*, 157, 112023, 2022.
21. Đorđević, V., Balanč, B., Belščak-Cvitanović, A., Lević, S., Kalušević, A., Kostić, I., Nedović, V., Trends in encapsulation technologies for delivery of food bioactive compounds. *Food Eng. Rev.*, 7, 452, 2015.
22. Frey, C., Fluid bed coating-based microencapsulation, in: *Microencapsulation in the Food Industry*, R. Sobel (Ed.), pp. 83–115, Academic Press, United States, 2023.
23. Oehme, A., Valotis, A., Krammer, G., Zimmermann, I., Schreier, P., Preparation and characterization of shellac-coated anthocyanin pectin beads as dietary colonic delivery system. *Mol. Nutr. Food Res.*, 55, S75, 2011.

24. Soh, S.H. and Lee, L.Y., Microencapsulation and nanoencapsulation using super-critical fluid (SCF) techniques. *Pharmaceutics*, 11, 21, 2019.
25. Sauer, D., Cerea, M., DiNunzio, J., McGinity, J., Dry powder coating of pharmaceuticals: A review. *Int. J. Pharm.*, 457, 488, 2013.
26. Niu, F., Haslam, J., Rajewski, R., Subramaniam, B., A fluidized-bed coating technology using near-critical carbon dioxide as fluidizing and drying medium. *J. Supercrit. Fluids*, 66, 315, 2012.
27. Matos, R.L., Lu, T., McConville, C., Leeke, G., Ingram, A., Analysis of curcumin precipitation and coating on lactose by the integrated supercritical antisolvent-fluidized bed process. *J. Supercrit. Fluids*, 141, 143, 2018.
28. Moncalvo, F., Martinez-Espinoza, M.I., Cellesi, F., Nanosized delivery systems for therapeutic proteins: Clinically validated technologies and advanced development strategies. *Front. Bioeng. Biotechnol.*, 14, 89, 2020.
29. Craik, D.J., Fairlie, D.P., Liras, S., Price, D., The future of peptide-based drugs. *Chem. Biol. Drug Des.*, 97, 366, 2013.
30. Thakur, N., Raigond, P., Singh, Y., Mishra, T., Singh, B., Lal, M.K., Dutt, S., Recent updates on bioaccessibility of phytonutrients. *Trends. Food Sci. Technol.*, 97, 366, 2020.
31. Nel, A.E., Mädler, L., Velegol, D., Xia, T., Hoek, E., Somasundaran, P., Thompson, M., Understanding biophysicochemical interactions at the nano–bio interface. *Nat. Mater.*, 8, 543, 2009.
32. Wongrattanakamon, P., Yooin, W., Sirithunyalug, B., Nimmanpipug, P., Jiranusornkul, S., Tentative peptide–lipid bilayer models elucidating molecular behaviors and interactions driving passive cellular uptake of collagen-derived small peptides. *Molecules*, 26, 710, 2021.
33. Moradi, S., Nowroozi, A., Shahlaei, M., Shedding light on the structural properties of lipid bilayers using molecular dynamics simulation: A review study. *RSC Adv.*, 9, 4644, 2019.
34. Machtakova, M., Wirsching, S., Gehring, S., Landfester, K., Thérien-Aubin, H., Controlling the semi-permeability of protein nano capsules influences the cellular response to macromolecular payloads. *J. Mater. Chem. B*, 9, 8389, 2021.
35. Hadjiev, N.A. and Amsden, B.G., An assessment of the ability of the obstruction-scaling model to estimate solute diffusion coefficients in hydrogels. *J. Controll. Release*, 199, 10, 2015.
36. Ranjbar, M., Khazaeli, P., Pardakhty, A., Tahamipour, B., Amanatfard, A., Preparation of polyacrylamide/polylactic acid co-assembled core/shell nanofibers as designed beads for dapsone *in vitro* efficient delivery. *Artif. Cells Nanomed. Biotechnol.*, 47, 917, 2019.
37. Piacentini, E. and Drioli, E., Encapsulation efficiency, in: *Encyclopedia of Membranes*, L. Giorno (Ed.), pp. 706–707, Springer, Berlin, 2016.
38. Latorres, J.M., Aquino, S., da Rocha, M., Wasielesky Jr., W., Martins, V.G., Prentice, C., Nanoencapsulation of white shrimp peptides in liposomes:

Characterization, stability, and influence on bioactive properties. *J. Food Process. Preserv.*, 45, e15591, 2021.

39. Haggag, Y.A., Matchett, K.B., Falconer, R.A., Isreb, M., Jones, J., Faheem, A., McCarron, P., El-Tanani, M., Novel Ran-RCC1 inhibitory peptide-loaded nanoparticles have anti-cancer efficacy *in vitro* and *in vivo*. *Cancers*, 11, 222, 2019.
40. Cheng, H., Zhang, X., Qin, L., Huo, Y., Cui, Z., Liu, C., Mao, S., Design of self-polymerized insulin loaded poly (n-butylcyanoacrylate) nanoparticles for tunable oral delivery. *J. Controll. Release*, 321, 64, 2020.
41. Dua, K., Bebawy, M., Awasthi, R., Tekade, R.K., Tekade, M., Gupta, G., Hansbro, P.M., Application of chitosan and its derivatives in nanocarrier based pulmonary drug delivery systems. *Pharm. Nanotechnol.*, 5, 243, 2017.
42. Shanmugam, H., Rengarajan, C., Nataraj, S., Sharma, A., Interactions of plant food bioactives-loaded nano delivery systems at the nano-bio interface and its pharmacokinetics: An overview. *Food Front.*, 3, 256, 2022.
43. Inchaurraga, L., Martínez-López, A.L., Martin-Arbella, N., Irache, J.M., Zein-based nanoparticles for the oral delivery of insulin. *Drug Deliv. Transl. Res.*, 10, 1601, 2020.
44. Banun, V.J., Rewatkar, P., Chaudhary, Z., Qu, Z., Janjua, T., Patil, A., Popat, A., Protein nanoparticles for enhanced oral delivery of coenzyme-Q10: *In vitro* and *in silico* studies. *ACS Biomater. Sci. Eng.*, 6, 2846–2856, 2021.
45. Erel-Akbaba, G., Akbaba, H., Keselik, E., Bahceci, S.A., Senyigit, Z., Temiz, T.K., Octaarginine functionalized nanoencapsulated system: *In vitro* and *in vivo* evaluation of bFGF loaded formulation for wound healing. *J. Drug Deliv. Sci. Technol.*, 71, 103343, 2022.
46. Hong, S., Dia, V.P., Baek, S.J., Zhong, Q., Nanoencapsulation of apigenin with whey protein isolate: Physicochemical properties, *in vitro* activity against colorectal cancer cells, and bioavailability. *LWT*, 154, 112751, 2022.

2

Smart Nanomaterials for Antiseptic Application

Sribharani Sekar[1], Jayaraman Pitchaimani[2] and A. Tamilselvi[1]*

[1]PG & Research Department of Chemistry, Thiagarajar College, Madurai, India
[2]Department of Chemistry, Sri Sai Ram Institute of Technology, West Tambaram, Chennai, India

Abstract

Skin infections are the most common in humans or animals mainly caused by infectious bacterial, fungal, and other pathogens. Also, humans and animals suffer from many dermatological diseases such as impetigo, abscess, and staph infection. Frequent cleaning of the skin with antiseptic soap or shampoo/lotion is recommended to suppress the spread of microbes. There are several biocides were used in hospitals to prevent nosocomial infections. Besides, the quotidian use of antiseptics in our daily routine. But we know that not all microorganisms are resistant to these antiseptic agents. Hence, there is a huge urge for a broad range of studies on novel antiseptics. Over the past few decades, nanotechnology has delivered a great effect on public use in various aspects. Unlike bulk materials, nanomaterials have a high surface-to-volume ratio, reflecting a unique action in fighting against harmful microbes. Due to the broad and potential use of nanomaterials, they have been the subject of significant investigation in various fields like medicine, food, and surface. Hence, we have discussed the formulation and the important application of nanomaterials as antiseptic agents. Especially, the role of metal, metal oxides, and non-metallic nanoparticle and their antiseptic property against infectious microorganisms.

***Keywords*:** Nanomaterials, antiseptic agent, metallic, non-metallic and metal-oxide nanoparticles, antimicrobes, mechanistic action

**Corresponding author*: tamilselvi_chem@tcarts.in

Sathish-Kumar Kamaraj, Arun Thirumurugan, Muthuchamy Maruthupandy, Mercedes Guadalupe López Pérez and Shanmuga Sundar Dhanabalan (eds.) Nanoscience and Nanotechnology for Smart Prevention, Diagnostics and Therapeutics: Fundamentals to Applications, (23–34) © 2024 Scrivener Publishing LLC

2.1 Introduction

The discovery of potential vaccines after World War II lead to a considerable decline in infectious diseases, such as plague, cholera, diphtheria, and measles. However, the onset of AIDS in 1980s, severe acute respiratory syndrome (SARS) in 2000, and the recent pandemic COVID 19 (a variant of SARS) alarms us that we need to overlook to infectious diseases as they cause millions of death worldwide. Most often, these diseases are caused by microbial pathogens, such as bacteria, viruses, fungi, or parasites. Antimicrobial agents play a key role against these pathogens and are classified as i) antibiotics, ii) antiviral, iii) antifungal, and iv) antiseptic. While the others are administrated orally, an antiseptic agent or biocide is a substance that are applied topically on living tissues like injured skin to inhibit the growth of harmful pathogens without killing them, thus preventing the spread of infectious diseases. Boric acid, iodoform, and Dettol (a mixture of chloroxylenol and terpineol) are a few examples of good antiseptics. However, the use of these antiseptics at physiological concentrations resulted in adverse side effects with notable toxicity [1]. To overcome this, several strategies like synthesizing antimicrobial substances from natural sources, and improvisation of existing antiseptic agents were carried out.

Nanomaterials, being small in size, well synchronize with cellular systems [2]. The revolution in nanotechnology has unveiled many nanosized antiseptic agents by unlocking new nanomaterials and processes at the nanoscale. Due to the high surface area to volume ratio, the nanomaterials exhibit improved chemical and antimicrobial activity as compared to other materials [3]. Nanomaterials can be classified into i) inorganic nanomaterials (Au, Ag, Cu, Zn, and Fe), ii) organic nanomaterials (dendrimers, cyclodextrin), iii) composite-based nanomaterials (Cu-doped TiO_2 NPs), and iv) carbon nanomaterials (fullerene, single, and multi-walled carbon nanotube). This chapter discusses the role of metal, metal oxide, and non-metallic nanoparticles as antiseptic agents. The mechanistic actions of nanoparticles against microbes are also discussed.

2.2 Metallic Nanoparticles

2.2.1 Gold Nanoparticles

Gold has been a fascinating material since its discovery in 4th century [4]. Colloidal gold is very well known for its curative property for various diseases like small box, measles and skin ulcer [2]. Well-controllable geometry, ease of surface functionalization with biomolecules, biocompatibility, size-dependent properties, surface plasmon resonance [5], facile synthesis, and low toxicity make them to hold top in the demand list. Gold NPs were used as antiseptics in diverse fields, few of them are listed below.

(i) ***Cleansing of contact lens:*** As mentioned earlier, the high surface area to volume ratio of NPs accommodates more functional groups [3], thereby enhancing their activity against pathogens. For example, the topical use of antiseptic agents alone, such as chlorhexidine digluconate, pentamidine isethionate on contact lens at recommended concentrations had resulted side effects in patients with mature cataracts and amoebic infections by free-living pathogen acanthamoeba with toxicity rate of 40%. But after the functionalization of chlorhexidine digluconate with gold nanoparticles exhibited enhanced anti-acanthamoeba effects with reduced toxicity of 60%. This is due to the fact that the NPs surface allows the incorporation of more antiseptic agents and thereby enabling multivalent interaction against amoeba viability as compared to free antiseptic molecules [1].

(ii) ***Skin care:*** Skin acts as a defense against pathogens. But when the skin is diseased like wounded, NPs loaded with drug can be penetrated through lesions and hair follicle openings. Reports showed that the NPs enter into the lipid matrix of the skin and slowly released the drug to achieve efficacy against pathogens. Figure 2.1 shows the permeation of NPs into skin [6].

Psoriasis: Psoriasis is a chronic skin disorder resulted by inflammatory factors with increased production of cytokines, often characterized by red patches. TNFα and IL-12 are the key inflammatory mediators in human psoriasis. The topical application of Cornus mas complexed Ag and Au NPs based ointment on patients exhibited that the NPs, especially Au NPs—being smaller in size than Ag NPs allowed easy penetration and significantly reduced IL-12 and TNFα cells by specific inhibition of NF-κB activity in psoriatic skin lesions [7].

Dermatitis: Inflammatory skin disease (ISD) or dermatitis is a serious social problem faced by the people as a consequence of inflammatory reaction in skin triggered by pollution, climate change. This includes atopic/eczema, allergic, contact dermatitis often characterized by redness,

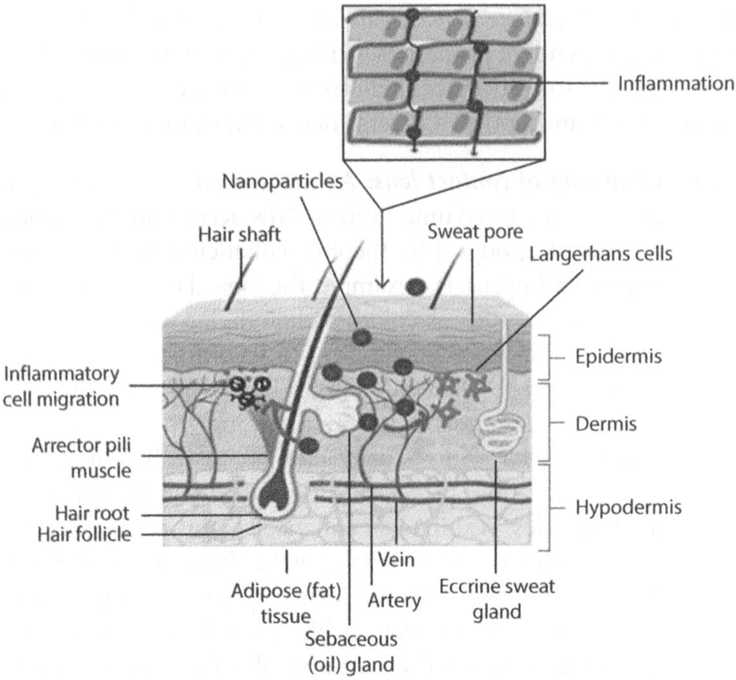

Figure 2.1 Permeation of nanoparticles into skin [6].

itchiness, and rash. Recent study show that EJ conjugated gold NPs serves as an anti-dermatitis candidate and successfully suppressed the skin-dermatitis in human by inhibiting the T+I-stimulated inflammatory chemokines (RANTES, TARC, and IL-8) and intracellular reactive oxygen species (ROS). The underlying mechanism of this inhibition effect is associated with the downregulation of the signaling pathways, such as mitogen-activated protein kinase and NF-κB in human epidermal keratinocytes (HaCaT cells) [8].

Skin cancer or Melanoma: Melanoma is a serious health problem. Common treatment methods include surgery, radio, and chemotherapy cause adverse side effects on patients. Topical administration of small molecular drugs do not possess specific target to cancer cells. Thanks to nanotechnology for designing nanoparticles whose surface readily conjugates with drug molecules and provides specific target to the cancer cells. For instance, 5-fluorouracil (5-FU) is a well-known anticancer drug, but this drug alone does not inhibit the tumor cells as this has poor skin permeability due to its hydrophilicity. To overcome this problem, first 5-FU was loaded into CTAB-capped Au NPs by taking the advantage of electrostatic interaction between cationic CTAB and anionic 5-FU. Next, the incorporation of 5-FU/AuNP- CTAB into vanishing cream and pluronic gel was tested against A431 tumor cells in mice. Indeed, the topical application of nano-formulated gel and cream, effectively inhibited the growth of tumor cells. This is attributed to the synergistic effect of smaller size of NPs and increase of lipophilicity (characteristic requirement for transport of drug into cell) of 5-FU by ion pairing during its complexation with CTAB increased the penetration rate of anticancer drug [9].

Wound dressing: Wound infection is caused by bacteria, such as *Staphylococcus aureus* (*S. aureus*—Gram-positive bacteria) and *E. coli* (Gram-negative bacteria). In purpose of multivalent interaction with both Gram-positive and Gram-negative bacteria, biocompatible chitosan (CS), and 2-mercapto-1-methylimidazole (MMT) were synchronously coated onto the surface of Au NPs. Blending with gelatine and crosslinking with tannic acid followed by freeze-drying fabricated CS-Au@MMT/gelatine dressing. The *in vivo* wound healing ability of this dressing was tested against *E. coli,* and *S aureus* in rabbits with dorsal wounds. After 10 days, it has been reported that the wounds treated with dressing form scabs with decrease in wound size. On day 16, the skin was rejuvenated and was similar to the normal skin [10]. In addition, the microscopic characterization showed that the dressing effectively killed *E. coli and S. aureus* by exhibiting red fluorescence—a characteristic of dead bacteria through the synergistic effect of Au NPs, MMT, and CS.

2.2.2 Silver Nanoparticles

Silver's toxicity to human cells is significantly lower than that of bacteria, it has long been recognized that it has antibacterial properties and has been used in many different applications. A wide range of antibacterial, antifungal, and antiviral activities can be found in silver nanoparticles. Invading bacterial cell walls with silver nanoparticles can alter the structure of cell membranes and even cause cell death. Their effectiveness is a result of both their enormous surface area-to-volume ratio and their nanoscale size [11]. Hundreds of harmful microbes are destroyed by silver ions, while the local microflora is unaffected. Silver nanoparticles, which have lyophobic qualities and are stabilized by a variety of mostly organic molecules, are frequently employed as hydrosols in practical applications. The silver nanoparticle-coated paper plays a crucial role in food preservation in addition to the uses of silver nanoparticles in the fields of medicine and environmental protection. This is because it acts as a reservoir for the slow release of ionic silver from the surface to the bulk to prevent microbial growth in the food, as well as the growth of pathogens on the surface. Because silver nanoparticles have such potent antimicrobial properties, the paint coating industry has shown a lot of interest in creating antibacterial coatings for surfaces to safeguard both human health and the environment.

2.2.2.1 Various Silver-Based Nanomaterials for Antiseptic Application

Seabuckthorn (SBT) extracts can be used to make silver nanoparticles that are endowed with natural biocides quickly and easily [12]. This approach made it possible to create silver nanoparticles (SBT@AgNPs) that were supported by paper and included biocides. Evaluation of the S0000000BT@AgNPs' antibacterial capabilities revealed significant suppression of both Gram-positive and Gram-negative bacterial strains, with antibacterial activity remaining above 99% even after a year in storage. Studies on the antioxidant characteristics revealed that SBT@AgNPs were effective at scavenging DPPH radicals. Overall, the Seabuckthorn biocides performed better when encapsulated in SBT@AgNPs in terms of their antibacterial and antioxidant properties. The authors assert that compared to two commercially available medicated dressing materials, the created substance is more stable, highly antiseptic, and effective against bacteria. Additionally, SBT@AgNPs may be made quickly and cheaply using a process that has great reproducibility and stability, and it can be tailored to work with a number of various surfaces and materials [12].

To generate silica nanoparticles with silver at their core and crust (SiNP-AgCC) around a template of silver nanoparticles suggested by Haider *et al.* single-pot protocol. $AgNO_3$ was used to create the silver nanoparticles, which had average diameters between 10 and 20 nm. A Stöber modified approach was then used to add TEOS. The silver nanoparticles are encased in the silica matrix that results from the condensation of TEOS in this manner. The residual silver ions in the solution are then reduced and deposited on the surface of the silica matrix by the addition of formalin to the process. Because the silica scaffold stops silver from leaking from the interior, this approach allows for the selective reduction of only the silver ions on the surface. The surface-formed silver nanoparticles had an average diameter of 2 to 5 nm. The results of the antibacterial tests showed that the substance had a high level of activity against bacterial colonies because those colonies were quickly eliminated upon initial contact with SiNP–AgCC. Being able to exhibit bactericidal capabilities even at lower concentrations speaks to the material's durability. The scientists suggested that SiNP-AgCC can be used in polymer nanocomposite water treatment membranes and for biomedical purposes due to the material's biocompatibility and ease of manufacture [13].

Silver nanoparticles (AgNPs, with average diameters ranging from 1 to 3 nm) were created (Lara *et al.*) using a microwave-assisted method, and their effectiveness against *Candida auris* biofilms—a well-known pathogenic fungus that can survive for several days attached to various surfaces—was examined. A powerful inhibitory capacity against the fungus, the development of biofilms, and even against pre-formed biofilms was discovered during antibiotic experiments on *C. auris*. The microbes could be killed by the nanoparticles by damaging their cell walls, according to a thorough imaging investigation. The authors used AgNPs to clean a number of medical surfaces, and they came to some incredibly encouraging discoveries. Silicone elastomers and bandage dressings treated with the substance demonstrated effective preventative action against bacterial biofilm development and contamination. The same materials were also repeatedly and completely washed in a phosphate buffer solution, and they continued to exhibit fungicidal activity against *C. auris* for up to six cycles of washing. The authors' substance is efficient for treating medical-based surfaces in a preventative manner [14].

A substance that combines silver and kaolin appears to hold promise for application in animal feed as an antibacterial. The *in vitro* genotoxicity experiments performed, which examined the silver-kaolin combination, produced unambiguously negative results. The quantities examined in the assays ranged from 0.083 to 27.6 g/mL when just the AgNPs content of the silver–kaolin formulation material was taken into consideration. The negative outcomes could be explained by the relatively low AgNP concentrations that were in direct contact with the cells. However, some of the studies analyzed by Rodrguez-Garraus and colleagues that did produce favorable results tested these amounts within the ranges that did. This may imply, as do other studies using related materials, that AgNPs' attachment to kaolin reduces their genotoxicity. It should be emphasized that the substance in the mouse lymphoma assay (MLA) and micronucleus (MN) tests were not entirely removed after washing; hence, the significance of the negative results is increased. The material was still present in all subsequent incubation phases. The silver–kaolin formulation also demonstrated its *in vitro* antimicrobial activity at concentrations much lower than those examined in our studies, with minimum inhibitory concentrations of 3.9 to 15.6 μg/mL and minimum bactericidal concentrations of 7.8 to 250 μg/mL, respectively, against a number of bacterial strains. In the *in vitro* genotoxicity studies, the test substance did not result in gene mutations, chromosomal abnormalities, or DNA damage (i.e., SBs, ALS, or oxidized bases) [15].

$Al_2(SO_4)_3.18H_2O$ was used in a hydrothermal procedure to create mesoporous alumina microspheres, known as porous aluminum oxide (POA), which were then thermally decomposed. Silver nanoparticle sorption in the POA matrix produced silver nanocomposites of POA (Ag/POAs) with significant biological activity. Silver nanoparticles are produced utilizing photochemical reduction during the synthesis of Ag/POAs, preventing the agglomeration of these particles. POA was shown to have less bactericidal activity than Ag/POAs. *Escherichia coli*'s capacity to form colonies was completely destroyed in 1 day on Ag/POAs at a concentration of 0.241 ppm silver nanoparticles. At all concentrations of silver nanoparticles, *Staphylococcus epidermidis* showed better tolerance, and its development was halted at a concentration of 0.374 ppm. The bactericidal action of Ag/POAs on bacteria in drinking water was discovered to be quite successful; at a concentration of 0.108 ppm silver nanoparticles, the growth of bacteria was totally lost in 1 day [16].

In order to create AgNPs, ZnONPs, and ZnO-AgNPs, Bidens pilosa's seed, leaf, and root extracts were used by Kyomuhimbo *et al.* Although the temperature had an impact on nanoparticle yield, the source of a certain plant extract source-root, leaf, or seed—could not be identified. The growth of *E. coli*, *S. aureus*, and *C. albicans* was inhibited in the medium treated with AgNPs, ZnONPs, and ZnO–AgNPs. Regardless of the extract employed, ZnO–AgNPs with a composition of 0.8 Ag/0.2 ZnO exhibited the best antibacterial activity. The ZnO–AgNPs made from the root extract had the lowest minimal inhibition concentration of the three extracts. In order to create a hand-sanitizing antiseptic, the active ingredient was the root extract ZnO–AgNPs (0.2:0.8). Strong antibacterial activity was displayed by the newly synthesized antiseptic. The hand antiseptic showed antibacterial efficacy against 14 mm *E. coli*, 20.5 mm *S. aureas*, and 24.5 mm *C. albicans* in inhibition zones [17].

2.2.3 Nonmetallic Nanomaterials as Antiseptic

Single-wall carbon nanotubes (SWCNTs) with a water-based povidone-iodine (PVPI) complex have been developed as nanotextured bandage with sustained release of antiseptic iodine by Simmons *et al.* [18]. This nanomaterial has been shown to have the antiseptic properties against *E. coli*. Moreover, the PVPI has been found to be low-toxic; hence, the electric pulse sent through the material helps quick healing and recovery of the damaged tissues.

Brooker *et al.* have designed anti-inflammatory fibrous sheets to treat Psoriasis and atopic dermatitis [19]. The synthesized fibrous sheet was composed of crosslinked poly(propylene sulphide) (PPS) encapsulated with polyethylene oxide (PEO) fibers. PPS-NP/PEO fibers are an example of soft organic nanoparticle obtained by electrospinning. It was observed that the antioxidant and anti-inflammatory property of the PPS was retained after encapsulation. Moreover, the fibrous sheet was observed to have great cytocompatibility.

2.2.4 Ionic Systems as Antiseptics

Ionic systems, such as ionic liquids and ionic surfactants, have been observed as antiseptics. Surfactants having the hydrophobic and hydrophilic ends, well-known surface active compounds, have been investigated to possess the antiseptic properties. In 2018, Marek *et al.* have shown that the pyridine-based cationic surfactants are effective against gram-positive and gram-negative bacteria. They have concluded that the hydroxyl methyl group on the pyridinium ion was not playing any role on its antimicrobial activity.

In 2012, Venugopalan *et al.* have synthesized 1-alky-3-methylimidazolium ions and shown this ionic liquid to have the antimicrobial and antibiofilm behavior to a cationic surfactant [20]. Ionic liquids, particularly, quaternary ammonium salts (QAS) are proven to have the antiseptic properties.

In 2021, Farisca *et al.* critically reviewed that the ionic systems and nanomaterials showing the antiseptic and disinfectant properties [21]. In this review, the authors concluded that the combination of surfactants with the nanomaterials would be more effective active surface agents that could be better antiseptic and disinfectant.

In 2017, Zein El Abedin and his co-workers have studied the physicochemical properties of organic and inorganic electrolytes on surface properties of C12 and C16 alcohol-based bis-sulfosuccinate anionic gemini surfactants [22].

2.3 Mechanism of Antimicrobial Action

The exact mechanism of antimicrobial action of nanoparticles was not completely understood but studies suggest that, first NPs are penetrated/adsorbed at the microbial cell surface through their functional group interaction with microbes. As a result of this interaction, the metabolic process is reduced by reducing the adenosine triphosphate (ATP) synthase activities. In next step, they declined the tRNA binding with ribosome followed by DNA and protein damage cause inhibition of various pathogens [23]. Scheme below (Figure 2.2) portraits the general antimicrobial mechanism of nanoparticles against bacteria, fungi [24].

NPs disturb the microbial cell wall protein and cytoplasm through their enormous interaction with microbes owing to their high surface area to volume ratio. In the case of Au NPs, its binding with thiol groups of nicotinamide adenine denucleotide (NADH) dehydrogenases produces oxidative stress by the release of ROS, which eventually led to cell death. Tunable size of NPs offers different doses of administration depending on the thickness of the microbial cell wall. Greater efficacy was observed with decrease in average size of NPs as the penetration can be made easily.

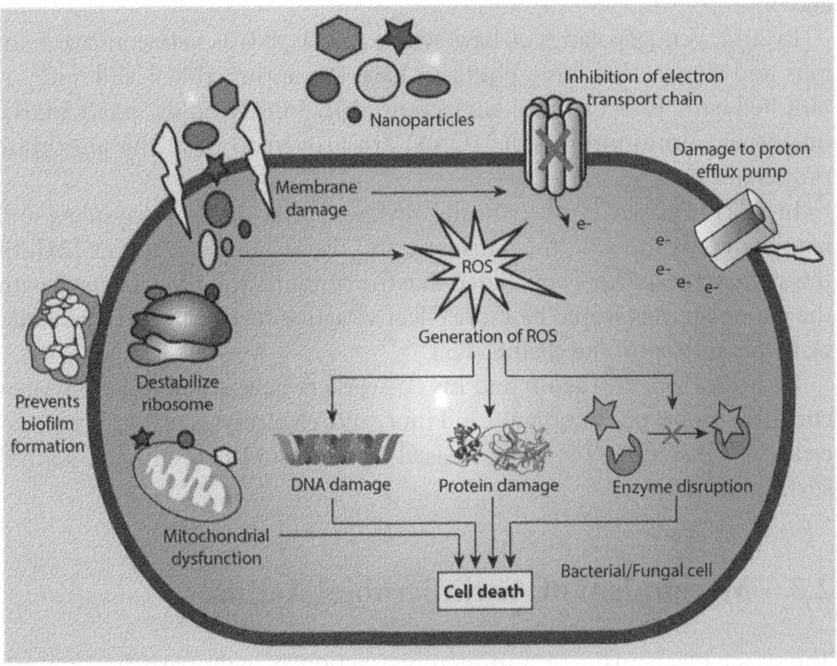

Figure 2.2 Schematic representation of antimicrobial mechanism of nanoparticles against microbes [24].

References

1. Aqeel, Y., Siddiqui, R., Anwar, A., Shah, M.R., Khan, N.A., Gold Nanoparticle Conjugation Enhances the Antiacanthamoebic Effects of Chlorhexidine. *Antimicrob. Agents Chemother.*, 60, 1283–8, 2015.
2. Yeh, Y.C., Creran, B., Rotello, V.M., Gold nanoparticles: preparation, properties, and applications in bionanotechnology. *Nanoscale.*, 4, 1871–80, 2012.
3. Azam, A., Ahmed, F., Arshi, N., Chaman, M., Naqvi, A., One step synthesis and characterization of gold nanoparticles and their antibacterial activities against E. coli (ATCC 25922 strain). *Int. J. Theor. Appl. Sci.*, 1, 1–4, 2009.
4. Sangeetha, G., Usha, N., Nandhini, R., Kaviya, P., Vidhya, G., Chaithanya, B., A Review on Properties, Applications and Toxicities of Metal Nanoparticles. *Int . J. App. Pharm.*, 12, 5, 58–63, 2020.
5. Link, S. and El-Sayed, M.A., Spectral Properties and Relaxation Dynamics of Surface Plasmon Electronic Oscillations in Gold and Silver Nanodots and Nanorods. *J. Phys. Chem. B*, 103, 8410–26, 1999.
6. Kakkar, V., Kumar, M., Saini, K., An Overview of Atopic Dermatitis with a Focus on Nano-Interventions. *Eur. Med. J.*, 3, 44–54, 2018.
7. Crisan, D., Scharffetter-Kochanek, K., Crisan, M., Schatz, S., Hainzl, A., Olenic, L. *et al.*, Topical silver and gold nanoparticles complexed with Cornus mas suppress inflammation in human psoriasis plaques by inhibiting NF-κB activity. *Exp. Dermatol.*, 27, 1166–9, 2018.
8. Xu, X.Y., Moon, S.-K., Kim, J.-K., Kim, W.J., Kim, Y.-J., Kim, H., Structural properties and anti-dermatitis effects of flavonoids-loaded gold nanoparticles prepared by Eupatorium japonicum. *Front. Pharmacol.*, 13, 1–16, 2022.
9. Safwat, M.A., Soliman, G.M., Sayed, D., Attia, M.A., Fluorouracil-Loaded Gold Nanoparticles for the Treatment of Skin Cancer: Development, *in Vitro* Characterization, and *in Vivo* Evaluation in a Mouse Skin Cancer Xenograft Model. *Mol. Pharm.*, 15, 2194–205, 2018.
10. Lu, B., Ye, H., Shang, S., Xiong, Q., Yu, K., Li, Q. *et al.*, Novel wound dressing with chitosan gold nanoparticles capped with a small molecule for effective treatment of multiantibiotic-resistant bacterial infections. *Nanotechnology*, 29, 425603, 2018.
11. Yin, I., Zhang, J., Zhao, I., Mei, M.L., Li, Q., Chu, C.-H., The Antibacterial Mechanism of Silver Nanoparticles and Its Application in Dentistry. *Int. J. Nanomed.*, 15, 2555–62, 2020.
12. Kalaiyarasan, T., Bharti, V.K., Chaurasia, O.P., Correction: One pot green preparation of Seabuckthorn silver nanoparticles (SBT@AgNPs) featuring high stability and longevity, antibacterial, antioxidant potential: a nano disinfectant future perspective. *RSC Advances.*, 10, 42961–3, 2020.
13. Haider, M.S., Shao, G., Ahmad, A., Imran, S.M., Abbas, N., Abbas, G. *et al.*, Facile, single-pot preparation of nanoporous SiO2 particles (carrier) with AgNPs at core and crust for controlled disinfectant release. *J. Saudi Chem. Soc.*, 23, 828–35, 2019.

14. Lara, H.H., Ixtepan-Turrent, L., Yacaman, M.J., Lopez-Ribot, J., Inhibition of Candida auris Biofilm Formation on Medical and Environmental Surfaces by Silver Nanoparticles. *ACS Appl. Mater. Interfaces*, 12, 21183–91, 2020.
15. Rodriguez-Garraus, A., Azqueta, A., Laborda, F., Gimenez-Ingalaturre, A.C., Ezquerra, A., Lostao, L. *et al.*, *In Vitro* Genotoxicity Evaluation of an Antiseptic Formulation Containing Kaolin and Silver Nanoparticles. *Nanomaterials (Basel, Switzerland)*, 12, 6, 914–931, 2022.
16. Gorbunova, M., Lemkina, L., Lebedeva, I., Kisel'kov, D., Chekanova, L., Synthesis and potential applications of silver—porous aluminium oxide nanocomposites as prospective antiseptics and bactericides. *J. Mater. Sci.: Mater. Med.*, 28, 40, 1–8, 2017.
17. Kyomuhimbo, H.D., Michira, I.N., Mwaura, F.B., Derese, S., Feleni, U., Iwuoha, E.I., Silver–zinc oxide nanocomposite antiseptic from the extract of Bidens pilosa. *SN Appl. Sci.*, 1, 681, 2019.
18. Simmons, T.J., Lee, S.H., Park, T.J., Hashim, D.P., Ajayan, P.M., Linhardt, R.J., Antiseptic single wall carbon nanotube bandages. *Carbon*, 47, 1561–4, 2009.
19. Brooker, C., d'Arcy, R., Mele, E., Willcock, H., Designing responsive dressings for inflammatory skin disorders; encapsulating antioxidant nanoparticles into biocompatible electrospun fibres. *Soft Matter.*, 17, 3775–83, 2021.
20. Venkata Nancharaiah, Y., Reddy, G.K.K., Lalithamanasa, P., Venugopalan, V.P., The ionic liquid 1-alkyl-3-methylimidazolium demonstrates comparable antimicrobial and antibiofilm behavior to a cationic surfactant. *Biofouling*, 28, 1141–9, 2012.
21. Faísca, F., Filipe, L., Petrovski, Z., Santos, M.M., Gago, S., Branco, L.C., Ionic Systems and Nanomaterials as Antiseptic and Disinfectant Agents for Surface Applications: A Review. *Surfaces*, 4, 169–90, 2021.
22. El-Shamy, A.M., Zakaria, K., Abbas, M.A., Zein El Abedin, S., Anti-bacterial and anti-corrosion effects of the ionic liquid 1-butyl-1-methylpyrrolidinium trifluoromethylsulfonate. *J. Mol. Liq.*, 211, 363–9, 2015.
23. Shamaila, S., Zafar, N., Riaz, S., Sharif, R., Nazir, J., Naseem, S., Gold Nanoparticles: An Efficient Antimicrobial Agent against Enteric Bacterial Human Pathogen. *Nanomaterials (Basel, Switzerland)*, 6, 4, 71–80, 2016.
24. Sharmin, S., Rahaman, M.M., Sarkar, C., Atolani, O., Islam, M.T., Adeyemi, O.S., Nanoparticles as antimicrobial and antiviral agents: A literature-based perspective study. *Heliyon*, 7, e06456, 2021.

3

Surface Plasmon-Based Diagnostic Technology

Sopan Nangare, Mahendra Mahajan and Pravin Patil*

Department of Pharmaceutical Chemistry, H. R. Patel Institute of Pharmaceutical Education and Research, Shirpur, Dist: Dhule (Maharashtra State), India

Abstract

Despite significant progress in whole domains, preliminary stage evaluation of health-related problems remains a major challenge for professionals. In the health care line of work, *in vitro* diagnostic tools are essential for early and precise diagnosis. Herein, the measurement of biomarker concentration of respective diseases is the primary method for identifying health conditions. In this shade, various varieties of biosensors have been unveiled with numerous benefits for diagnosis. Among the several types of biosensors, surface plasmon resonance (SPR)-guided optical biosensors have garnered a lot of emphasis for recognizing target biomarkers. Principally, it has assorted advantages, including high sensitivity, low cost, high accuracy, label-free detection, faster recognition, etc. Therefore, we have covered SPR-based diagnostic technology, including the principle and sorts of SPR-based biosensors. Then, advanced nanomaterials, such as carbon and metal-based nanomaterials for the fabrication of surface SPR biosensors, are also reported. The relevance of biotransducers including bioreceptor types and the principle of bioconjugates in the design of SPR biosensors are highlighted. Furthermore, applications of SPR-based biosensors for *in vitro* diagnosis of health concerns have been demonstrated. Along with this, the present obstacles and opportunities in SPR-based detection are also described. The combination of plasmonic metal-based nanomaterials and non-plasmonic carbon-based nanomaterials for bioreceptor immobilization leads to superior function. Furthermore, it has a reduced limit of detection and offers a broad linear concentration range for biomarkers over other traditional biosensors. As well, SPR-based diagnosis complies with the basic quality assurance requirements of sensitivity, selectivity, stability, repeatability, etc.

**Corresponding author*: rxpatilpravin@yahoo.co.in

Sathish-Kumar Kamaraj, Arun Thirumurugan, Muthuchamy Maruthupandy, Mercedes Guadalupe López Pérez and Shanmuga Sundar Dhanabalan (eds.) Nanoscience and Nanotechnology for Smart Prevention, Diagnostics and Therapeutics: Fundamentals to Applications, (35–60) © 2024 Scrivener Publishing LLC

Essentially, the use of SPR biosensors in real-time clinical sample analysis has been proven. In the future, this book chapter will provide detailed insights for aspiring researchers interested in SPR-based biosensing for *in vitro* diagnostic applications.

Keywords: Surface plasmon resonance, biosensor, *in vitro* diagnosis, nanomaterials, biotransducer, sensitivity

3.1 Introduction to Surface Plasmon-Based Diagnostic Technology

3.1.1 Concept of Surface Plasmon

Since its beginnings, the employment of surface plasmon-based plasmonic sensors has drawn tremendous emphasis in the science discipline [1]. A surface plasmon-based highly sensitive sensory system is primarily focused on changing the dielectric environment at chosen plasmonic metal and dielectric contacts [2]. Contact with an external electromagnetic field and a metal surface also caused surface electrons to collectively oscillate, a phenomenon known as surface plasmon resonance (SPR). The preference for plasmonic nanoparticles in this line is heavily accounted for in sensor development. In this, the electron density of such plasmonic nanomaterials at the interface between chosen particles and the dielectric medium allows interaction with incoming light [3]. In brief, surface plasmons (SPs) are coherent conduction electron oscillations on the surfaces of objects triggered by quantized energy. In this situation, SPs are categorized into two types: localized surface plasmons (LSPs) and propagating surface plasmon polaritons (PSPPs or SPPs). These two forms of SPs are the primary components of plasmonic sensors. Electron oscillations emerge on its surface whenever electromagnetic waves are transmitted to appropriate plasmonic materials, with the wavelength of the waves being higher than the size of the nanomaterial. Localized surface plasmons resonance (L-SPR) is a phenomenon that emerges when an arriving light's electric field interacts with surface electrons comprising collective oscillations [1]. Figure 3.1 depicts the presentation of SP-based diagnostic technology. Here, SPR-mediated sensory systems have been widely exposed for biomedical applications that mostly address the biosensing of various life-threatening health problems since their beginnings. Interestingly, the application of a cutting-edge SPR-based biosensing platform provides an ideal replacement for traditional approaches because no labeling chemicals are required. Furthermore, the application of SP sensing brings several perks such as quick response,

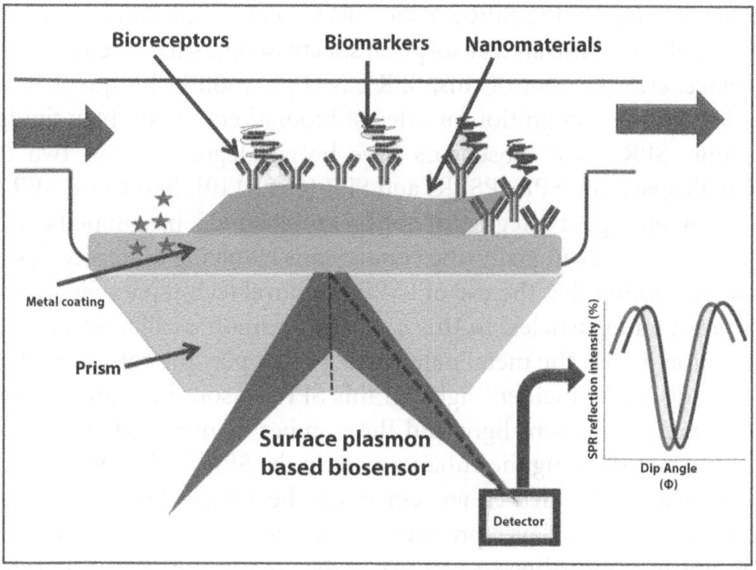

Figure 3.1 Presentation of SP-based biosensor for diagnostic applications.

limited sample demand, and the possibility to examine the putative interaction involved in molecules [4]. Furthermore, SPR-based diagnostics is simple to implement and may be employed for distant monitoring applications [1]. Notably, SPR-mediated sensors are a form of an affinity-based optical sensor. Moreover, it provides real-time observation, as well as excellent responsiveness. It offers high specificity, as target molecules attach to the modified surface of the sensor in the absence of a labeling agent. As a result, it delivers highly selective objective analyte assessment in given complex samples with the appropriate limit of detection (LOD) [5].

3.1.2 Types of SP-Based Diagnostic Technology

The biosensor is extensively employed in the biomedical area for diagnostics. Principally, the biosensor is an advanced analytics gadget. It is employed to analyze biological samples. It aids in understanding bio-composition, architecture, and activity by transforming a biochemical stimulus into an electrical pulse [6]. Out of several categories of biosensors, SPR-based biosensors are incredibly powerful instruments for biomedical applications. It can be utilized for measuring biomolecular dynamics, chemical identification, and biomarker identification [7]. The scientific community has documented many sorts of plasmon-based diagnostic techniques.

Notably, it includes the SPR sensor, surface-enhanced Raman scattering, surface-enhanced infrared absorption spectroscopy, surface-enhanced fluorescence, etc., [8]. Out of this, SPR based plasmon technique is broadly revealed for the recognition of interest biomarkers in complex fluids. At this time, SPR-based biosensors have been categorized into two types mainly Propagating SPR (PSPR) and SPR (LSPR) [9]. In outline, SPR sensors are an emergent category of optical biosensors. It has piqued the curiosity of researchers in analyzing connections involving biomolecules [10]. It is worth noting that the use of LSPR is a novel technique that associates with metal nanoparticles. In this case, the plasmon oscillation undergoes localization within the metal nanoparticles that play the role of oscillating dipole induced by incident light. In this SPR sensor, there are two major factors namely incident light and the number of nanoparticles available on the surface showing the enhancement in the SPR field [6, 9]. Moreover, sensitivity and SPR filed enhancement can be achieved using an amalgamation of metal thin layer-produced SP waves and L-SPR as compared to the P-SPR [6]. According to a survey of the literature, SPR imaging (SPRi) is increasingly being used for the recognition of particular analytes without the necessity for labeling agents. It has been noted that the employment of SPRi aids in the monitoring of the change in optical reflectivity caused by the immobilization of chosen bioreceptors on the surface of plasmonic metals [11]. Overall, SPR is a fantastic sophisticated approach for detecting clinical analytes with various objectives. In this case, it is advantageous to measure the particular attachment between the analyte and receptor using refractive index (RI) changes. As a consequence, it is advantageous to reduce the time required to achieve the result compared to the previously published approaches. As per the literature, SPR has various advantages including label-free detection and a simple and cost-effective technique. There is always the option of increasing the sensitivity and selectivity of the SPR sensor by tweaking the RI using a cutting-edge method. Furthermore, the identification of biomarkers needed a modest sample size due to the few non-specific interactions involved in the detection procedure. Overall, SPR is a superior alternative to recently employed *in vitro* diagnostic methods for medical diagnosis of several health conditions [12]. Several reports have been reported the preference of SPR based sensor for interaction analysis between ligand receptor, antibody to antigen, enzyme to substrate, etc. As well, it has been suggested for analysis of conformational alteration in protein or small molecules. Despite this merits, the preference of SPR is suffering from major limitations. Mainly, it includes the simultaneous detection of diverse target interactions. To achieve this task, SPR imaging

(SPRi) is reported by scientific fraternity. In this shade, the multiplexed assay have been design using SPRi based sensor [13].

3.2 Nanomaterials for the Design of Surface Plasmon-Based Biosensor

Innovative SPR-based identification is an optical method that may reveal chemical and physical changes that occur during the metal/dielectric interaction. This optical sensing approach relies on SPR. It has been touted as affording highly sensitive and label-free determination of a target analyte [14]. Several forms of nanomaterials that cover carbon-based nanomaterials and metal or their derivatives have been discovered to adjust the sensitivity of plasmonic sensors (Figure 3.2) [6, 9]. In particular, metal nanoparticles and their oxides have been regularly documented in SPR-based biosensors for the measurement of biomarkers since their inception [6]. It provides detection limits for analytes up to femtomolar (fM) mass concentration. Unfortunately, analytes with low molecular mass and highly diluted concentrations mainly limit their adoption in biosensing. Simply stated, there is a strong desire to employ cutting-edge technology to overcome the sensitivity barriers of SPR biosensors. The best option in SPR biosensors is the utilize nanomaterials that offer SPR signal enhancement. Interestingly, the preference of metal particles, such as silver nanoparticles (AgNPs) and gold nanoparticles (AuNPs), are an exceptional option for SPR biosensors. In addition, different types of nanomaterials such as nanorods (NRs), quantum dots (QDs), and magnetic nanoparticles have been disclosed in the sensitivity enhancement of SPR biosensors [15]. It has been reported that the utilization of metal nanoparticles over other reported nanomaterials in SPR-based sensing systems exhibited high suitability. Here, the size, assembly, structure, shape, and composition of a metal nanoparticle can be monitored to tune the desirable optical merits [16, 17]. A literature survey claimed the shape of metal nanoparticles plays an imperative function to decide the properties such optical, physical, and chemical properties [17]. As per data, AuNPs are extensively used among metal nanoparticles to strengthen SPR sensitivity. Furthermore, the synthesis of AuNPs is a simplistic process whereas it shows the potential to conjugate with biological components to create bioconjugates (biotransducer) [6, 17]. Recently, AgNPs have been discovered as a component for the creation of bioconjugates for increased sensitivity and selectivity in the measurement of target analytes. Importantly, AgNPs have responsiveness to minor adjustments in their dielectric environment. Furthermore, using an appropriate functionalizing agent, the surface of AgNPs may be chemically changed, resulting in an improvement in sensor presentation in terms of sensitivity and selectivity [16]. The preference of platinum nanoparticles and platinum-mediated nanocomposites in the creation of SPR-based sensors has been broadly documented in recent years, with platinum providing a rise in sensor sensitivity. Furthermore, the composite with platinum metal nanoparticles can afford a new option for recognition of interest analyte using SPR biosensor with numerous advantages such as fast recognition, high responsiveness to low analyte quality, high specificity in complexed samples, and larger surface area for receptor immobilization [6, 18]. In addition, various publications have discussed the usage of metal oxides in diagnostic applications. Interestingly, metal oxides may enable small size, specific surface exposure, maximum surface area to volume ratio, and other benefits that increase analyte selectivity and sensitivity [19]. Specifically, tin oxide, zinc oxide, cerium oxide, copper oxide, and other metal oxides can be employed in the construction of SPR biosensors [6]. Lately, the uses of carbon-based nanomaterials or nanostructures have been revealed for boosting the sensitivity of SPR biosensors [15]. Interestingly, it includes graphene and graphene-based nanomaterials (graphene oxide: GO), carbon nanotubes (CNTs), QDs, carbon fiber, fullerenes, etc., [6]. This carbon-based advanced nanoparticle has several benefits over previously reported nanomaterials. It offers a large surface area for functionalization, biocompatibility, a high surface area to volume (SA: V) ratio, electrical conductivity, and other benefits [20]. Literature reported that GO is now being used in the creation of SPR-based biosensors. The oxidation state of GO has a significant impact on reflection potential, which facilitates the recognition of tiny mass molecules [9]. Interestingly, as a non-plasmonic nanomaterial, GO plays a critical function almost similar to metal nanoparticles. As a result, the conjugation of plasmonic and non-plasmonic materials for the creation of SPR-based sensor systems can furnish better performance than the previously disclosed SPR sensor methods [21]. The preference for CNTs in SPR biosensor innovation is also gaining recognition in the biomedical field. Numerous varieties of CNTs for example single/double/multiwall CNTs are reported. Importantly, it has a customizable surface shape and structure, as well as a large surface area with a variety of surface functions. Furthermore, the unique physical and chemical properties of CNTs provide exceptional sensitivity and selectivity. Furthermore, surface customized CNTs are superior to the naked form of CNTs in terms of improved SPR sensor performance [18].

3.3 Biotransducers in Surface Plasmon-Based Biosensor

Bioreceptors in SPR-based sensor systems permit the selective detection of analytes of interest in provided clinical samples. As a result, the exact identification of many health-related disorders is conceivable by employing an SPR-based sensor. These receptors are found on the surface of metal chips as well as nanomaterials, depending on the intention and approach used in the SPR assay (Figure 3.2). This receptor interacts with a certain analyte, resulting in the SPR response [6]. The literature survey assured that the two types of biorecognition materials namely catalytic and affinity-based bioreceptors in SPR medicated biosensors. Here, catalytic receptors offer the biochemical reaction between analyte and receptors whereas affinity receptors illustrate a high specificity among analyte and receptors. Example of affinity receptors includes aptamers, antibody, nucleic acid, and receptors.

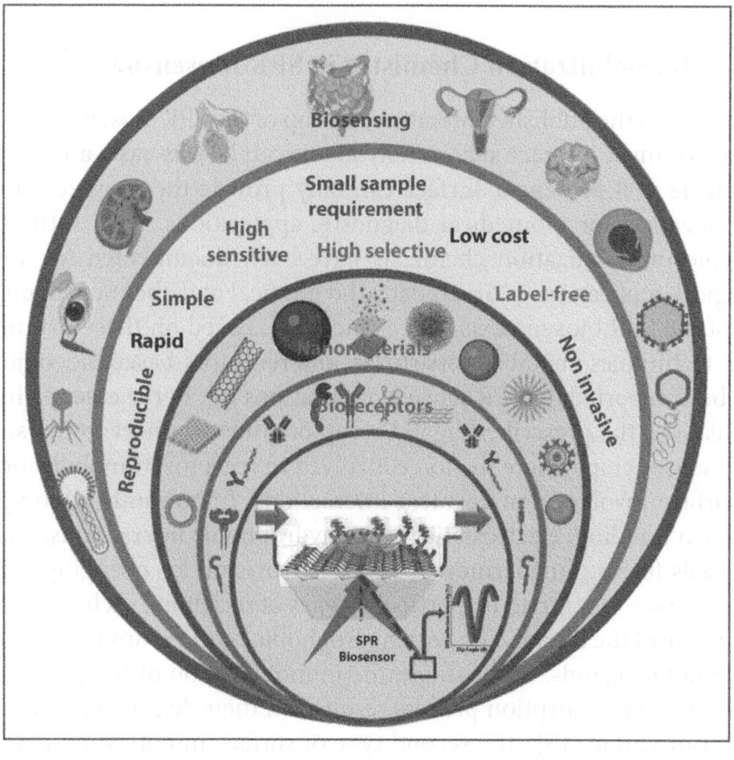

Figure 3.2 Presentation of SPR-based biosensor for *in vitro* diagnosis of different diseases/disorders.

Similarly, the catalytic receptors include the such as cells, microbes, tissues, organelles, and enzymes [20, 22]. The utilization of antibodies in the construction of SPR-based biosensors has been known for a long time among many types of bioreceptors. In essence, the antibody is prepared up of two binding sites with a 'Y' shape. In specific, antigen binding sites (Fab) of antibodies must be uncovered in case the SPR biosensor after immobilized on the sensor chip surface [23]. Interestingly, strong specificity between antigen and antibody is required, which may bestow an enhanced presentation of SPR biosensors in terms of selectivity followed by sensitivity [6, 24]. The utilization of aptamers in SPR biosensors has gotten a lot of attention as a bioreceptor for diagnostic applications [6]. A review of the literature indicates that aptamers have nearly equivalent characteristics to other bioreceptors. Importantly, it addresses the restrictions of antibodies' insufficient affinity to target [25]. Aptamers endow with binding to molecules of diverse mass and form. Overall, the application of aptamers in SPR-based biosensors delivers a great presentation regarding affinity, specificity, and sensitivity [6, 20].

3.3.1 Immobilization Chemistry in SPR Biosensor

According to the published literature, the top of the SPR biosensor at which molecules must interact successfully is critical. In this case, a robust and repeatable SPR biosensor surface can only provide the prospect of commercialization for biomedical diagnostic applications. Hence, the many probable immobilization chemicals have been documented for various biological applications [26]. To date, the physical and chemical manipulation of the SPR biosensor surface is included. Based on these two modulation techniques, metal nanoparticles, and receptors-based bioconjugates have been reported for SPR sensing applications [6]. In the case of physical modulation, the prescribed ligand/receptors must connect to the surface of the sensor chip, where the force involved in receptor immobilization on the surface involves non-covalent interactions. Importantly, it comprises hydrogen bonding, electrostatic forces, hydrophobic interactions, and van der Waals forces. Furthermore, there is a provision for changing the surface of the sensor chip to fix particular types of receptors, such as adjusting the surface of the sensor chip with hydrophobic head groups to immobilize hydrophobic ligands. Surprisingly, the immobilization of receptors on the surface via the adsorption process resulted in their delayed removal from the sensor surface [5]. The second type of surface modification is chemical modulation. In this case, carboxy-methylated dextran (CM5) is extensively reported for the immobilization of bioreceptors. In addition to this,

J1 (unmodified gold surface), F1 (short dextran), C1 (flat carboxymethylated), B1 (low charge carboxymethylated dextran), L1 (lipophilic dextran), HPA (hydrophobic monolayer), NTA (nickel chelation), etc. have been accounted for surface modification of SPR sensor [26]. Various elements of receptor immobilization have been observed. Herein, thiol coupling is being extensively exploited in several SPR-based sensors. In these circumstances, the thiols group containing receptors (R-SH) is required to immobilize on the surface of the metal layer (Au), which resulted in the creation of a boding between R-S-Au. Ligands lacking thiol functionality can also be employed with R-SH tuning on the ligand's surface [6, 27]. Covalent immobilization is another example of receptor immobilization on the surface of the sensor termed a self-assembled monolayer (SAM). It offers the platform for the immobilization of biological molecules for the development of affinity SPR sensors [5]. If the covalent immobilization fails, there is an alternate capture approach. Here is a fundamental requirement for optimal ligand affinity. In this scenario, such selected ligands must be covalently immobilized on the surface of the sensor. Chosen molecules can be immobilized on the surface of the ligand (termed a capturing ligand). In this context, several sorts of designs have been recorded, including antibody and antigen interactions, as well as biotin-streptavidin as an immobilization method [5, 28]. The immobilization of receptors on the surface of the SPR biosensor can be chosen based on its appropriateness and the expertise of respective researchers.

3.4 Applications of Surface Plasmon-Based Diagnostic Technology

According to the literature, SPR-based analyte detection is frequently utilized in biomedical applications that can contribute to the assessment of certain illnesses [14]. Furthermore, significant efforts have been made to design modern SPR biosensors for detecting biomarkers in clinical investigations. Here, research studies support the uses of SPR-mediated biosensors for the recognition of biomarkers as analytes at the required quantities in a suitable buffer, spike sample, and clinical samples. As an output, it has been used to recognize proteins, hormones, nucleic acids, enzymes, antibodies, and other molecules [29]. In brief, Alzheimer's disease and associated health conditions are challenging to manage due to a lack of effective diagnosis. As a result, the recognition of beta-amyloid ($A\beta_{1\text{-}42}$) using a GO-based SPR biosensor was reported. In summary, the

SPR-based biosensor has been created via layer-by-layer assembly (LbL) coated on AgNPs and decorated GO nanosheet. It has been done to immobilize beta-amyloid antibodies on the surface of chitosan-based LbL assembly termed bioconjugation. In this case, GO nanosheets offer a high surface area for bioconjugate immobilization (Figure 3.3) and, eventually, the lowest detection limit was obtained to be 1.21 fg/mL. The proposed affinity SPR biosensor has been built on a mix of non-plasmonic and plasmonic nanoparticles that increase the representation in terms of sensitivity and selectivity of the SPR biosensor. Furthermore, the measurement of $A\beta_{1\text{-}42}$ in Alzheimer's disease-induced animal fluids supported the sensor's real-time applicability. It shows good stability and repeatability for the detection of $A\beta_{1\text{-}42}$. As a whole, this revolutionary step might usher in a new era of SPR-mediated diagnostics of a variety of life-threatening disorders [21].

Detection of tau proteins such as total tau protein and phosphorylated tau protein is crucial for the diagnosis of AD. Therefore, Nu and colleagues

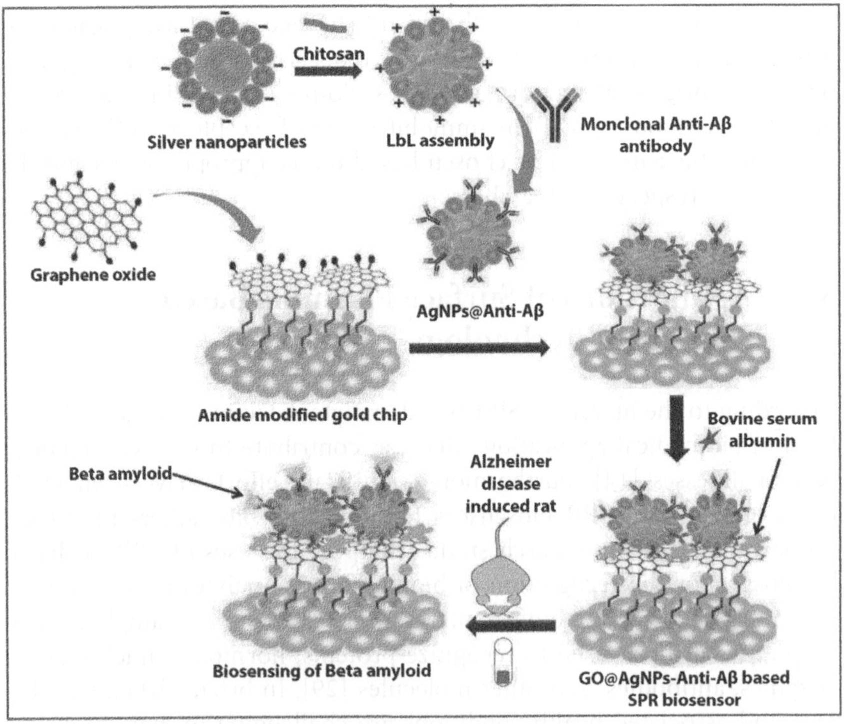

Figure 3.3 Detection of $A\beta_{1\text{-}42}$ using chitosan containing LbL coated AgNPs-GO nanosheet-based SPR biosensor [reprinted (adapted) with permission from Ref. [21]. Copyright © 2022 Elsevier].

designed the SPR fiber sensors for the recognition of tau protein and phosphorylated tau protein in clinical samples. In brief, 11-mercaptoundecanoic acid (MUA) functionalization of gold-coated fiber core has been achieved followed by immobilization of specific antibodies on the modified surface via carbodiimide chemistry. This present study validates that the level of both biomarkers total tau protein (six times) and phosphorylated tau protein (three times) has been found high than the health group. As a response, it demonstrates the 2.4 and 1.6 pg/mL of the detection limit for total tau protein and phosphorylated tau protein, accordingly. Hence, these novel SPR fiber sensors can be used for the detection of total tau protein and phosphorylated tau protein in clinical samples of patients including blood and cerebrospinal fluid [30]. The detection of microRNAs (miRNA) using the appropriate approach has garnered a lot of attention in the biomedical sector for cancer diagnosis. A hybridization-enzymolysis recycling process-mediated SPR biosensor for miRNA finding has been disclosed. The sensitivity of the intended sensor has been increased by employing streptavidin-modified AuNPs and duplex-specific nuclease (DSN). Furthermore, the bio-DNA-bio (biotinylated DNA) probe has been preferred for the sensor design. In which, the 3′ and 5′ termini of DNA have been biotinylated (Figure 3.4). Based on a biotin-streptavidin affinity in the absence of miRNA, this bio-DNA-bio indicates the binding affinity towards the streptavidin-modified sensor chip. As a result, it allows the capture of streptavidin-modified AuNPs with identical responses and produces the SPR signal. Interestingly, the presence of DSN allows for the digestion of bio-DNA-bio hybridized with miRNA by the injection of miRNA in this system. Fortunately, the last miRNA strand is unharmed. It allows for hybridization-enzymolysis recycling. As a result of one miRNA's hydrolysis, it generates multiple bio-DNA fragments. Simultaneously, bio-DNA competes with undigested bio-DNA-bio to fix with a streptavidin-modified sensor. It has been found that the low concentration of streptavidin altered the attachment of AuNPs due to digestion and competitive binding. Besides, it also gives SPR lesser signals. The LOD and linearity range were determined to be 1 fM and 1 fM to 15 fM, respectively. As a result, the suggested SPR biosensor may be utilized to diagnose cancer and other diseases [31].

A review of the literature revealed that carbohydrate Antigen 153 (CA 153) is a useful biomarker in the diagnosis of breast cancer (BC). In the event of post-surgery, serological levels of CA153 greater than 30 U/mL (high) illustrate disease progression or recurrence. As a result, precise identification of CA 15-3 in clinical samples is critical in the diagnosis of BC. Ribeiro and colleagues developed a highly specific and sensitive SPR

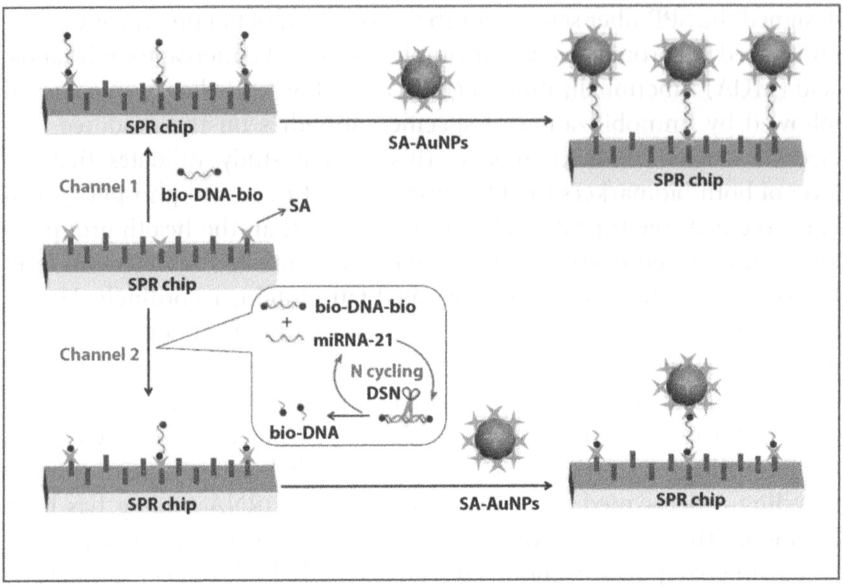

Figure 3.4 Detection of miRNA using hybridization-enzymolysis recycling process-mediated SPR biosensor [reprinted with permission from Ref. [31]. Copyright © 2021 Elsevier].

biosensor for assessing CA 15-3 levels in samples. In summary, the sensor chips with the gold-modified surface have been changed with mercapto-succinic acid (MSA) as a SAM that helps the immobilization of CA 15-3 antibodies. The detection method has been used in combination with SPR and electrochemical responses (eSPR). Initially, direct SPR has been employed to detect CA 15-3, which has a poor detection limit of 21 U/mL. Notably, the second-designed eSPR biosensor has the lowest detection limit for CA 15-3 at 0.0998 U/mL. The electrochemical performance of this biosensor has been validated at the modified gold surface of SPR. It demonstrates that the interaction of CA 15-3 with immobilized antibodies results in a decrease in the peak current of the redox probe, which aids in increased sensitivity. Furthermore, the selectivity investigation confirmed the anti-interference capability of eSPR biosensors, while detection in human serum samples confirmed clinical usefulness [32]. In 2018, Xu and colleagues discovered an inimitably responsive recognition of carcinoembryonic antigen (CEA) using SPR with an adjustable flexible liquid core-coupling sensing approach. Briefly, SPR biosensors have been fabricated using a liquid core with an SPR sensor using a regular chip. The antibodies have been immobilized on the surface of the MUA-modified gold-coated sensor chip via carbodiimide chemistry. The injection of sample protein on the surface of the sensorchip shows the change in sensorgram due to the high bio-affinity of antigen to antibody. Then, it has been observed that the designed SPR biosensor shows the potential to improve sensitivity by changing the liquid core RI. It claims that the sensitivity increases by 7.8 times and the limit of detection were found to be low up to nanogram (ng) (LOD = 0.5 ng/mL). Additionally, the high sensitivity of the sensor for CEA detection has been demonstrated through its real-time application. Finally, the proposed method will be beneficial for medical diagnosis, food hygiene, and pollution monitoring [33]. In 2019, Wu *et al.* designed a super sensitive SPR probe based on Ti_3C_2-MXene and multi-walled carbon nanotube (MWCNT) polydopamine (PDA) - AgNPs for the recognition of CEA with improved detection limit. The Ti_3C_2-MXene/AuNPs composites have been designed with *staphylococcal* protein A (SPA). After that, MWCNTs-PDA-AgNPs-polyclonal anti-CEA antibody (MWPAg-Ab_2) and monoclonal anti-CEA antibody (Ab_1) have been immobilized. The analysis assured that the proposed sensor has a linear response range of 2×10^{-16} to 2×10^{-8} M ($R^2 = 0.9989$), with a detection limit of 0.07 fM (Femtomolar: fM; S/N=3). Similarly, the selectivity of the designed SPR biosensor has been tested in the existence of interferences that assured the anti-interference potential of the SPR biosensor for CEA. In addition, this biosensing approach has shown excellent reproducibility and specificity for CEA (in serum samples). Hence, it validates that the designed SPR-based affinity sensor is an attractive and conceivable strategy for sensing CEA in clinical body fluid samples from serum samples. Therefore, it will be useful in the future for rapid diagnosis and recognition of cancer [34]. Klestova and colleagues developed an SPR-based affinity biosensor for the recognition of infectious bronchitis coronavirus (IBV). In this study, an Au-coated sensor chip has been functionalized using a "SAM" and subsequently, antigen samples have been injected. Following that, antibody-containing samples have been assessed to demonstrate antigen-antibody affinity. In this case, SAM is critical in SPR biosensors to increase sensitivity in the presence of IBV. As a result, such a new method in SPR biosensors may be used for pathogen identification [35]. Recent investigations have indicated that the measurement of Immunoglobulin G (IgG) can be used to make a diagnosis of an infectious viral disease. Yang and colleagues reported the detection of IgG using an SPR biosensor. Here, SPR biosensors have been created utilizing graphene nanomaterials and Au film. Rabbit anti-human IgG has been subsequently placed in the active layer of graphene nanomaterials. The integration of IgG through the sample port reveals the reaction owing to antigen-antibody affinity. The sensitivity of SPR biosensors has been increased by the inclusion of graphene nanomaterials, metal nanoparticles, and affinity bioreceptors. As a result, it has a detection range of 0 to 250 ng/mL. The analysis of IgG in human serum supports real-time applicability of designed SPR biosensors. As a consequence, expected SPR biosensors might offer up new avenues for the identification of viral infections such as the Hepatitis B virus, dengue fever virus, and others [36]. Scientists have difficulties in detecting SARS-CoV-2 biomarkers because of a paucity of sensing techniques. Lewis and colleagues created a portable, enzyme/antibody-free L-SPR-based biosensor that uses aptamer as a bioreceptor to detect S1 spike protein. The sensing platform has been coated with nanogold and then modified with biotin-streptavidin to fix the chosen aptamers (biotinylated) before being used in S1 spike protein detection. The envisaged L-SPR biosensor delivers great selectivity for S1 spike protein in the presence of other interfering compounds. Possibly, it is because of the high bio-affinity between S1 spike protein and aptamers. As a result, the suggested biosensor has a detection limit of up to 0.26 nM and a linear concentration range of 1 to 100 nM. Furthermore, it has high stability, remarkable reproducibility, and real-time application. This developed SPR-based approach can be utilized in the future to analyze S1 spike protein in artificial saliva and other clinical samples [37]. An appropriate sort of design is required to provide quick, simple, selective, and sensitive detection of analytes of interest. An overall paper shows the creation of an SPR-based biosensor for detecting biomarkers of various health problems. It ensured the SPR biosensor's high sensitivity, simplicity, real-time measurement, cost-effectiveness, high selectivity, dependability, and so on. Furthermore, SPR-based sensing devices are portable and provide quick detection. It may be utilized as a smart biosensing instrument for illness detection, providing a both quantitative and qualitative evaluation of biomarkers in complicated samples. The gold nanorod (GNRs)-based SPR biosensor-mediated sandwich assay has been used to detect norovirus (NoV) capsid protein (Figure 3.5). In summary, the DNA aptamer sequence-I has been immobilized on a gold surface modified with MUA. Here, the MUA surface has been activated using carbodiimide chemistry. After that, the activated amine functionality aids in the formation of bioconjugates with aptamer II and a gold surface modified with MUA. Similarly, the GNRs surface has been changed, and aptamer immobilization achieved. To produce the sandwich approach, the NoV capsid protein has been injected through the sample injection port, followed by the secondary conjugate of aptamer I-GNRs. As a result, it has a wide dynamic range of 70 to 200 aM and a low limit of detection of 70 aM. Furthermore, the analysis of NoV capsid protein in blood samples verifies the real-time

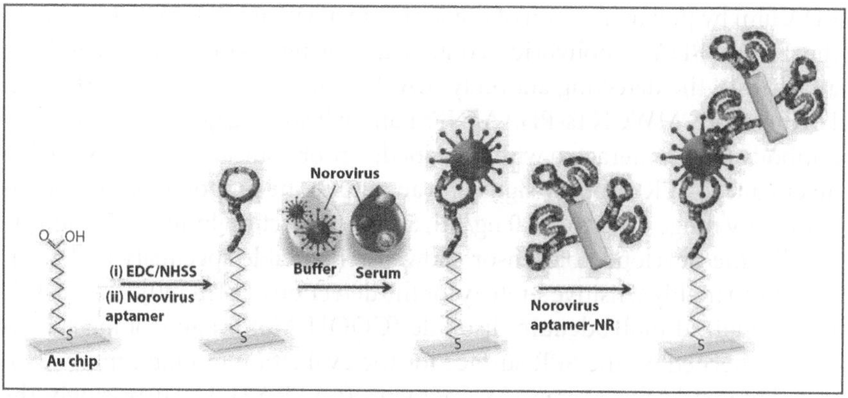

Figure 3.5 Detection of NoV capsid protein using aptamer-based SPR biosensor [reprinted (adapted) with permission from Ref. [38]. Copyright © 2018 Elsevier].

applicability in clinical applications. Surprisingly, large variations in RI caused by the complex formation of GNRs-aptamer I and gold aptamer II/NoV protein give increased sensitivity for SPR biosensors. Furthermore, due to the preferences of aptamers I and II, it provides a high specificity to NoV capsid protein. Overall, the method of GNRs-mediated SPR biosensors for the detection of NoV capsid protein with high sensitivity and selectivity appears promising [38].

Recently, SARS-CoV-2 spike antigen detection has been performed utilizing an SPR biosensor with modest changes to laser heterodyne feedback interferometry. This modification to the SPR biosensor allows for the recognition of minute changes in RI in the presence of various concentrations of interest analyte. Initially, the chip surface has been functionalized with poly-L-lysine (PLL), which aids in the immobilization of anti-SARS-CoV-2 spike monoclonal antibodies. Following the injection of the SARS-CoV-2 spike antigen, the antibody exhibits a high affinity for the antigen, whilst the change in the system's RI is seen as a reaction to quantify the quantity in samples. Furthermore, it demonstrates SARS-CoV-2 spike antigen detection with good selectivity, reproducibility, and label-free detection. As a result, the newly developed SPR biosensor may be employed for COVID-19 *in vitro* diagnosis [39]. Cardiovascular disease (CVD) diagnosis using respective biomarkers is an interesting area in biomedical fields. Out of several CVD biomarkers, cardiac troponin-I (cTnI) is the main biomarker that needs to be detected in patient samples. Therefore, Chen *et al.* fabricated an SPR biosensor for the detection of cTnI by designing iron oxide (Fe_3O_4) and PDA-based Fe_3O_4@PDAs. Subsequently, 'Au' has been coated with a

PDA film by polymerization of dopamine in a native solution. Additionally, the Fe_3O_4@PDA nanohybrids contain a sensing platform that has been attached to the detecting antibody (dAb) termed aFe_3O_4@PDA-dAb-cTnI. Furthermore, MWCNTs-PDAAgNPs nanohybrids engaged with secondary antibody (Ab2) interacted with cTnI loaded probe surface which resulted in an enhanced SPR response signal. As an output, the proposed biosensor has a linearity range of 15 to 2500 ng/mL and the detection limit is 3.75 ng/mL. Finally, the developed biosensor showed remarkable specificity and accuracy and a highly sensitive strategy for the detection of cTnI [40]. The COOH functionalized molybdenum disulfide (COOH-MoS_2) nanocomposite has been patterned on the SPR surface for the evaluation of lung cancer-associated biomarker cytokeratin 19 fragments (CYFRA21-1). Surprisingly, the COOH-MoS_2 sensor device has a large surface area, a large number of carboxylic sites for antibody attachment, and great affinity. Interestingly, the COOH-based GO sheet modification provides excellent conductivity and electron transfer rate. As a result, it strengthens affinity more than the typical SPR sensor. Moreover, it has a broad linear range and a low detection limit of 0.05 pg/mL to 100 ng/mL, and 0.05 pg/mL, respectively. It also facilitates real CYFRA21-1 analysis in serum samples. As a result, the developed COOH-MoS_2 nanocomposite-based SPR biosensor may be employed as a point-of-care diagnostic device for the detection of CYFRA21-1 in clinical samples [41]. Identification of human chorionic gonadotropin (hCG) protein was done in 2017 utilizing a combination of GO nanosheets and a specific peptide aptamer-driven SPR biosensor. Specified peptide aptamers have been anchored on GO nanosheets utilizing carboxylic functionality. It has a detection limit of 0.065 nM for hCG and a high binding constant of $6.37\times10^{12}M^{-1}$. It is due to the desired combination of GO and peptide aptamers, which affords high specific binding and increased sensitivity. Overall, this GO-peptide-based SPR biosensor has great specificity, exceptional stability, high sensitivity, the lowest detection limit, and superb affinity. As a result, tailored SPR biosensor will broaden the scope of clinical diagnostic applications [42]. Zhang and colleagues successfully detected IgG using an advanced SPR biosensor. In summary, an SPR biosensor based on GO and GNRs was invented for identifying bovine IgG. The synthesized GNRs were mounted on a surface gold-modified SPR sensor chip by covalent attachment with the Au-coated sensor chip's disulfide functionality (Figure 3.6). Moreover, GO nanosheets were formed on the surface by electrostatic interaction between positive charge CTAB-GNRs and negative charge COOH-GO. Ultimately, covalent interaction between antibodies with amine functionality and GO-associated COOH groups fixed the antibody on the surface. It provides binding with antigen and antibody after

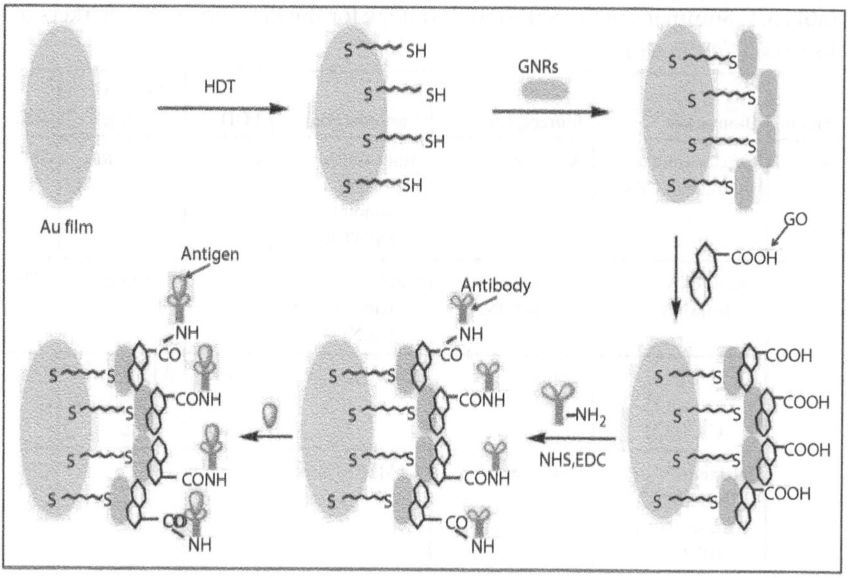

Figure 3.6 Detection of IgG using GO and GNRs-based SPR biosensor [Reprinted (adapted) with permission from Ref. [43]. Copyright © 2013 Elsevier; HDT: 1,6-hexanedithiol].

Table 3.1 Summary of the SPR-based strategy for the detection of biomarkers of diseases.

Sr. no.	Biomarker	Bioreceptor	Nanomaterial	LOD	Linearity range	Ref.
1.	Beta-amyloid ($A\beta_{1-42}$)	Antibody	Graphene oxide (GO), silver nanoparticles (AgNPs)	1.21 fg/mL	2 fg/mL to 400 ng/mL	[21]
2.	Tau protein and phosphorylated tau protein	Antibody	Gold (Au) coated fiber core	2.4 and 1.6 pg/mL	10 to 2360 pg/mL of total tau and 10 pg/mL to 4360 pg/mL	[30]
3.	Carcinoembryonic antigen (CEA)	Antibody	Gold (Au) film	0.5 ng/mL	—	[33]

(*Continued*)

Table 3.1 Summary of the SPR-based strategy for the detection of biomarkers of diseases. (*Continued*)

Sr. no.	Biomarker	Bioreceptor	Nanomaterial	LOD	Linearity range	Ref.
4.	Carcinoembryonic antigen (CEA)	Antibody	Multi-walled carbon nanotube (MWCNT)	0.07 fM	2×10^{-16} to 2×10^{-8} M	[34]
5.	MicroRNA (miRNA)	Deoxyribonucleic acid (DNA)	Gold nanoparticles (AuNPs)	1 fM	1 to 15 fM	[31]
6.	Carbohydrate Antigen 15–3 (CA 15–3)	Antibody	Gold (Au) film	0.0998 U/mL	0.10 U/mL to 500 U/mL	[32]
7.	Infectious bronchitis coronavirus (IBV)	Antibody	Gold (Au) film	—	—	[35]
8.	Immunoglobulin G (IgG)	Antibody	Graphene	—	0 to 250 ng/mL	[36]
9.	S1 spike protein	Aptamer	Nanogold	0.26 nM	1 to 100 nM	[37]
10.	Norovirus (NoV) capsid protein	Aptamer	Gold Nanorods (GNRs)	50 aM	70 to 200 aM	[38]
11.	Severe acute respiratory syndrome coronavirus 2 (SARS-CoV-2)	Antibody	Gold (Au) film	0.08 pg/mL	0.01 to 1000 ng/mL	[39]
12.	Troponin-I (cTnI)	Antibody	Iron oxide (Fe_3O_4) nanoparticles	3.75 ng/mL	15 to 2500 ng/mL	[40]
13.	CYFRA21-1	Antibody	COOH-Molybdenum disulfide (MoS_2)	0.05 pg/mL	0.05 pg/mL to 100 ng/mL	[41]
14.	Human chorionic gonadotropin (hCG) protein	Peptide aptamer	Graphene oxide (GO) sheets	0.065 nM	0.065 to 8.32 nM	[42]
15.	Bovine IgG	Antibody	Gold nanorods (GNR), graphene oxide (GO)	—	0.075 to 40 μg/mL	[43]

μg, microgram; pg, picogram; nM, nanomolar; ng, nanogram; aM, attomolar; fM, femtomolar; fg, femtogram.

the addition of antigen due to high affinity, which is reflected in RI change. Because of the link between localized SPR and propagating plasmon, GNRs immobilization on an Au film-coated SPR sensor kit culminated in a significant change in SPR reflectivity. It has a 32-fold lower detection limit for bovine IgG than the previously published approach. As well, it also has a broad linear range of 0.075 to 40 μg/mL. In this case, GO provides a large surface area with rich carboxylic functionality for the immobilization of antibodies, which is significant in sensitivity modulation. In this case, the use of GO nanoparticles in SPR sensors allows for the adjustment of surface plasmon polariton propagation and hence a change in RI. Overall, this sensor provides an entirely novel avenue for monitoring bovine IgG [43]. Table 3.1 summarizes the SPR-based strategy for the detection of biomarkers of diseases.

3.5 Current Challenges and Prospects

In the last few years, SPR-mediated biosensors have emerged as a new and intriguing foundation for biomedical diagnostic applications due to their multiple advantages. Essentially, it has a quick response time, a small sample size, and an extremely sensitive identification of the analyte of interest. This SPR-based technique allows for the measurement of changes in the detecting medium's RI. Moreover, SPR-based biosensors have been described for obtaining information on receptor–analyte interactions as well as making high-profile discoveries [44]. Despite significant advancements in biosensor technology, the operation of designing an SPR sensory system is tricky and necessitates expertise for a variety of purposes. It involves the thickness of metal layers, the concentration of bioreceptors on the surface sensor chip, and so on [20]. There is potential to address this issue in the future by optimizing metal layer thickness and immobilizing a precise known amount of bioreceptors. Furthermore, the simulation technique has been proposed to tackle the problem with SPR biosensors. Furthermore, there is a requirement to adjust the metal and dielectric layer in SPR, which might enable strengthening the demonstration of the sensor for biomarker detection. The demand for optimal metal and other nanomaterial layers, as well as their composition, will provide innovative opportunities for SPR biosensors. In the future, there will be a prerequisite to lower both the total cost and the timeframe of design of the sensor and detection of the analyte. In most situations, small molecule analysis reveals the signal-to-noise ratio, which is the key issue in validating real single and actual noise. The mass of analysts is widely recognized to exhibit the shift in reflected light angle. As a result, the tiny molecules may provide a poor response (sensorgram), making it difficult to obtain an accurate result [9]. There are significant changes in the biological functions of biomolecules that must be controlled during bioreceptor immobilization. The identification of biomarkers in a highly viscous material is severe trouble in SPR biosensors. The flow cells may become clogged, resulting in damage. In addition, SPR-based diagnosis involves the most challenging part namely binding kinetics analysis. The majority of the literature on simple bimolecular interaction binding mode fitting lacks these data. Overall, there are a few crucial considerations to consider during the actual use of SPR biosensors. It covers reagent analytical quality, baseline stability, non-specific interaction monitoring, and instrument cleaning. Furthermore, suitable washing processes, the use of a reference, the avoidance of the avidity effect, the optimal flow rate, the possibility for bioreceptor immobilization on the surface, and numerous replicates are all significant considerations in SPR biosensors. SPR biosensors will accomplish high sensitivity and selective detection with control over the aforementioned characteristics. As a result, in the future, it will provide new opportunities for the early accurate detection of serious health-related concerns. Presently, SPR-based biosensors have been preferred over radioimmunoassay, enzyme-linked immunosorbent assay (ELISA), calorimetry, and other methods. It might be because the SPR sensor provides real-time kinetic data, whereas the majority of the previously described sensor types do not. Furthermore, fast contact between two molecules failed a signal that the SPR biosensor can overcome. SPR biosensors are more helpful than ELISA for analyzing both slow and strong interactions [6]. As a result, in the upcoming days, sophisticated nanomaterials combined with SPR biosensors are a preferable choice for diagnostic applications.

3.6 Concluding Remarks

Over the previous two decades, SPR-mediated sensors have advanced significantly. It results in a wide range of modified SPR sensor platforms. Importantly, SPR biosensor exposes biomolecular identification of components and various assessment formats. It has been observed that the SPR biosensor technology has increasing interest in commercialization, even though most are currently on the market. It demonstrates the involvement in the detection and identification of biological and chemical contaminants. The designed ultramodern SPR-based biosensors demonstrate the capability of detecting tiny analytes, and the detection level is relevant in terms of medical values. In SPR-based diagnostics, sophisticated nanomaterials such as metal nanoparticles, metal oxides, carbon-based nanomaterials, and others are frequently employed in the construction of SPR sensors. As an output, it provides increased performance for detecting target analytes with high sensitivity and selectivity. More than typical forms of SPR biosensors, the combination of affinity receipts and nanoparticles provides a synergetic presentation to SPR sensors. Of several nanomaterials, the application of Au-based nanomaterials and GO is excellent for diagnostic applications. In addition, the use of an antibody as an affinity biosensor has been widely documented in SPR-based analysis. To summarize, SPR-based detection is an optical approach that can give insights into chemical and physical changes that occur at the metal/dielectric interaction. As well, this SPR-based optical sensing approach has been suggested as providing very sensitive and label-free detection of a specific analyte, such as disease biomarkers that can aid in the diagnosis of specific illness conditions. Furthermore, SPR is an effective technique for investigating the interaction between binding surfaces and analyte molecules. In the future, the right application of SPR-based biosensor technology will be employed as an extraordinary diagnostic tool defeating other approaches. A unique integrated SPR with modern technologies, such as smartphones, may be built in the next few days. Such an advanced design of SPR biosensors will serve as a new alternative for biomolecule detection with high responsiveness and specificity in delivered non-diluted complicated samples from clinics. Furthermore, the perfect combination of optimal surface chemistry and nanomaterials, as well as chemical analysis, system engineering, and microfluidics exploration can assist in achieving the goal of clinical detection. Finally, the strong performance of SPR biosensors in the diagnosing industry presented here ensures that the SPR biosensor will have a significant impact on the biomedical profession within a couple of years.

Conflict of Interest

The authors state that they have no conflicts of interest.

Acknowledgments

Sopan Nangare is grateful to the Indian Council for Medical Research (ICMR) for giving him a Research Associate Fellowship to work on surface plasmon resonance-based biosensors. The authors are also grateful to Principal Dr. Sanjaykumar Bari, H. R. Patel Institute of Pharmaceutical Education and Research in Shirpur for providing the necessary facilities to finalize this book chapter.

References

1. Divya, J., Selvendran, S., Raja, A.S., Sivasubramanian, A., Surface plasmon based plasmonic sensors: A review on their past, present and future. *Biosens. Bioelectron.*, X, 100175, 2022.
2. Tiwari, K. and Sharma, S.C., Surface plasmon based sensor with order-of-magnitude higher sensitivity to electric field induced changes in dielectric environment at metal/nematic liquid-crystal interface. *Sens. Actuator. A: Phys.*, 216, 128–135, 2014.
3. Bhagat, B., Mehta, K., Sinha, T., Baruah, P., Mukherjee, K., Recent Advances and Opportunities of Plasmonic Sensors, in: *Plasmon-enhanced Light-Matter Interactions*, vol. 297-330, 2022.
4. Gomes, J.C.M., Souza, L.C., Oliveira, L.C., SmartSPR sensor: Machine learning approaches to create intelligent surface plasmon based sensors. *Biosens. Bioelectron.*, 172, 112760, 2021.
5. Mahmoudpour, M., Dolatabadi, J.E.N., Torbati, M., Homayouni-Rad, A., Nanomaterials based surface plasmon resonance signal enhancement for detection of environmental pollutions. *Biosens. Bioelectron.*, 127, 72–84, 2019.
6. Nangare, S.N. and Patil, P.O., Affinity-based nanoarchitectured biotransducer for sensitivity enhancement of surface plasmon resonance sensors for in vitro diagnosis: A review. *ACS Biomater. Sci. Eng.*, 7, 1, 2–30, 2020.
7. Wijaya, E., Lenaerts, C., Maricot, S., Hastanin, J., Habraken, S., Vilcot, J.-P., Boukherroub, R., Szunerits, S., Surface plasmon resonance-based biosensors: From the development of different SPR structures to novel surface functionalization strategies. *Curr. Opin. Solid State Mater. Sci.*, 15, 5, 208–224, 2011.
8. Shrivastav, A.M., Cvelbar, U., Abdulhalim, I., A comprehensive review on plasmonic-based biosensors used in viral diagnostics. *Commun. Biol.*, 4, 1, 1–12, 2021.
9. Patil, P.O., Pandey, G.R., Patil, A.G., Borse, V.B., Deshmukh, P.K., Patil, D.R., Tade, R.S., Nangare, S.N., Khan, Z.G., Patil, A.M., Graphene-based nanocomposites for sensitivity enhancement of surface plasmon resonance sensor for biological and chemical sensing: A review. *Biosens. Bioelectron.*, 139, 111324, 2019.
10. Nor, S.N.S., Rasanang, N.S., Karman, S., Zaman, W.S.W.K., Harun, S.W., Arof, H., A Review: Surface Plasmon Resonance-Based Biosensor for Early Screening of SARS-CoV2 Infection. *IEEE Access*, 10, 1229–1244, 2022.

11. Wang, B. and Park, B., Immunoassay biosensing of foodborne pathogens with surface plasmon resonance imaging: A review. *J. Agric. Food Chem.*, 68, 46, 12927–12939, 2020.
12. Fu, E., Chinowsky, T., Nelson, K., Yager, P., SPR Imaging for Clinical Diagnostics, in: *Handbook of Surface Plasmon Resonance*, pp. 313–332, 2008.
13. Bellassai, N., D'Agata, R., Jungbluth, V., Spoto, G., Surface plasmon resonance for biomarker detection: advances in non-invasive cancer diagnosis. *Front. Chem.*, 7, 570, 2019.
14. Kadhuma, F., Kafia, S., Karamb, A., Al-Zukya, A., Al-Kadhemya, M., Al-Salehc, A., Simulation of surface plasmon resonance (SPR) layers of gold with silicon nitride as a Bi-layer biosensor. *Dig. J. Nanomater. Biostructures*, 17, 623–633, 2022.
15. Mohammadzadeh-Asl, S., Keshtkar, A., Dolatabadi, J.E.N., de la Guardia, M., Nanomaterials and phase sensitive based signal enhancement in surface plasmon resonance. *Biosens. Bioelectron.*, 110, 118–131, 2018.
16. Haes, A.J. and Duyne, R.P.V., Preliminary studies and potential applications of localized surface plasmon resonance spectroscopy in medical diagnostics. *Expert Rev. Mol. Diagn.*, 4, 4, 527–537, 2004.
17. Holzinger, M., Le Goff, A., Cosnier, S., Nanomaterials for biosensing applications: A review. *Front. Chem.*, 2, 63, 2014.
18. Maduraiveeran, G., Sasidharan, M., Ganesan, V., Electrochemical sensor and biosensor platforms based on advanced nanomaterials for biological and biomedical applications. *Biosens. Bioelectron.*, 103, 113–129, 2018.
19. Lan, L., Yao, Y., Ping, J., Ying, Y., Recent advances in nanomaterial-based biosensors for antibiotics detection. *Biosens. Bioelectron.*, 91, 504–514, 2017.
20. Nangare, S. and Patil, P., Nanoarchitectured bioconjugates and bioreceptors mediated surface plasmon resonance biosensor for *in vitro* diagnosis of Alzheimer's disease: Development and future prospects. *Crit. Rev. Anal. Chem.*, 52, 5, 1139–1169, 2022.
21. Nangare, S. and Patil, P., Chitosan mediated layer-by-layer assembly based graphene oxide decorated surface plasmon resonance biosensor for highly sensitive detection of β-amyloid. *Int. J. Biol. Macromol.*, 214, 568–582, 2022.
22. Nguyen, H.H., Lee, S.H., Lee, U.J., Fermin, C.D., Kim, M., Immobilized enzymes in biosensor applications. *Materials*, 12, 1, 121, 2019.
23. Liu, J.K., The history of monoclonal antibody development–progress, remaining challenges and future innovations. *Ann. Med. Surg.*, 3, 4, 113–116, 2014.
24. Hufton, S.E., Moerkerk, P.T., Meulemans, E.V., de Bruıne, A., Arends, J.-W., Hoogenboom, H.R., Phage display of cDNA repertoires: The pVI display system and its applications for the selection of immunogenic ligands. *J. Immunol. Methods*, 231, 1-2, 39–51, 1999.
25. Kedzierski, S., Khoshnejad, M., Caltagirone, G.T., Synthetic antibodies: the emerging field of aptamers. *Bioprocess. J.*, 11, 4, 46–49, 2012.

26. Pattnaik, P., Surface plasmon resonance. *Appl. Biochem. Biotechnol.*, 126, 2, 79–92, 2005.
27. Lin, S., Zhang, B., Skoumal, M.J., Ramunno, B., Li, X., Wesdemiotis, C., Liu, L., Jia, L., Antifouling poly (β-peptoid). *Biomacromolecules*, 12, 7, 2573–2582, 2011.
28. Bergström, G. and Mandenius, C.-F., Orientation and capturing of antibody affinity ligands: Applications to surface plasmon resonance biochips. *Sens. Actuators B Chem.*, 158, 1, 265–270, 2011.
29. Masson, J.-F., Surface plasmon resonance clinical biosensors for medical diagnostics. *ACS Sens.*, 2, 1, 16–30, 2017.
30. Nu, T.T.V., Tran, N.H.T., Nam, E., Nguyen, T.T., Yoon, W.J., Cho, S., Kim, J., Chang, K.-A., Ju, H., Blood-based immunoassay of tau proteins for early diagnosis of Alzheimer's disease using surface plasmon resonance fiber sensors. *RSC Adv.*, 8, 14, 7855–7862, 2018.
31. Huang, Y., Sun, T., Liu, L., Xia, N., Zhao, Y., Yi, X., Surface plasmon resonance biosensor for the detection of miRNAs by combining the advantages of homogeneous reaction and heterogeneous detection. *Talanta*, 234, 122622, 2021.
32. Ribeiro, J.A., Sales, M.G.F., Pereira, C.M., Electrochemistry-Assisted Surface Plasmon Resonance Biosensor for Detection of CA 15–3. *Anal. Chem.*, 93, 22, 7815–7824, 2021.
33. Xu, J. and Chen, Y., Surface plasmon resonance sensing with adjustable sensitivity based on a flexible liquid core coupling unit. *Talanta*, 184, 468–474, 2018.
34. Wu, Q., Li, N., Wang, Y., Xu, Y., Wei, S., Wu, J., Jia, G., Fang, X., Chen, F., Cui, X., A 2D transition metal carbide MXene-based SPR biosensor for ultrasensitive carcinoembryonic antigen detection. *Biosens. Bioelectron.*, 144, 111697, 2019.
35. Klestova, Z., Voronina, A., Yushchenko, A.Y., Vatlitsova, O., Dorozinsky, G., Ushenin, Y.V., Maslov, V., Doroshenko, T., Kravchenko, S., Aspects of "antigen–antibody" interaction of chicken infectious bronchitis virus determined by surface plasmon resonance. *Spectrochim. Acta A Mol. Biomol.*, 264, 120236, 2022.
36. Yang, Z.-W., Pham, T.-T.-H., Hsu, C.-C., Lien, C.-H., Phan, Q.-H., Single-Layer-Graphene-Coated and Gold-Film-Based Surface Plasmon Resonance Prism Coupler Sensor for Immunoglobulin G Detection. *Sensors*, 22, 4, 1362, 2022.
37. Lewis, T., Giroux, E., Jovic, M., Martic-Milne, S., Localized surface plasmon resonance aptasensor for selective detection of SARS-CoV-2 S1 protein. *Analyst*, 146, 23, 7207–7217, 2021.
38. Kim, S., Lee, S., Lee, H.J., An aptamer-aptamer sandwich assay with nanorod-enhanced surface plasmon resonance for attomolar concentration of norovirus capsid protein. *Sens. Actuators B Chem.*, 273, 1029–1036, 2018.

39. Dai, Z., Xu, X., Wang, Y., Li, M., Zhou, K., Zhang, L., Tan, Y., Surface plasmon resonance biosensor with laser heterodyne feedback for highly-sensitive and rapid detection of COVID-19 spike antigen. *Biosens. Bioelectron.*, 206, 114163, 2022.
40. Chen, F., Wu, Q., Song, D., Wang, X., Ma, P., Sun, Y., Fe3O4@ PDA immune probe-based signal amplification in surface plasmon resonance (SPR) biosensing of human cardiac troponin I. *Colloids Surf. B: Biointerfaces*, 177, 105–111, 2019.
41. Chiu, N.-F. and Yang, H.-T., High-sensitivity detection of the lung cancer biomarker CYFRA21-1 in serum samples using a Carboxyl-MoS2 functional film for SPR-based immunosensors. *Front. Bioeng. Biotechnol.*, 8, 234, 2020.
42. Chiu, N.-F., Kuo, C.-T., Lin, T.-L., Chang, C.-C., Chen, C.-Y., Ultra-high sensitivity of the non-immunological affinity of graphene oxide-peptide-based surface plasmon resonance biosensors to detect human chorionic gonadotropin. *Biosens. Bioelectron.*, 94, 351–357, 2017.
43. Zhang, H., Daqian, S., Shang, G., Hanqi, Z., Jia, Z., Ying, S., Enhanced wavelength modulation SPR biosensor based on gold nanorods for immunoglobulin detection. *Talanta*, 115, 857–862, 2013.
44. Antiochia, R., Bollella, P., Favero, G., Mazzei, F., Nanotechnology-based surface plasmon resonance affinity biosensors for *in vitro* diagnostics. *Int. J. Anal. Chem.*, 2016, 1–15, 2016, https://doi.org/10.1155/2016/2981931.

4

Nanoprobes for Glutathione Investigation and Real-Time Quantitative Imaging

Janani Archana K. and Karthikeyan Balasubramanian*

Nanophotonics Laboratory, Department of Physics, National Institute of Technology, Tiruchirappalli, India

Abstract

Numerous illnesses and ailments have been linked to the pathogenesis of oxidative stress and antioxidant deficiencies. Glutathione, a potent antioxidant thiol molecule, functions as a free radical scavenger, protecting the cells from reactive oxygen species and playing an indispensable role in maintaining the redox balance. An alteration in the glutathione levels renders the cells more susceptible to tumor progression. As a result, it is pivotal to investigate an efficacious method for determining the glutathione concentration to identify and diagnose its associated processes and diseases. The utilization of nanomaterials has tremendous potential, particularly in medicine and industry. Propelled by the recent advancements in nanotechnology, plenty of works have been designed using various nanomaterials for the detection, disease monitoring and diagnosis of glutathione. Employing nanomaterials for fluorescence sensing is a commendable technique with potent advantages like increased stability, sensitivity, lower detection limit, and enhanced performance. Nanometal surface energy transfer and Forster resonance energy transfer-based sensors are emerging techniques for detecting levels of metabolites and their changes in living cells. Herein, the present chapter is precise on the latest advances in monitoring glutathione concentration using the nanometal surface energy transfer, Forster resonance energy transfer methods and real-time imaging of glutathione using various nanohybrid probes.

***Keywords*:** Nanomaterials, glutathione, real-time imaging, fluorescence sensing, cancer diagnosis

**Corresponding author*: bkarthik@nitt.edu

Sathish-Kumar Kamaraj, Arun Thirumurugan, Muthuchamy Maruthupandy, Mercedes Guadalupe López Pérez and Shanmuga Sundar Dhanabalan (eds.) Nanoscience and Nanotechnology for Smart Prevention, Diagnostics and Therapeutics: Fundamentals to Applications, (61–90) © 2024 Scrivener Publishing LLC

Abbreviations

GSH	Glutathione
GSSG	Glutathione disulphide
NADPH	Nicotinamide Adenine Dinucleotide Phosphate
NADH	Nicotinamide Adenine Dinucleotide
FRET	Forster Resonance Energy Transfer
NSET	Nanometal Surface Energy Transfer
SV	Stern-Volmer
D	Donor
A	Acceptor
[Q]	Concentration
UCNP	Upconverting nanoparticles
NIR	Near infrared
$NaYF_4$	Sodium yttrium fluoride
Yb	Ytterbium
Tm	Thulium
MOF	Metal-organic framework
N-CQD	Nitrogen-doped carbon quantum dot
NW	Nanowire
LOD	Limit of detection
OPA	Ortho-phthalaldehyde
GQD	Graphene quantum dot
CQD	Carbon quantum dot
NMM	*N-Methylmaleimide*
BSO	Buthionine sulfoximine
LPA	Lipoic acid
VS_2	Vanadium disulphide
TMD	Transition metal dichalcogenides
CVD	Chemical vapour deposition
PEI	Polyethyleneimine
FITC	Fluorescein isothiocyanate
MB	Methylene blue
FS	FITC-doped SiO_2 nanospheres
FSM	FITC-doped SiO_2 nanospheres & MnO_2
FSMP	FITC-doped SiO_2@MnO_2 core-shell nanocomposite
PEG	Polyethylene glycol
Au NP	Gold nanoparticles
Ce	Cerium
PBS	Phosphate-buffered saline

CP	Chromogenic probe
g-C_3N_4	Graphitic carbon nitride
MES	*2-(N-morpholino) ethanesulfonic acid*
$KMnO_4$	Potassium permanganate
PBEC	Poly[3-{2,5-bis(2-ethyl-hexyloxy)}-phenyl]-vinyl]-9-ethyl-carbazole
CPN	Conjugated polymer nanoparticles
PSMA	Poly (styrene-co-maleic anhydride)
CNC	Copper nanoclusters
Lys	Lysine
BSA	Bovine serum albumin
NC	Nanocluster
NP	Nanoparticle
CP	Conjugated polymer
SPR	Surface plasmon resonance

4.1 Introduction

Over the last few decades, various engineered nanomaterials have been researched and designed for multiple applications. A great deal of attention has been generated about their promising applications as agents for diagnosis and therapy. By tailoring the structure and composition of nanomaterials to improve the sensitivity for a particular target, by effectively utilizing the molecular-recognition events to use their optical, magnetic, and electrical properties susceptible to the analyst of interest. Various classes of nanoparticles are being investigated and studied for applications in biomedical, with a special emphasis on antioxidant agents based on nanoparticles.

The detection technique is critical in biosensing, where the characteristics of the choice of material in the sensing units and the analyte must be perceived. When designing biosensing platforms, key factors to keep in mind include the need for low costs and the prospects for real-time monitoring, notably for point-of-care uses where easiness is essential. Extremely sensitive and accurate nanomaterials-based biosensors and imaging probes have enabled the development of novel technologies for the early detection and diagnosis of disease-related biomarkers.

Thiols (R-SH) are ubiquitous in the physiological fluid. Glutathione, a natural tripeptide, nonprotein thiol antioxidant, is abundant in living organisms. Their reactions to oxidative stress allow researchers to use them in diagnostics from which the pathophysiology of various circumstances could be indirectly supervised. Interestingly, over the last decade, the roles of glutathione in apoptosis, amino acid transportation, bioreduction, nitrogen and sulfur metabolism, enzyme activity, xenobiotics, and controlling oxidative stress in redox homeostasis have been observed. Hence, glutathione is regarded as a potent, significant self-defence molecule. Neurological diseases, such as Parkinson's and Alzheimer's, arthritis, diabetes mellitus, epilepsy, and atherosclerosis are caused by the alterations or significantly reduced concentrations of glutathione.

Variations in glutathione levels are like a two-edged sword for cells [1], as depicted in Figure 4.1. GSH metabolism performs both advantageous and detrimental roles. At one point, glutathione helps regulate redox homeostasis, enhancing antioxidant capability, immune response regulation, deoxyribonucleotide synthesis, cell proliferation regulation, and electrophilic xenobiotic detoxification. On the contrary, glutathione is also linked to the onset of cancer, its growth and metastatic spread. For instance, an increase in the glutathione level guards cancer cells in colon, bone marrow, lung, breast, and throat cancer. Besides that, glutathione levels in cancer cells can impact resistance to drugs and clinical outcomes.

As a consequence of its importance, glutathione has been the focus of various studies aimed at developing chemosensors and fluorescent

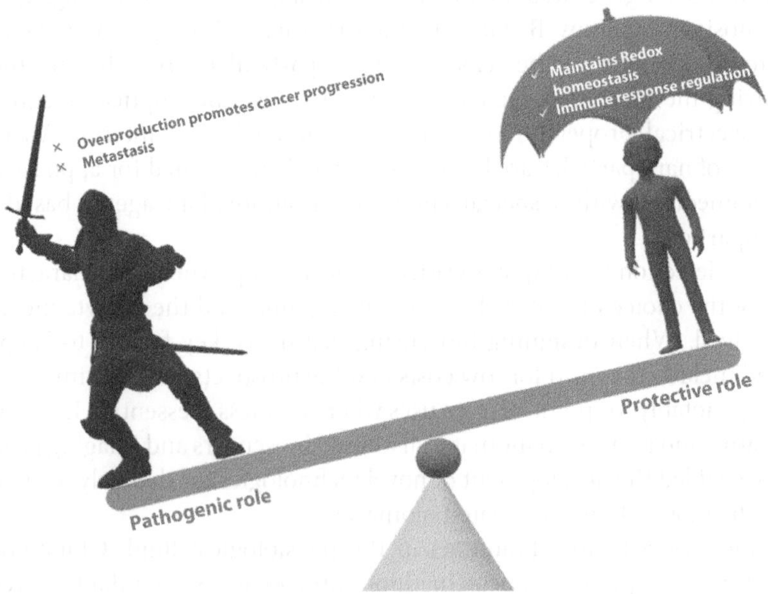

Figure 4.1 An illustration of glutathione metabolism as a regulator of cellular processes and its attribution to pathological processes.

imaging probes based on nanomaterials. It is prophesied that the insightful use of nanostructures will enhance the functional capability of biomolecular devices with significant sensitivities and limits of detection. Also, in clinical diagnosis, real-time monitoring of glutathione assists us in comprehending its role in cancer and contributes to efficient cancer treatment and drug design.

Biosensors are organized into various classes based on receptors and transducers. Depending on the receptors involved, a biosensor can be classified into enzyme, antibody, whole-cell, peptide, and nucleic acid-based biosensors. Depending on the transducer, biosensors are classified into optical, calorimetric, electrochemical, piezoelectric, and acoustic biosensors. During the past decade, optical biosensors have achieved exponential growth by attracting researchers since they outperform traditional analytic approaches in terms of cost-effectiveness, label-free detection, and highly selective and specific recognition of analytes. Further, optical biosensors are classified into absorption, fluorescence, phosphorescence, surface plasmon resonance, fibre optic, surface-enhanced Raman scattering (SERS), colorimetric and resonance dispersion-based biosensors.

Numerous conventional methods are available to quantify glutathione, including colorimetric, capillary electrophoresis, electrochemical, high-performance liquid chromatography, and flow injection analysis [2]. Although different techniques are available, fluorescence is advantageous and highly commendable owing to its merits. It includes an excellent limit of detection, fast response, ease of detection, excellent sensitivity and selectivity, low cost, noninvasive, reliable, ease of operation, and instrumentation. Nanomaterial-based fluorescent probes have rapidly progressed in biomedical applications during the last few decades. Also, these probes make biological imaging, detection of analytes, and drug delivery faster and more efficient. The bloom of nanotechnology has paved the way in the fluorescence field by transitioning from the traditional labeled method to the label-free detection of analytes.

A vast array of nanomaterials has been used for fluorescence imaging. Fluorescence imaging has gained popularity since they are versatile, selective and sensitive, contrast-rich. A clear understanding of the two different types of fluorescent imaging is obligatory. The first kind is mainly focused on imaging with certain intrinsically fluorescent species (e.g., green fluorescent protein extracted from jellyfish; NADH in living tissues). The second method encompasses the use of nanoparticles and the addition of synthetic fluorescent probes, nanomaterials, and labels.

The present chapter aims to elucidate the recent advancements in the design of nanomaterials as effective and selective fluorescent probes for glutathione sensing. Moreover, an overview of the myriad of nanomaterials used for quantitative real-time imaging of glutathione in live cells has been discussed. A brief discussion on energy transfer-based sensing platforms using nanomaterials has been carried out. This chapter also aims to impart to the readers the understanding of nanomaterials as quenchers, fluorophores, and fluorescence signal enhancers and their potential in fluorescence-based sensing and imaging platforms.

4.2 Glutathione—A Potent, Master Antioxidant

In humans, glutathione, a nonenzymatic antioxidant composed of γ-l-glutamyl-L-cysteinyl-glycine, is the most abundant sulfhydryl reagent, essential for various biological functions, including eradicating free radicals, repairing oxidative damage, signaling molecule for the activation of gene expression and helping maintain redox homeostasis. This antioxidant is distinguished by the presence of a thiol group that is reactive and γ-glutamyl bond. Cysteine's sulphydryl group facilitates the antioxidant capability of glutathione. A wide range of oxidants is scavenged directly by glutathione, namely nitric oxide, superoxide anion, carbon radicals, and hydroxyl radicals. Also, catalytically, GSH detoxifies peroxynitrite, hydroperoxides, and lipid peroxides.

In the liver, GSH is generated intracellularly using the amino acids glutamate, cysteine, and glycine and then transported to various organelles. GSH concentration in human blood is in the millimolar range, whereas in the plasma, the concentration is in the micromolar range. Typically, in cells, glutathione can be found in the cytosol (1–11 mM), nuclei (3–15 mM), and mitochondria (5–10 mM) [3].

Reduced and oxidized forms are the two different states of existence of glutathione. The active form is the reduced glutathione (GSH), which is a monomer and is predominant under physiological circumstances. The other form is glutathione's oxidized state, a disulfide dimer named glutathione disulphide (GSSG). The cellular prooxidants like reactive aldehydes and free radicals oxidize thiol in GSH to form GSSG. The reduction of GSSG back to GSH is facilitated by the enzyme glutathione reductase, which replenishes GSH for antioxidant activity. GSH and GSSG work in closer liaison with redox-active compounds like NADPH to retain cellular redox homeostasis.

The concentration of GSH is 10 to 100 times greater than GSSG. Generally, the ratio of GSH to GSSG varies from 100:1, and any variation in this level indicates redox homeostasis and oxidative stress. This ratio is

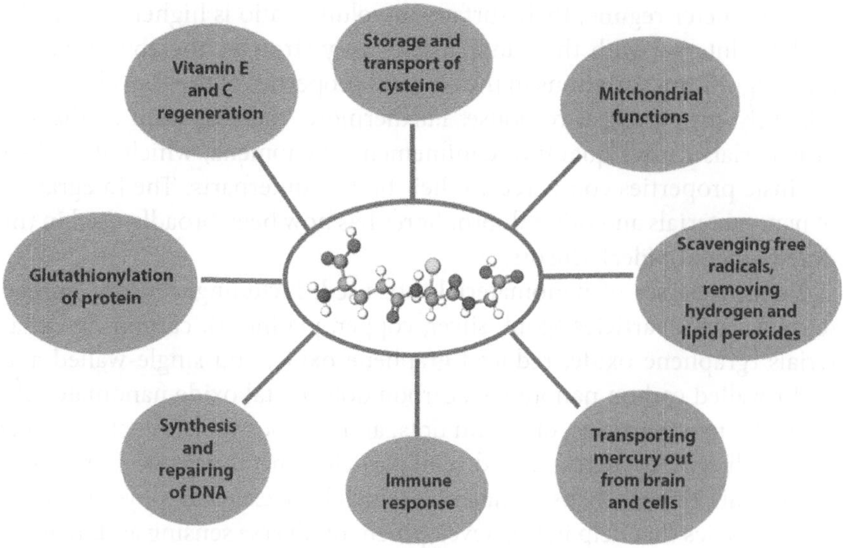

Figure 4.2 Schematic of the vital physiological roles of glutathione.

monitored by the enzyme named glutathione reductase. Under normalcy, the concentration of GSSG is maintained at less than 5% of the overall GSH concentration in the cell.

GSH has a tremendous ability to maintain and regulate various cellular processes. It is indeed difficult to understate the importance of glutathione, whose key roles are depicted in Figure 4.2. Distinctly, sufficient glutathione is essential for defending the body from toxic elements to sustain health and support a longer lifespan.

4.3 Biosensing of Glutathione Using a Variety of Nanomaterials

4.3.1 Nanomaterials: A Game Changer in the Past Decade

Nanotechnology has sparked much curiosity because of the advancements in the synthesis procedures, which help us synthesize nanomaterials in varied, desirable morphology and physicochemical properties. The nanoscale regime is typically less than 100 nm in size and can be synthesized from organic and inorganic materials. Based on the dimensions, the nanomaterials are classified into 0D, 1D, 2D, and 3D. Nanomaterials offer lesser size when compared to their bulk counterparts. As the particle size is in the nanometer regime, their surface-to-volume ratio is higher; hence, the analytes interact with the nanoparticles very strongly, thereby providing more significant variations in the physical properties of the transducer and relatively strong signal response. Furthermore, the reduction in the size of materials causes quantum confinement phenomena, which alter their intrinsic properties compared to their bulk counterparts. The integration of nanomaterials and other fluorophores has now been broadly used in the development of ideal sensors.

Various classes of nanomaterials are used, including, but not limited to, metal nanoparticles (gold, silver, copper, platinum), carbon nanomaterials (graphene oxide, reduced graphene oxide, and single-walled and multi-walled carbon nanotubes), carbon dots, metal oxide nanomaterials, magnetic nanoparticles, quantum dots, and composite nanomaterials like core-shell nanoparticles, doped nanoparticles, metal-organic framework (MOFs) are the widely used materials which have excellent physicochemical properties that help in the development of diverse sensing and imaging platforms.

Conventional organic fluorophore dyes are a vast class of molecules with smaller sizes and are easily modifiable, convenient, and cost-effective. However, their low photobleaching resistance, shorter-lived lifetimes, and limited chemical stability limits their use in fluorescence applications. The above trouble has driven the onset of nanomaterials with high quantum yields for energy transfer-based applications. Fluorescent nanomaterials overcame these challenges by offering a longer lived excited state lifetime, excellent quantum yield, and are highly photostable.

4.3.2 Fluorescence-Based Biosensors for Glutathione Sensing

4.3.2.1 Understanding Fluorescence

Fluorescence spectroscopy has evolved into the most widely used technique in many fields, including sensing and diagnostics. Fluorescence is caused by electrons freely moving over a fluorophore and emitting absorbed energy. Initially, electrons will be in the ground state. Upon excitation, when a photon of particular energy hits the sample, the electron will absorb the photon's energy and reaches the higher vibrational level of an excited state. The molecule loses its energy through collision and returns to the lowest vibrational level in the ground state. Fluorescence is a recursive process since a single molecule is excited and detected repetitively. A molecule's lifetime between excitation and return to the ground state is typically 10^{-9} s.

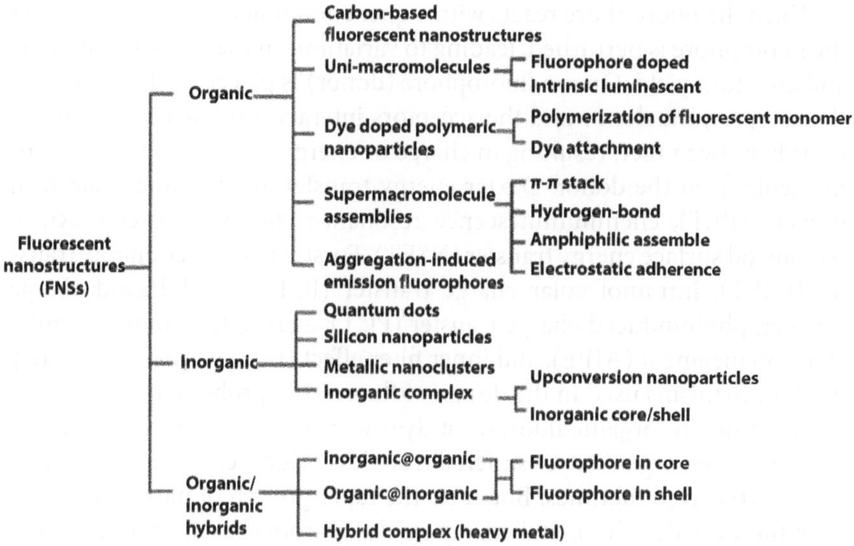

Figure 4.3 Fluorescent nanostructures categorized based on chemical composition and structure. Reprinted (adapted) with permission from reference [4]. Copyright (2014) Elsevier.

Molecules that emit fluorescence are termed fluorophores. Generally, fluorescent nanostructures are classified into inorganic, organic, and organic–inorganic hybrid based on their structure and composition, as shown in Figure 4.3. The shape and size of the nanomaterials, synthesis methodology and fluorophore environment during synthesis greatly influence the position of the absorption and emission peaks. Also, factors, such as pH, temperature, viscosity, and solvent effects, significantly influence the emission peaks. Fluorescence spans the electromagnetic spectrum from ultraviolet (190–400 nm), visible (400–800 nm), and near-infrared regimes (800–1400 nm).

4.3.2.2 Fluorescence Sensing Strategy

Different sensing methodologies have been implemented with the help of several sensing parameters like fluorescence intensity, lifetime, emission wavelength, and anisotropy. Among these, sensors operating on the variations in fluorescence intensity are the most extensively used since they are the easiest, simplest, and most sensitive to the changes in the environment of the fluorophore for detecting various analytes.

When the fluorophore reacts with a quencher molecule, the emission of the fluorophore is perturbed, leading to variations in the intensity, lifetime, and quantum yield. Once a fluorophore (donor) is placed in the vicinity of the acceptor, the donor and the acceptors interact if the necessary conditions have been met, resulting in charge or energy transfer to the acceptor molecule from the donor. Dexter energy transfer, photoinduced electron transfer (PET), chemiluminescence resonance energy transfer (CRET), nanometal surface energy transfer (NSET), Forster resonance energy transfer (FRET), intramolecular charge transfer (ICT), metal–ligand charge transfer, photoinduced charge transfer (PCT), aggregation-induced emission enhancement (AIEE), and inner filter effect (IFE) are the most widely used mechanisms used in the design of fluorescent probes currently.

Traditionally, organic fluorescent dyes were used as donors and acceptors. However, organic dyes suffered from less brightness, photobleaching, and shorter-lived lifetimes. But, after the rapid growth of nanotechnology over the past decade, metal nanoparticles, carbon-based materials, and inorganic quantum dots have opened new avenues for fluorescent sensing systems relying on charge and energy transfer. The selection of a donor with a high quantum yield is essential for attaining good FRET/NSET efficiency. In this chapter, we have primarily focused on the energy transfer-based (FRET and NSET) sensors for glutathione sensing and imaging.

4.3.2.3 Turn-Off and Turn-On Sensing

Generally, two detection strategies based on the fluorescence intensity for sensing are used; "turn-on" (enhancement) or "turn-off" (quenching). Turn-on fluorescent probes cause an increase/enhancement in the fluorescence emission intensity of the fluorophore without modifying the fluorophore's spectra. Fluorescence-sensing platforms are typically focused on the electronic interaction or energy transfer that occurs when analytes are introduced, resulting in the variation in fluorescence signals corresponding to the analyte. In the present chapter, systems based on energy transfer have been given prime importance.

In the presence of an analyte of interest, fluorescence sensors generally use fluorescent nanomaterials, which act as donors or acceptors and either enhance or quench the fluorescence intensity. Quenching refers to the decrease in the fluorescence intensity of the fluorophore without altering its spectrum. The molecule that absorbs energy from the fluorophore and causes a decrease in the intensity of the fluorophore is the quencher. Molecular interactions like the formation of ground-state complexes, rearrangements in the molecular environment, energy transfer, and excited state reactions [5]. Static and dynamic quenching mechanisms are widely being studied. The Stern–Volmer (SV) plot has been used to understand the quenching mechanism, which provides information about the excited state deactivation process. An SV plot is plotted with F_0/F on the y-axis and the concentration of the fluorophore on the x-axis. Here, F_0 and F are the fluorescence emission intensities in the absence and presence of the quencher.

In the case of dynamic quenching, the fluorophore and the quencher undergo collision. Upon collision between the fluorophore and the quencher, the fluorophore in the excited state gets deactivated, resulting in energy transfer or charge transfer. Diffusion plays a vital role in this quenching because the fluorophore and the quencher have to diffuse and collide. Changes in the fluorescence lifetime values characterize dynamic quenching.

In static quenching, a complex is formed between the quencher and the fluorophore, which results in the formation of a ground-state complex, that is nonfluorescent. The ground-state complex is formed in the ground-state itself before even the excitation happens. The formation of ground-state complexes will affect the absorption spectra of the fluorophore and the excited state lifetime remains unchanged. Hence, static quenching can be characterized by recording the absorption spectra.

4.3.2.3.1 Forster Resonance Energy Transfer

A nonradiative energy transfer process, FRET, is where the energy from an excited donor (fluorophore) in the excited electronic state is transferred to an acceptor via intermolecular dipole-dipole interactions. This technique is now widely used in various fields, including biosensing, protein folding and unfolding studies, cell biology, single-molecule experiments, monitoring density of biological membranes and DNA mechanical movements. Due to these underlying conveniences, FRET is also known as the "spectroscopic ruler." In FRET, the fluorophore is an oscillating dipole capable of transferring its energy to another dipole with the same resonance frequency.

The efficiency (E) of energy transfer, given in (4.1), varies with the inverse sixth power of the distance between D and A.

$$E = \frac{R_0^6}{R_0^6 + r^6} \tag{4.1}$$

Here, R_0 is the distance at which the energy transfer is 50%, and r is the distance between D and A. The distance between D and A can be found using the relation,

$$R_0 = 0.211\{k^2\, \phi_D\, J(\lambda) n^{-4}\}^{(1/6)} \tag{4.2}$$

where k is the orientation factor of the dipoles, n is the refractive index of the medium, ϕ_D is the quantum yield of the donor, and J(λ) is the extent of spectral overlap between the acceptor's absorption and the donor's emission spectrum. A larger value of J(λ) signifies an excellent spectral overlap between the D and A.

Generally, FRET results in the reduction of emission intensity of fluorophore, accompanied by a decrease in the excited state lifetime of the fluorophore. For efficient FRET to take place, the following conditions must be satisfied: i) spectral overlap—there must be an overlap between the absorption spectra of the acceptor and emission spectrum of the donor (Figure 4.4), ii) the relative orientation of the transition dipole of the donor and acceptor should be parallel, and iii) the distance between the D and A should be between 1 and 10 nm.

A FRET probe must be designed meticulously. The fluorophore's emission and the acceptor's absorption should significantly overlap. In the conventionally used fluorophores, many shortcomings like auto-fluorescence and lesser lifetime in the excited states were observed. Extraordinary advantages like long-range working distance, and reduction of crosstalk between D and A, were rendered when conventional fluorescent dyes were

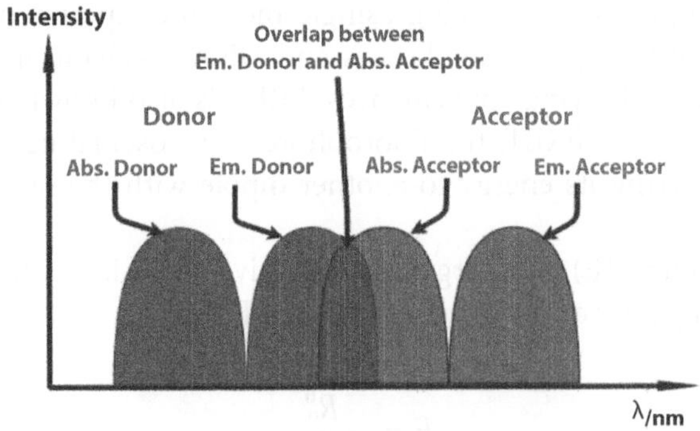

Figure 4.4 Spectral overlap of the acceptor's absorption and donor's emission spectra. Reprinted (adapted) with permission from reference [6]. Copyright (2020) Royal Society of Chemistry.

replaced with nanomaterials. Moreover, when nanomaterials were used for FRET-based sensing, the distance between the D and A could be controlled easily.

4.3.2.3.2 Nanometal Surface Energy Transfer

The Nanometal surface energy transfer process describes the nonradiative energy transfer process from a dipole to a metallic surface. NSET is the interaction between the acceptor's free conduction electrons with the donor dipole's electromagnetic field. In NSET, the acceptor has a metallic surface with an isotropic distribution of dipole vectors, facilitating energy transfer efficiently. The oscillating dipole and the electrons interact stronger when they are near and perpendicular to the metallic surface. Metallic nanoparticles as acceptors with appropriate Fermi levels can aid in collecting dipoles and generating additional energy oscillations, leading to a larger energy transfer efficiency [7].

Applying Fermi's golden rule in the dipole approximation of energy transfer, where the product of interaction elements of the donor (F_D) and acceptor (F_A) are related to the energy transfer rate (k_T), as shown in Table 4.1. The distance between the donor and acceptor purely depends on their geometric arrangement [8]. From Table 4.1, it is clear that as there is an increase in dimension, there is a decrease in the power of the distance factor [9].

In NSET, the characteristic distance between the donor and acceptor is given by

$$d_0 = [(\frac{0.225 * C^3 * \phi_D}{\omega_F * \omega_D^2 * k_F})]^{(1/4)} \tag{4.3}$$

Table 4.1 Dipole arrangement and interaction elements for FRET and NSET.

<table>
<tr><th rowspan="2">Dipole arrangement</th><th rowspan="2">Interaction element (F)</th><th colspan="2">Energy transfer rate $k_T = F_D * F_A$ (R-Separation distance between dipoles)</th></tr>
<tr><th>FRET</th><th>NSET</th></tr>
<tr><td>3D dipole volume</td><td>F~ Constant</td><td rowspan="3">• Point dipole-point dipole interaction
• $k_T \sim (1/R^3)*(1/R^3)$
• $\mathbf{k_T \sim (1/R^6)}$</td><td rowspan="3">• Point dipole-surface dipole interaction
• ·$k_T \sim (1/R^3)*(1/R)$
• $\mathbf{k_T \sim (1/R^4)}$</td></tr>
<tr><td>2D dipole surface</td><td>F~ 1/R</td></tr>
<tr><td>Point dipole</td><td>F~ $1/R^3$</td></tr>
</table>

where ϕ_D is the quantum yield of the donor, ω_F is the Fermi frequency of the metal, and k_F is the Fermi wavevector of the acceptor and ω_D is the donor's electronic transition frequency. The efficiency (E) of energy transfer varies with the inverse fourth power of the distance between D and A. FRET and NSET differ by the operating distance range between the donor and acceptor moieties (for FRET, distance dependency is $1/R^6$ and for NSET, it is $1/R^4$). FRET is effective only when the distance between the donor–acceptor pairs is between 10 nm, whereas NSET can efficiently detect twice the distance compared to FRET. Thus, energy transfer between a donor dipole and a metal surface opens up a new framework for developing biosensors that can attain energy transfer efficiency double the FRET range.

4.3.3 Fluorescence Imaging

Molecular imaging encompasses modalities such as magnetic resonance imaging, computed tomography, scanning probe microscopy and positron emission tomography. These techniques are efficient only for imaging the macroenvironment (at the tissue level). Bioimaging is acquiring images of biological materials using nanomaterials, fluorescent tags and probes to have sectional views of cells, tissues and compartments in the cells.

Fluorescence imaging is a versatile technique owing to its excellent sensitivity, ease of operation, absence of ionizing radiation, real-time monitoring, noninvasiveness, ability to span the imaging from visible to the near-infrared range and high spatiotemporal resolution. Using the visible wavelength range (400–950 nm) has very minimal tissue penetration, limiting its intravital imaging capability. To overcome this limitation, the fluorescence imaging technique has been developed in the near-infrared region (1000–1700 nm), which offers high-fidelity imaging, lesser background fluorescence than visible light and better penetration [10]. For fluorescence imaging, the probes used should have a suitable excitation and emission wavelength to mitigate the background noise, high quantum efficiency and high photostability. Enhanced brightness, even distribution and inertness to their microenvironment [11] are the advantages of using nanoparticles in fluorescence imaging.

4.3.4 Outlook of Different Nanoprobes for Glutathione Sensing and Imaging

4.3.4.1 Graphene and Carbon Quantum Dots-Based Materials as Donors

Graphene quantum dots and carbon quantum dot materials have emerged as splendid materials due to their excellent photoluminescent property, low cost, solubility, biocompatibility and simple synthesis methods. Since they are small and have tunable size, GQDs exhibit quantum confinement effects, exciton confinement and edge effects where the excitation spectrum can be tailored. GQDs are created by clipping large graphene structures, where their physical and chemical properties are identical to graphene. GQDs have excellent tunable optical properties, including a longer lifetime in the excited state, photostability, and high brightness. GQDs provide a major sensing area for adsorption for other molecules through π–π stacking interaction, which allows easy adherence for acceptors on its surface [12]. CQDs are carbon-based fluorescent nanostructures with quasi-spherical shapes. They have distinctive properties, including high photostability, cost-effective synthesis procedures, aqueous stability and bio-compatible and tunable composition. Generally, the emission spectrum of CQDs is in the UV region extending to the visible region. On the surface of GQDs and CQDs, they have an abundance of O_2-containing functional groups. Compared to GQDs, CQDs exhibit relatively more defects and poor crystallinity due to lower sp^2 carbon [13, 14].

Lakshita *et al.* [15] have synthesized nitrogen-doped carbon quantum dots (N-CQDs) and MnO_2 nanowires using hydrothermal synthesis. For FRET to happen, the absorption spectrum of the acceptor should overlap with the donor's emission spectrum. N-CQDs blue emission at 450 nm had a good overlap with the broad absorption spectra of MnO_2 nanowires ranging from 300 to 500 nm. At different concentrations of MnO_2 NWs, the emission of N-CQDs is quenched, signifying the FRET between N-CQDs (donor) and MnO_2 NWs (acceptor). GSH was added in the nanomolar range at different concentrations ranging from 20 to 500 nM, where GSH showed a "turn-on" response to varying concentrations of GSH with a good LOD of 7.52 nM (Figure 4.5). MnO_2 NW is converted into Mn^{2+} and GSH is oxidized to GSSG due to its reducing capability. This nanoprobe has been used to quantify GSH in real-time human plasma, where 60 μL aliquot of human plasma was added to the nanoprobe solution resulting in

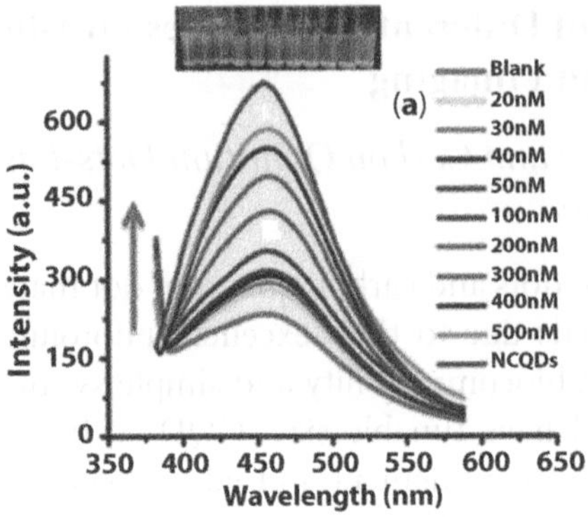

Figure 4.5 Turn-on fluorescence sensing of GSH with N-CQDs and MnO_2 NW probes. Reprinted (adapted) with permission from reference [15]. Copyright (2021) American Chemical Society.

a final volume of 3 mL where they have determined the concentration of GSH in human plasma to be ranging from 19.85 to 29.86 nM.

Cai *et al.* [16] have synthesized C dots-MnO_2 colloidal solutions via *in situ* synthesis. MnO_2 nanosheets quenched the emission of carbon dots through FRET. This nanohybrid probe showed a "turn-on" response while varying the concentration of GSH from 0 to 250 μM with a LOD of 300 nM. Since this nanohybrid probe exhibited good sensitivity, it was employed to detect GSH in human serum samples, demonstrating strong potential for glutathione sensing. In the human serum sample, the concentration of GSH was found to be 1.304 μM.

Yan *et al.* [17] have synthesized graphene quantum dots and ultra-thin MnO_2 nanosheets. MnO_2 nanosheets quench the emission of GQDs through FRET. It is found that MnO_2 nanosheets act as recognition units for GSH. This nanocomposite probe was employed for the detection of GSH. A gradual restoration of the emission intensity was observed with the increasing GSH concentration with a good LOD of 150 nML^{-1}. Since GSH has reducing capability, it decomposes MnO_2 nanosheets into Mn^{2+}, thereby facilitating the recovery of emission intensity of GQDs. The nano-composite comprising GQDs and MnO_2 nanosheets was used to monitor intracellular GSH. In this study, MCF-7 cells were used where MCF-7 is an extensively researched breast adenocarcinoma epithelial cancer cell line. MCF-7 cells have been pretreated with thiol-reactive scavenger (*N*-methylmaleimide), inhibitors (Buthionine sulfoximine) and enhancers (lipoic acid) to assess the response of the nanocomposite probe to variation

Figure 4.6 Images from confocal fluorescence microscope of MCF-7 cells that had been pretreated with NMM, BSO, and LPA, followed by incubation with GQDs and MnO_2 nanohybrid probe. Reprinted (adapted) with permission from reference [17]. Copyright (2016) American Chemical Society.

in the intracellular GSH levels. Using these cancer cells, control experiments were performed where these cells were incubated with GQD alone and GQD along with BSO and NMM for another 2 hours and 4 hours. In the case of cells treated with GQDs, a continuous blue signal was observed, which signifies that the GQDs are transferred to cancer cells.

In contrast, for cells treated with GQDs with NMM and BSO, the emission signal was similar to that of the control, indicating that the emission intensity had not been mediated either by the inhibitors or scavengers. Figure 4.6 clearly shows that the cancer cells incubated with LPA show a good fluorescence enhancement, while cells treated with NMM and BSO show a reduction in the fluorescence signal. Thus, this study reveals that the nanohybrid probe of GQDs and MnO_2 is appropriate for GSH monitoring *in vitro*.

4.3.4.2 Metal-Oxide Material as Donors

Metals such as iron, zinc, cerium, magnesium, nickel, zirconium, titanium, iron, and copper were discovered to have captivating properties like nontoxicity, high chemical stability, and strong adsorption capacity. Metal oxides have exceptional optical properties due to a larger surface-to-volume ratio, strong adsorption tendency, and electron and phonon confinement, making them a potential candidate for biosensing platforms [18]. Also, they are morphologically versatile, cost-effective synthesis, and the synthesis methodology is simple, which makes them a fascinating material for biosensors. In transition metal ions, unfilled d-shells are prevalent, which have wider bandgaps, allow reactive electronic transitions, and increased dielectric constants, which makes them have exceptional optical properties [19].

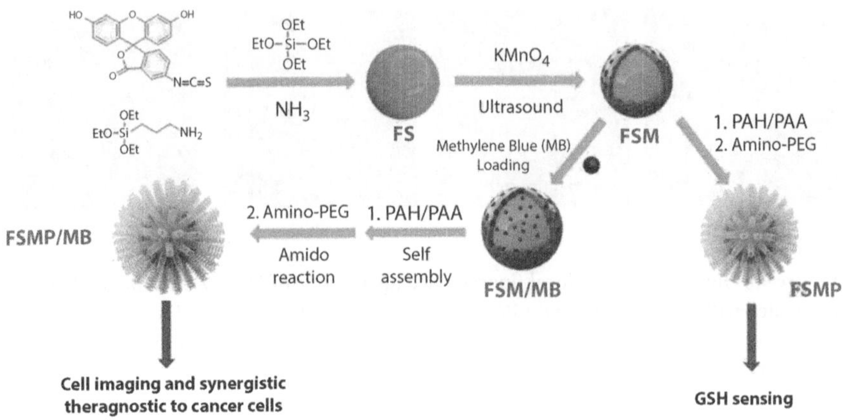

Figure 4.7 Pictorial representation of the synthesis of fluorescein isothiocyanate (FITC)-doped SiO_2 nanosphere for monitoring GSH levels, cell imaging and theragnostic to cancer cells. Reprinted (adapted) with permission from reference [20]. Copyright (2021) American Chemical Society.

Zhang *et al.* [20] have designed core-shell architecture consisting of a SiO_2@MnO_2 nanocomposite probe to detect GSH and cancer cell theranostics. The Modified Stober method has been used to synthesise FITC-doped SiO_2 nanospheres (Figure 4.7). MnO_2 quenches the fluorescence of FITC-doped Si nanoparticles through FRET. A turn-on response is observed upon varying the concentration of GSH from 0 to 450 μM, which is attributed to the redox decomposition of MnO_2 to Mn^{2+} by GSH with a LOD of 2.1 μM. The proposed nanocomposite probe has been used for sensing GSH in human serum samples, which provided exemplary results, denoting the reliability and feasibility of the proposed probe. As a result, the crafted nanoprobe can be used as an efficient tool to monitor levels of GSH. The porous MnO_2 has been loaded with MB to understand the prepared nanocomposite's therapeutic capability. PEG has been loaded on the outer layer to ensure that the photosensitizer is internalized by tumor cells. Once FSMP has entered the tumor cells. The high concentration of GSH in cells will degrade the MnO_2 layer, thereby creating Mn^{2+} and restoring the emission of FITC. This nanocomposite probe has been applied as a theranostic to cancer cells.

4.3.4.3 *Metal Nanoparticles as Donors*

Noble metal nanoparticles have gained significant attention owing to their exceptional plasmonic properties that can be employed to modify the optical signal of molecules placed in their vicinity. Noble metals such as gold, silver, copper and platinum are the most prevalent and broadly examined owing to their unique conductivity, biocompatibility and very large surface-to-volume ratio. Compared to their bulk counterparts, noble metal nanoparticles exhibit distinct physical, chemical and optical properties at the nanoscale. Exploiting different physicochemical characteristics of metal nanoparticles like SPR, emission quenching/enhancement, surface-enhanced Raman scattering (SERS), and a wide range of biosensing platforms for detecting enzymes, nucleic acids and other biological molecules have been developed. Using various synthesis procedures, metal nanoparticles can be fine-tuned in size and shape according to the intended applications. When exposed to optical energy, metal nanostructure develop a high plasmon field around them. The fluorescence emission varies extensively when a fluorophore is brought alongside a metal nanostructure with a strong surface plasmon field. The collective excitation of the conduction electrons governs the optical properties of metal nanoparticles, and these collective oscillations are known as localized surface plasmon resonance. In this section, importance has been given to metal nanoparticles that act as donors for GSH detection.

Wang *et al.* [21] proposed a simple one-step chemical reduction method to produce fluorescent Cu NCs using BSA as a template. The nanocomposite was prepared by reducing $KMnO_4$ in MES buffer by ultrasonication comprising Cu NCs, and MnO_2 nanosheets were prepared by *in situ* synthesis (Figure 4.8). In the presence of MnO_2 nanosheets, the emission of Cu NCs is quenched owing to FRET, where Cu NCs act as a donor and MnO_2 nanosheets as acceptor. In the vicinity of GSH, MnO_2 nanosheets have been reduced to their lower valence state, Mn^{2+}. GSH was added to the composite where the nanosheets have been reduced to Mn^{2+} and GSH has been oxidized to GSSG via the thiol-disulfide exchange. The concentration of GSH has been varied from 0 to 600 μM, where the fluorescence recovered gradually with LOD of 0.1 μM. The proposed nanocomposite was used to determine GSH levels in human serum samples, which showed satisfactory results in sensing GSH in serum samples.

Li *et al.* [22] have developed a GSH sensor based on Cu nanoclusters (CNCs) and MnO_2 nanospheres. A complex of CNCs and MnO_2 nanospheres was assembled by injecting CNCs onto MnO_2 nanospheres via electrostatic interaction between two components. Here, MnO_2 nanospheres are found to quench the fluorescence emission of CNCs via FRET. This nanocomplex probe was employed for sensing GSH. Once GSH is added, the emission spectra significantly recover (turn on). The enhancement in emission is attributed to a two-stage enhancement process where

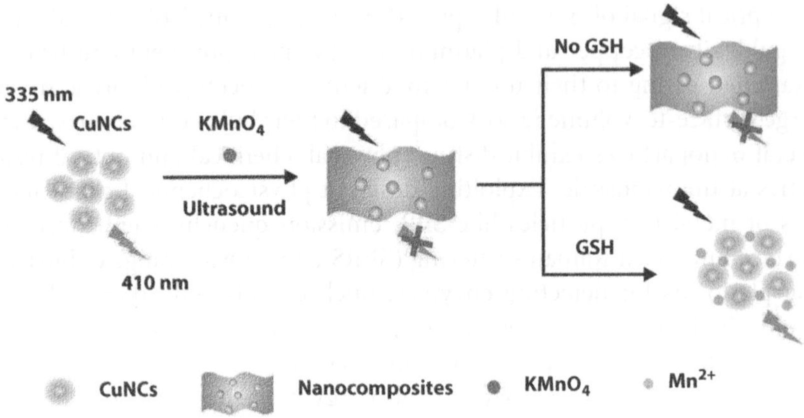

Figure 4.8 Graphical representation of GSH detection Cu nanoclusters and MnO_2 nanosheets. Reprinted adapted with permission from reference [21]. Copyright (2016), Royal Society of Chemistry.

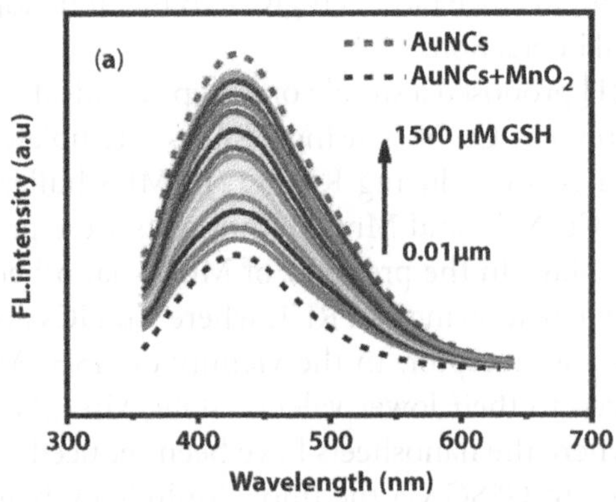

Figure 4.9 Fluorescence emission spectra of GSH detection with AuNCs@Lys-MnO_2 nanosheets. Reprinted adapted with permission from reference [23]. Copyright (2020), Elsevier.

GSH induces the decomposition of MnO_2; moreover, the complex and GSH will have host-guest interaction. The concentration of GSH has been varied from 0 to 2 mM and a relatively good LOD of 17 μM has been attained using this probe. The proposed nanocomplex probe has been used to detect GSH in human serum samples, showing acceptable and reliable results for real-time detection.

Fu *et al.* [23] have developed an AuNCs@Lys-MnO_2 nanohybrid probe-based system for the sensitive detection of GSH. Using lys as the ligand, Au NCs were synthesized using chloroauric acid as the precursor and hydrazine hydrate as the reducing agent (AuNCs@Lys). AuNCs@Lys-MnO_2 NSs and Au@Lys were combined using electrostatic adsorption. AuNCs@Lys exhibits blue emission when excited at 365 nm. The presence of MnO_2 NSs quenches the fluorescence of AuNCs@Lys. The emission of the nanocomposite probe gradually recovered upon adding GSH in different concentrations ranging from 0.5 to 1000 μM, as shown in Figure 4.9, with a LOD of 5 nM. The developed AuNCs@Lys-MnO_2 nanocomposite probe solution has also demonstrated feasibility for detecting GSH in spiked human serum.

4.3.4.4 Metal-Organic Framework as Donors

MOFs are porous structures which constitute the emerging class of nanomaterials where organic linkers link metal clusters together through coordination bonds. Due to the availability of plentiful metal clusters and organic linkers with controllable photoluminescence, MOF-based nanomaterials have been preferred for biosensing. Because of their high porosity and the presence of functional sites that use pore space to capture different species, MOFs are excellent candidates. The sensing capacity of a MOF-based sensor is impacted by the metal centres and materials used [24].

Wang *et al.* [25] have developed a ratiometric fluorescent sensor for sensitive GSH and dopamine sensing using UiO-66-NH_2 MOF. The prepared MOF UiO-66-NH_2 belongs to the family of luminescent MOF, with emission at 450 nm was synthesized by hydrothermal method. In the aqueous solution of PEI, dopamine self-oxidises itself and co-exists to form P_{DA-PEI}, resulting in a new emission peak at 530 nm. The emission of MOF at 450 nm is significantly quenched. Moreover, a significant spectral overlap between the absorption spectra of P_{DA-PEI} and the emission spectra of UiO-66-NH_2 has been observed. When PEI, GSH and dopamine co-existed, the emission at 450 nm is recovered and the emission at 530 nm is quenched. Thus, these results show that this ratiometric fluorescent probe can detect dopamine and GSH where the LODs are 0.68 and 0.57 μM. This nanohybrid probe was employed for detecting GSH in the human serum sample, which had good accuracy, employing that this MOF-based probe can be used for detecting GSH.

Gong *et al.* [26] have developed cerium MOF and gold nanoparticles for monitoring GSH levels. Electrostatic interaction between Ce-MOFs-Au NPs and FRET, the emission of Ce-MOF has been quenched by Au NPs. Once GSH has been added in varying concentrations (0–40 μM) to the nanocomposite probe, a "turn-on" response has been observed owing to the formation of Au-SH covalent bond between Au nanoparticles and GSH with an excellent LOD of 58 nM. The nanocomposite probe was used for monitoring GSH in human serum samples, showing excellent, reliable response for detecting, thus proving the practicability of using this probe.

4.3.4.5 Transition Metal Dichalcogenide Materials as Donors

Fascinated by graphene and graphene-like 2D materials, single or few-layered TMDs have gained attention in recent years as a newly evolving class of nanomaterials. TMDs are MX_2-type materials (M is the transition metal and X is the chalcogen) with outstanding chemical, optical, physical, electrical and magnetic properties and the configurable bandgap, along with their exceptional optoelectronic properties, makes them ideal for developing highly sensitive biosensors. In TMDs, vertically stacked layers are retained by weak van der Waals force. Every layer of TMDs is made up of dichalcogenide atoms and transition metals bonded covalently that are arranged in a hexagonal lattice [27]. They are synthesized via CVD, liquid phase exfoliation, wet-chemical synthesis, mechanical cleavage, and exfoliation techniques. The number of layers determines the properties of TMD materials [28]. The cytotoxicity of 2D TMDs is comparatively small to biological species, which is the demand for any biosensing application. The emphasis of this section will be on cutting-edge research efforts in the biosensing of GSH using 2D TMDs.

Du *et al.* [29] have synthesized water-soluble vanadium disulphide quantum dots using the hydrothermal method with excellent fluorescence emission. Owing to FRET, the emission of VS_2 quantum dots was quenched by MnO_2 nanosheets. Moreover, an appreciable spectral overlap has been observed with the emission spectrum of VS_2 quantum dots and the absorption spectrum of MnO_2 nanosheets, signifying the presence of FRET. With the addition of GSH from 0 to 500 μM to the nanohybrid probe comprising of VS_2 and MnO_2, a "turn-on" response has been observed where nanosheets have been reduced into Mn^{2+} with a LOD of 0.31 μM. They have also demonstrated the upconversion fluorescence property with NIR excitation of 780 nm, which again confirmed the 'turn-on' response of the nanohybrid probe (Figure 4.10). This probe was used for GSH detection in fresh human serum, which could detect the levels of GSH in serum. Using VS_2 QDs is advantageous because they provide excellent detection of GSH in both UV and NIR excitation.

Archana *et al.* [30] have used nanohybrid probes consisting of MoS_2 nanosheets and Cu nanoparticles synthesized by hydrothermal and chemical

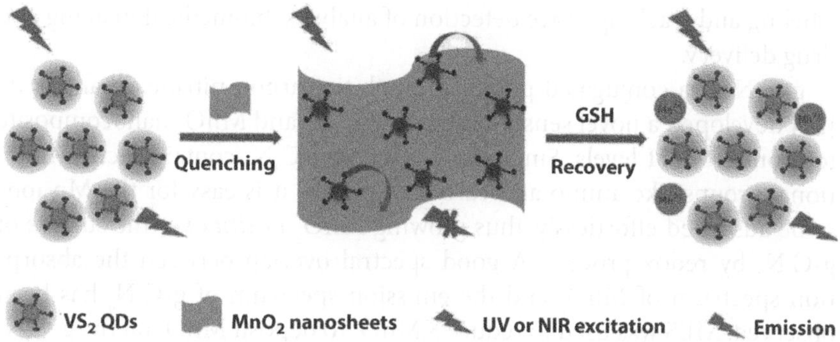

Figure 4.10 Graphical representation of GSH sensing using VS_2 quantum dots and MnO_2 nanosheets. Adapted with permission from reference [29]. Copyright (2018), Elsevier.

reduction method. In the presence of Cu nanoparticles, the emission of MoS_2 has been quenched significantly owing to the NSET process, where the energy has been transferred from MoS_2 to the conduction electrons of Cu nanoparticles. This nanohybrid probe was used for sensing GSH where the concentration of GSH has been varied from 50 to 350 nM. The nanohybrid probe exhibited a "turn-off" response upon adding different concentrations of GSH with an excellent detection limit of 2.55 nM. Here, GSH gets adsorbed on the surface of the nanohybrid probe. GSH being a thiol molecule gets conjugated with the sulfur vacancy of MoS_2, which reduces fluorescence intensity after adding GSH. Since this showed an excellent LOD, it can be a promising probe for detecting GSH.

4.3.4.6 *Polymer Nanoparticles as Donors*

In recent years, the interest towards CPs has expanded dramatically. In the initial days, researchers were fascinated towards developing optoelectronic devices based on CPs owing to their excellent electrical properties. Subsequently, it has been identified that CPs have impressive optical properties that accelerate their application in imaging, sensing and diagnosis. The use of CPs as fluorescent probes has drawn a great deal of interest in biology and medicine over the last couple of years. A significant benefit of using CPs is that their optical properties can be controlled owing to the ability to synthesise new polymer frameworks. Also, CPs are biocompatible with good quantum yields, water-soluble, surface modification, photostability, and less toxic. CPs, owing to their distinctive signal amplification characteristics, have a multitude of applications in phototherapy, cellular labeling and tracking, trace detection of analytes, biomedical imaging and drug delivery.

g-C_3N_4 is a conjugated polymer, graphitic carbon nitride. Zhang *et al.* [31] developed a novel sensor based on g-C_3N_4 and MnO_2 nanocomposite to monitor GSH levels. Since the surface of g-C_3N_4 contains active functional groups like amino and carboxyl groups, it is easy for the Mn ions to be adsorbed effortlessly, thus growing MnO_2 *in situ* over the surface of g-C_3N_4 by redox process. A good spectral overlap between the absorption spectrum of MnO_2 and the emission spectrum of g-C_3N_4 has been observed. MES was used to reduce $KMnO_4$ to deposit MnO_2 on the g-C_3N_4 nanosheet. At different concentrations of $KMnO_4$, the emission intensity of g-C_3N_4 reduced considerably, owing to FRET from g-C_3N_4 to MnO_2. The nanocomposite probe comprising g-C_3N_4 and MnO_2 was studied for GSH detection, which exhibited a 'turn-on' fluorescence response with a LOD of 0.2 μM. Here, GSH gets oxidized to GSSG, and MnO_2 gets reduced to Mn^{2+}. Thus, the proposed nanocomposite probe has been studied to track the intracellular GSH in living cells. Initially. HeLa cells were first incubated with nanocomposite. When studied under scanning confocal microscopy, the nanocomposite incubated HeLa cells show strong blue emission. But HeLa cells that were not incubated with nanocomposite showed no emission, as seen in Figure 4.11. To track the changes in the intracellular GSH level, the HeLa cells were treated for 30 minutes with 500 μM of NMM, a thiol-reactive scavenger, followed by a nanocomposite. As observed from the figure, it is evident that cells incubated with NMM showed no appreciable fluorescent signal since NMM reduces the GSH concentration. Thus, this demonstrates the use of g-C_3N_4 and MnO_2 nanocomposite for easy intracellular detection and monitoring.

Wang *et al.* [32] have constructed a new polymer PBEC using carbazole and hydroquinone as raw materials. To obtain conjugated polymer nanoparticles ($CPNs_{PBEC}$), the designed polymer PBEC with carboxylic acid groups were synthesized by PBEC, and PSMA using co-precipitation method and MnO_2 nanosheets were prepared by hydrothermal method. With the increasing concentrations of MnO_2, the fluorescence of $CPNs_{PBEC}$ has been quenched by MnO_2. There was a significant spectral overlap between the emission spectra of $CPNs_{PBEC}$ and the absorption spectra of MnO_2, signifying the presence of FRET. The concentration of GSH was varied from 0 to 90 μM, and in the presence of GSH, the fluorescence of $CPNs_{PBEC}$ was recovered with LOD 20.4 nM (Figure 4.12). Here, MnO_2 is reduced to Mn^{2+} and GSH is oxidized to GSSG. This nanocomposite probe comprising $CPNs_{PBEC}$ and MnO_2 has been used for detecting GSH

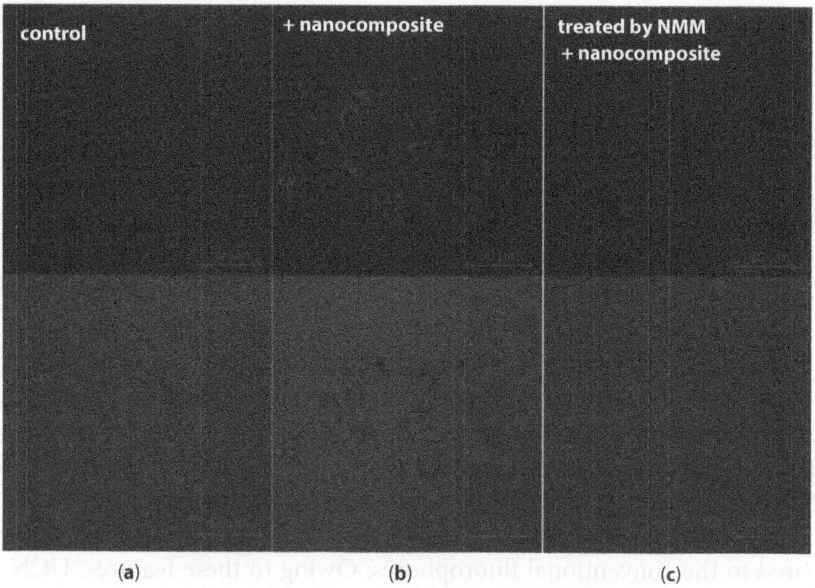

Figure 4.11 HeLa cells under confocal fluorescence microscopy (a) cell as control; (b) cells incubated with the g-C_3N_4–MnO_2 nanocomposite, (c) cells pretreated with NMM (500 μM) for 20 min, followed by nanocomposite. Reprinted (adapted) with permission from reference [31]. Copyright (2014) American Chemical Society.

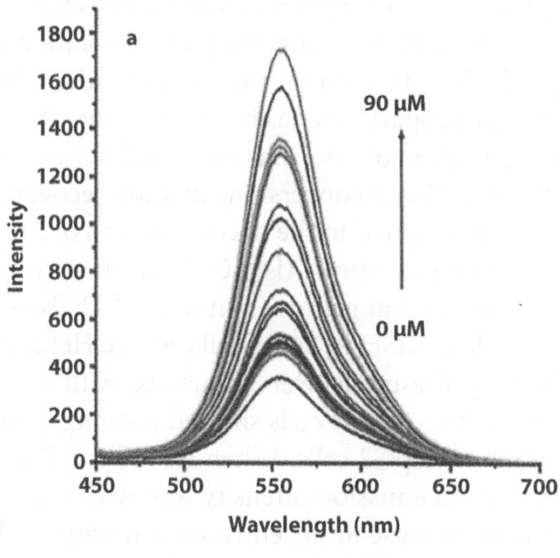

Figure 4.12 Reprinted (adapted) with permission from reference [32]. Copyright (2020) Elsevier.

in human serum. The nanoprobe showed an excellent response to GSH detection in human serum samples.

4.3.4.7 *Upconversion Nanoparticles as Donors*

Upconversion nanoparticles are currently gaining popularity due to their exceptional luminescence since they have the ability to offer visible light emission from the excitation of the near-infrared (NIR) band. Rare earth ions, metal ions and certain organic molecules are found to exhibit upconversion luminescence. UCNPs have garnered recognition owing to their interesting optical properties like high quantum yield, chemical stability, longer lifetimes in the excited states, sharp emission and absorption properties, extremely good photodegradation, and photobleaching resistance. The autofluorescence from biological samples is significantly less in the NIR region. Hence, designing biosensing and bioimaging systems with UCNPs offers good penetration depth, with lesser damage to tissues, compared to the conventional fluorophores. Owing to these features, UCNPs have gained significant interest in the fields of bioimaging and sensing in recent years.

Deng *et al.* [33] have synthesized Yb/Tm co-doped core-shell $NaYF_4$ UCNP. The growth of MnO_2 nanosheets was directed in the presence of MES buffer by the prepared core-shell $NaYF_4$:Yb/Tm@$NaYF_4$ nanoparticles resulting in the formation of MnO_2 nanosheets-modified UCNP as represented by the schematic in Figure 4.13(a). Here, the emission of core-shell NP overlaps with the absorption of MnO_2 nanosheets, confirming the FRET. Moreover, with NIR excitation, the particle solution exhibited emission quenching, which confirmed the presence of FRET. The aqueous solution of these hybrid nanoparticles has been tested for GSH monitoring.

When the concentration of GSH was increased, the emission turned on, with a LOD of 0.9 μM. The upconversion emission recovery after the addition of GSH has been assigned to the decomposition of MnO_2 nanosheets, where MnO_2 is reduced to Mn^{2+}. Also, GSH has been oxidized to GSSG via the thiol-disulfide exchange. MnO_2-modified UCNPs were tested to monitor the intracellular GSH in living cells, where HeLa and HepG2 cells were chosen. Initially, these cells were incubated with nanoparticles for 3 hours, and when analysed, these cells showed a strong emission, implying a high level of GSH in HepG2 cells. When NMM and BSO were added to these cells, a decrease in emission intensity was observed. With HeLa cells treated with LPA, an increase in the emission intensity has been observed. When NMM was added to HeLa cells treated with LPA, a decrease in the emission intensity was observed, as shown in Figure 4.13(b). These findings

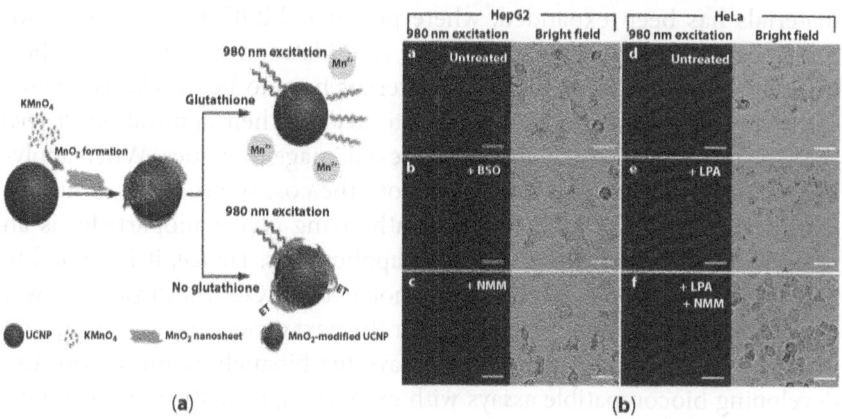

Figure 4.13 (a) Schematic representation of the experimental design of MnO_2 modified UCNPs (b) Upconversion emission and bright-field microscope images for intracellular detection of GSH in HepG2 and HeLa cell line (a) HepG2 cells labeled with nanocomposite (b) HepG2 cells treated with BSO followed by incubation with nanocomposite and (c) HepG2 cells treated NMM followed by incubation with nanoparticles (d) HeLa cells labeled with nanocomposite (e) HeLa cells pretreated with LPA followed by incubation with nanoparticles (f) HeLa cells pretreated with LPA, NMM followed by incubation with nanocomposite. Reprinted (adapted) with permission from reference [33]. Copyright (2011) American Chemical Society.

strongly suggest the practicability of tracking GSH intracellularly with the help of MnO_2-modified UCNPs.

4.4 Conclusions

The rapid advancement of nanotechnology in recent years has propelled biosensing into the age of nanoprobes, where diseases can be diagnosed at the cell and molecular levels more quickly, accurately and efficiently. Integrating nanomaterials into fluorescent sensors has greatly enhanced the functionality of fluorescent sensors in terms of sensitivity, photostability, shelf life and multiplexed capability. The architecture of fluorescence sensing and imaging varies according to the application, and numerous distinctive nanomaterial combinations are found in the literature. GSH is the focus of numerous studies dedicated to creating fluorescence-based sensors and imaging probes due to its significance. Remarkable advancements have been made in developing fluorescent sensors for sensing GSH based on energy transfer mechanisms. Though metal nanoparticles and carbon, graphene quantum dot-based materials are used, a repertoire of materials has been expanded, where polymer, MOF, and UCNP nanomaterials have emerged as a prevalent candidates in biosensing and bioimaging. Many UCNP-based nanomaterials have to be developed, which can span from visible to NIR regions to expand their application *in vivo* imaging technique, which also causes less damage to tissues. When applying nanoparticles for real-time detection, the cost of synthesis should be considered. E.g., The high cost of synthesizing gold nanoparticles is an underlying constraint for a variety of applications. Hence, it is pivotal to explore and develop cost-effective nanoparticles. Optical biosensors will undeniably be an effective technique for the next generation of sensing and imaging. Forthcoming nanoprobe assays for bioanalysis must prioritise developing biocompatible assays with excellent specificity and sensitivity, lower cost and excellent stability for *in vivo* and *in vitro* applications.

References

1. Tian, M., Liu, X.Y., He, H., Ma, X.Z., Liang, C., Liu, Y., Jiang, F.L., Real-Time Imaging of Intracellular Glutathione Levels Based on a Ratiometric Fluorescent Probe with Extremely Fast Response. *Anal. Chem.*, 92, 10068–10075, 2020.
2. Kannan, J.A. and Balasubramanian, K., Thermally influenced, optical and fluorescence properties of Zinc Oxide nanoparticles for glutathione sensing. *Appl. Phys. A Mater. Sci. Process.*, 126, 602, 2020.
3. Fiser, B., Szóri, M., Jójárt, B., Izsák, R., Csizmadia, I.G., Viskolcz, B., Antioxidant potential of glutathione: A theoretical study. *J. Phys. Chem. B.*, 115, 11269–11277, 2011.
4. Chen, M. and Yin, M., Design and development of fluorescent nanostructures for bioimaging. *Prog. Polym. Sci.*, 39, 2, 365–395, 2014.
5. Lakowicz, J.R., *Principles of fluorescence spectroscopy*, Third edition, Springer New York, NY, 2006.
6. Kaur, A., Kaur, P., Ahuja, S., Förster resonance energy transfer (FRET) and applications thereof. *Anal. Methods*, 12, 5532–5550, 2020.
7. Torabfam, M., Kurt, H., Bayazıt, M.K., Yüce, M., Plasmonic nanometal surface energy transfer-based dual excitation biosensing of pathogens. *Sens. Diagn.*, 1, 765–774, 2022.
8. Yun, C.S., Javier, A., Jennings, T., Fisher, M., Hira, S., Peterson, S., Hopkins, B., Reich, N.O., Strouse, G.F., Nanometal surface energy transfer in optical rulers, breaking the FRET barrier. *J. Am. Chem. Soc.*, 127, 3115–3119, 2005.
9. Persson, B.N.J. and Lang, N.D., Electron-hole-pair quenching of excited states near a metal. *Phys. Rev. B*, 26, 5409–5415, 1982.

10. Li, C., Chen, G., Zhang, Y., Wu, F., Wang, Q., Advanced fluorescence imaging technology in the near-infrared-II window for biomedical applications. *J. Am. Chem. Soc.*, 142, 14789–14804, 2020.
11. Wolfbeis, O.S., An overview of nanoparticles commonly used in fluorescent bioimaging. *Chem. Soc. Rev.*, 44, 4743–4768, 2015.
12. Shi, J., Tian, F., Lyu, J., Yang, M., Nanoparticle based fluorescence resonance energy transfer (FRET) for biosensing applications. *J. Mater. Chem. B*, 3, 6989–7005, 2015.
13. Ji, C., Zhou, Y., Leblanc, R.M., Peng, Z., Recent Developments of Carbon Dots in Biosensing: A Review. *ACS Sensors*, 5, 2724–2741, 2020.
14. Li, M., Chen, T., Gooding, J.J., Liu, J., Review of carbon and graphene quantum dots for sensing. *ACS Sensors*, 4, 1732–1748, 2019.
15. Dewangan, L., Korram, J., Karbhal, I., Nagwanshi, R., Satnami, M.L., N-Doped Carbon Quantum Dot-MnO2 Nanowire FRET Pairs: Detection of Cholesterol, Glutathione, Acetylcholinesterase, and Chlorpyrifos. *ACS Appl. Nano Mater.*, 4, 13612–13624, 2021.
16. Cai, Q.Y., Li, J., Ge, J., Zhang, L., Hu, Y.L., Li, Z.H., Qu, L.B., A rapid fluorescence "switch-on" assay for glutathione detection by using carbon dots-MnO2 nanocomposites. *Biosens. Bioelectron.*, 72, 31–36, 2015.
17. Yan, X., Song, Y., Zhu, C., Song, J., Du, D., Su, X., Lin, Y., Graphene Quantum Dot- MnO2 Nanosheet Based Optical Sensing Platform: A Sensitive Fluorescence "turn Off-On" Nanosensor for Glutathione Detection and Intracellular Imaging. *ACS Appl. Mater. Interfaces*, 8, 21990–21996, 2016.
18. Solanki, P.R., Kaushik, A., Agrawal, V.V., Malhotra, B.D., Nanostructured metal oxide-based biosensors. *NPG Asia Mater.*, 3, 17–24, 2011.
19. Yoon, Y., Truong, P.L., Lee, D., Ko, S.H., Metal-Oxide Nanomaterials Synthesis and Applications in Flexible and Wearable Sensors. *ACS Nanosci. Au.*, 2, 64–92, 2022.
20. Zhang, Y., Wang, D., Meng, Y., Lu, W., Shuang, S., Dong, C., Biodegradable Fluorescent SiO2@MnO2-Based Sequence Strategy for Glutathione Sensing in a Biological System and Synergistic Theragnostics to Cancer Cells. *ACS Sustain. Chem. Eng.*, 9, 2770–2783, 2021.
21. Wang, H.B., Chen, Y., Li, Y., Liu, Y.M., A sensitive fluorescence sensor for glutathione detection based on MnO2 nanosheets-copper nanoclusters composites. *RSC Adv.*, 6, 79526–79532, 2016.
22. Li, T., Wang, Z., Jiang, D., Wang, H., Lai, W.F., Lv, Y., Zhai, Y., A FRET biosensor based on MnO2 nanosphere/copper nanocluster complex: From photoluminescence quenching to recovery and magnification. *Sens. Actuators B: Chem.*, 290, 535–543, 2019.
23. Fu, W., Wang, H., Chen, Y., Ding, J., Shan, G., Fluorescence sensing analysis for rapid detection of serum glutathione based on degrading AuNCs@Lys-MnO2 fluorescence resonance energy transfer system. *Microchem. J.*, 159, 105556, 2020.

24. Mohan, B., Kumar, S., Xi, H., Ma, S., Tao, Z., Xing, T., You, H., Zhang, Y., Ren, P., Fabricated Metal-Organic Frameworks (MOFs) as luminescent and electro-chemical biosensors for cancer biomarkers detection. *Biosens. Bioelectron.*, 197, 113738, 2022.
25. Wang, N., Xie, M., Wang, M., Li, Z., Su, X., UiO-66-NH2 MOF-based ratiometric fluorescent probe for the detection of dopamine and reduced glutathione. *Talanta*, 220, 121352, 2020.
26. Gong, C., Li, Z., Liu, G., Wang, R., Pu, S., A sensitive fluorescence "turn on" nanosensor for glutathione detection based on Ce-MOF and gold nanoparticles. *Spectrochim. Acta - Part A Mol. Biomol. Spectrosc.*, 265, 120362, 2022.
27. Sarkar, D., Xie, X., Kang, J., Zhang, H., Liu, W., Navarrete, J., Moskovits, M., Banerjee, K., Functionalization of transition metal dichalcogenides with metallic nanoparticles: Implications for doping and gas-sensing. *Nano Lett.*, 15, 2852–2862, 2015.
28. Ping, J., Fan, Z., Sindoro, M., Ying, Y., Zhang, H., Recent Advances in Sensing Applications of Two-Dimensional Transition Metal Dichalcogenide Nanosheets and Their Composites. *Adv. Funct. Mater.*, 27, 1–18, 2017.
29. Hu, H., Zavabeti, A., Quan, H., Zhu, W., Wei, H., Chen, D., Ou, J.Z., Recent advances in two-dimensional transition metal dichalcogenides for biological sensing. *Biosens. Bioelectron.*, 142, 111573, 2019.
30. Janani, A.K. and Karthikeyan, B., Probing nanometal surface energy transfer between nanocopper and MoS2 for glutathione sensing. *J. Phys. D. Appl. Phys.*, 55, 265101, 2022.
31. Zhang, X.L., Zheng, C., Guo, S.S., Li, J., Yang, H.H., Chen, G., Turn-on fluorescence sensor for intracellular imaging of glutathione using g-C3N4 Nanosheet-MnO2 sandwich nanocomposite. *Anal. Chem.*, 86, 3426–3434, 2014.
32. Wang, Q.-B., Zhang, C.J., Lu, Q., Liu, Z.-E., Yao, J.-S., Zhang, X., A "turn-on" fluorescence platform of detection glutathione using MnO2 nanosheets quenched fluorescent conjugated polymer nanoparticles. *Dye. Pigment.*, 176, 108189, 2020.
33. Deng, R., Xie, X., Vendrell, M., Chang, Y.T., Liu, X., Intracellular glutathione detection using MnO2-nanosheet-modified upconversion nanoparticles. *J. Am. Chem. Soc.*, 133, 20168–20171, 2011.

5

Diagnosis of Physical Stimuli Response Enhances the Anti-Quorum Sensing Agents in Controlling Bacterial Biofilm Formation

Govindan Ramachandran[1], Balamurugan Palanisamy[1], Gnansekaran Chackaravarthy[1], Chenthis Kanisha Chelliah[2], Govindan Rajivgandhi[1,3*], Franck Quero[3] and Manoharan Natesan[1†]

[1]*Department of Marine Science, Bharathidasan University, Tiruchirappalli, Tamil Nadu, India*
[2]*Department of Nanotechnology, Noorul Islam Centre for Higher Education, Kumaracoil, Kanyakumari, Tamil Nadu, India*
[3]*Laboratorio de Nanocelulosa y Biomateriales, Departamento de Ingeniería Química, Biotecnología y Materiales, Facultad de Ciencias Físicas y Matemáticas, Universidad de Chile, Santiago, Chile*

Abstract

Biofilm is a self-assembled microbial community that attached each other in inherent surface and encapsulated by extracellular polymeric substances. The mechanism of biofilm is one of the important virulence factors in Gram-negative bacteria, which not only difficult to eradicate with existing antibiotics treatment but also enables bacteria to be a major problem in clinical settings. Importantly, the cell–cell communication process is helped to provide strong mechanical stability and also provide the nutrients to biofilm. In particular, the biofilm cells communicate each other through harmonic signaling molecule is referred as quorum sensing (QS), influenced by QS molecules. Especially, Gram-negative bacteria (GNB) biofilm formation depends on the cell signaling molecules, and it can alter and regulate the metabolic pathways due to the stimulation of toxin genes and other QS signaling molecules, such as N-acylhomoserine lactones (AHL) and

**Corresponding author*: rgovindan@ing.uchile.cl
†*Corresponding author*: biomano21@yahoo.com

Sathish-Kumar Kamaraj, Arun Thirumurugan, Muthuchamy Maruthupandy, Mercedes Guadalupe López Pérez and Shanmuga Sundar Dhanabalan (eds.) Nanoscience and Nanotechnology for Smart Prevention, Diagnostics and Therapeutics: Fundamentals to Applications, (91–116) © 2024 Scrivener Publishing LLC

oligopeptides. Both the QS molecules are autoinducers, and play an important role in the construction of biofilm behavior in bacteria due to the gene regulations. Therefore inhibition of QS using QS inhibitors is a best choice to eradicate the complete biofilm formation of pathogens, and also promising approach to control severe infections. Recent years, stimuli-responsive drug delivery nanocarriers are the emerging weapons to eradicate the QS genes in controlling of biofilm formation. In this inhibition mechanism, the stimuli-responsive drug delivery nanocarriers are constructed by various natural and chemical sources that enable a considerable alteration in response to intrinsic/chemical stimuli and extrinsic/physical stimuli including temperature, pH, NaCl, redox and enzymes and light sources, ultrasound, near infrared, magnetic sources, and also electrical field. Further, different strategies are fabricated to arrest the QS process of bacteria in clinical field, such as usage of liposoluble drugs, antibiotics, bioactive compounds, and choose the surface of the bacterial tissue for minimize the toxicity effect, and then controlled drug release in the targeted sites using different stimuli. All these factors are promising factors for improve the anti-QS drugs to inhibit the biofilm formation of gram negative bacteria. Hence, the current study of stimuli-responsive drug delivery nanocarriers is the best choice to control the drug release in target sites, and influence on biofilm eradication therapy.

Keywords: Quorum sensing, biofilm formation, physical stimuli response, nanocarrier, controlled drug delivery

5.1 Introduction

According to studies from the World Health Organization (WHO), antimicrobial resistance is a serious worrying danger on a global scale and has resulted in serious public health concerns in recent years. Antimicrobial resistance is the second most death-causing infectious disease in the world which challenges the search for new antibiotics in the treatment of multidrug-resistant infections [1]. Whereas the performance of antibiotics against microorganisms is substantially lowered by the counteraction of microbes against antibiotics is a serious issue that needs to be addressed which causes medical and socio-economic issues [2]. Biofilms are communities of microorganisms that adhere to surfaces and produce extracellular polymeric substances (EPS) [3]. These EPS play a crucial role in providing the stability and adhesion required for biofilm formation. The extracellular polymeric substance is a mixture of proteins, extracellular deoxyribonucleic acid, lipids, and polysaccharides. Biofilms play a major role by inactivating antibiotics by decomposing or absorbing through enzymes or they form a barrier preventing the penetration of the antibiotic. One of the major concerns of biofilm-forming bacteria is their number of folds differing from the normal microbes (i.e., 1000 folds larger), which makes the bacteria more resistant to antibiotics and causes recurrent infections and mortality [4]. Biofilms have become a significant concern in medicine and industry as they can cause infections, corrosion, and fouling. Current treatments for biofilms are inadequate due to their resilience and the development of antibiotic resistance. Therefore, there is a significant need for a new approach to preventing or eradicating biofilms [5].

Nanotechnology has emerged as a promising field for resolving the issue of biofilm formation. Nanoparticles, including both organic and inorganic varieties, have been studied extensively for their potential to disrupt or prevent biofilm formation. These materials can function as therapeutic agents or as means of delivering different antibiotics or bactericidal agents to biofilms. Nanomaterials are more suitable for the delivery of drugs in various disease treatments due to their significant features such as flexibility in size, shape, surface, charge, solubility, and prolonged release of drugs in required minimal concentration, which reduces the side effects and decreases the drug toxicity [6]. The important aspects to be considered for the construction of optimal drug delivery systems are (i) improved antimicrobial treatment effectiveness, (ii) increased local drug concentration at the target site, (iii) reduced antibiotic accumulation in uninfected host tissues, and (iv) decreased risks of toxicity. These nanoparticles can be designed to target specific bacterial strains or to disrupt the biofilm matrix. Furthermore, nanocarriers can be used to deliver antimicrobial agents directly to the biofilm, increasing their effectiveness [7].

5.1.1 Biofilm Formation and Quorum Sensing Mechanism

Quorum sensing (QS) is a mechanism used by bacteria to communicate with each other and coordinate their behavior, such as the expression of virulence factors or the formation of biofilms. It involves the production and detection of signaling molecules called autoinducers (AI), which accumulate in the extracellular environment and reach a threshold concentration that triggers a response in the cells [8]. The formation of biofilm occurs due to various factors such as nutrient deficiency, increased salt concentrations, and high pH levels [9]. The biofilm formation of microbes is a dynamic process that takes place in five different stages (i) adherence of bacteria to biotic and abiotic surfaces via van der Waals forces, electrostatic contacts, sedimentation, etc. (ii) Growth and multiplication of microorganisms on the substrate accompanied by EPS secretion, which increases bacterial adherence and promotes the creation of three-dimensional biofilm formations and microbe aggregation. (iii) Maturing stage of the biofilms where the microbial aggregations develop into huge clusters. (iv) The bacterial density in the biofilm has reached the maximum permissible level and the biofilm is fully developed. (v) The genes linked with fimbriae or flagella are activated, which stimulates the development of fimbriae or flagella, allowing the microbe to detach from the biofilm and migrate to new environments. Also, mature biofilm microbes are able to produce some hydrolytic enzymes, which aid in the bacteria's separation from the biofilm [10].

Antiquorum sensing activity refers to the ability of certain compounds or microorganisms to disrupt or inhibit the cellular communication process called quorum sensing (QS) in bacteria.

5.1.2 Stimuli–Response Systems

The concept of stimuli-responsive systems was initially proposed in the late 1970s with the use of thermosensitive liposomes for the local release of medications via hyperthermia. The basic concept of Stimuli-responsive materials is their smart response to the exogenous or endogenous stimulus which are responsive to environmental cues or pathological disorders, capable of changing their dynamic structure or shape in response to external stimuli in a variety of ways, acquiring additional special physical or chemical characteristics, and performing specified functions [11]. Recent studies reported the effectiveness of the drug delivery systems built from a combination of stimuli-responsive components had drastically improved the therapeutics and resolved the contradiction between therapy and biocompatibility. This innovation demonstrated excellent promise for application in the sectors of tissue engineering, antibacterial, and medication delivery [12].

The difference between the conventional dosage drugs and stimuli-responsive drugs was discussed in the Table 5.1 below.

A drug delivery system is a device designed to carry therapeutic medicines through biological membranes without any biomolecule compound loss to the target areas. Nanodrug delivery systems are the most popular among the various types of drug delivery systems due to their biocompatibility and capacity to deliver drugs effectively to the target region.

Table 5.1 Difference between conventional dosage drugs and stimuli-responsive drugs.

Conventional dosage drugs	**Stimuli-responsive drugs**
Short half-life	Long half-life and sustained drug release
Low therapeutic index	Broad therapeutic index
Unspecified target delivery	Specific target delivery
Low solubility	High solubility
Large-size compounds with limited access	Small size easy cell penetration
Frequent administration of higher dose concentration	High bioavailability
Fluctuations in drug concentrations	Controlled release

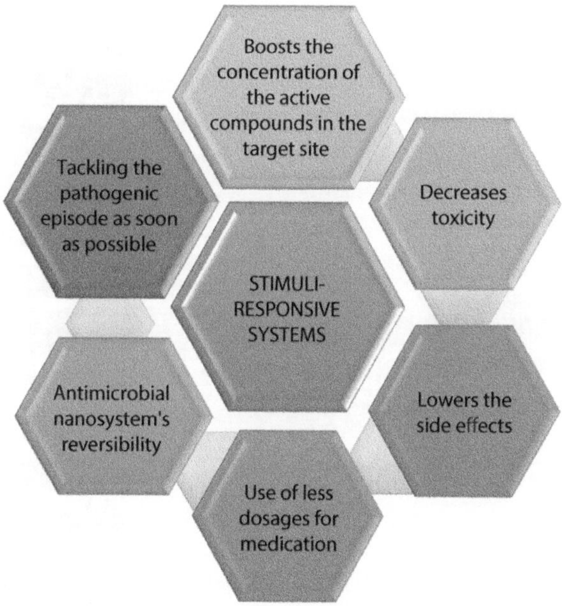

Figure 5.1 Advantages of stimuli-responsive smart drug delivery systems.

The prominence of stimuli-responsive DDSs has risen significantly for their efficient drug release at the target sites which can actively perceive and react to internal or external stimuli for the regulated release of the therapeutic agents [13]. The advantages of the stimuli-responsive smart drug delivery systems are given in Figure 5.1.

5.2 Types of Stimuli Response for Material Synthesis

The stimuli-responsive materials are classified into two major types one that responds to the physical or extrinsic stimuli (i.e. light, magnetic field, heat, etc.) and another type that responds to the microbial stimulus or intrinsic stimulus (i.e. pH, enzymes, redox, etc.). The extrinsic stimulus-responsive materials are constructed by altering their whole molecular structure whereas, the intrinsic stimulus-responsive materials respond to the stimuli through unique chemical bonds, such as those that are sensitive to acid and enzymes [14]. The various types of stimuli-responsive materials are available in Figure 5.2.

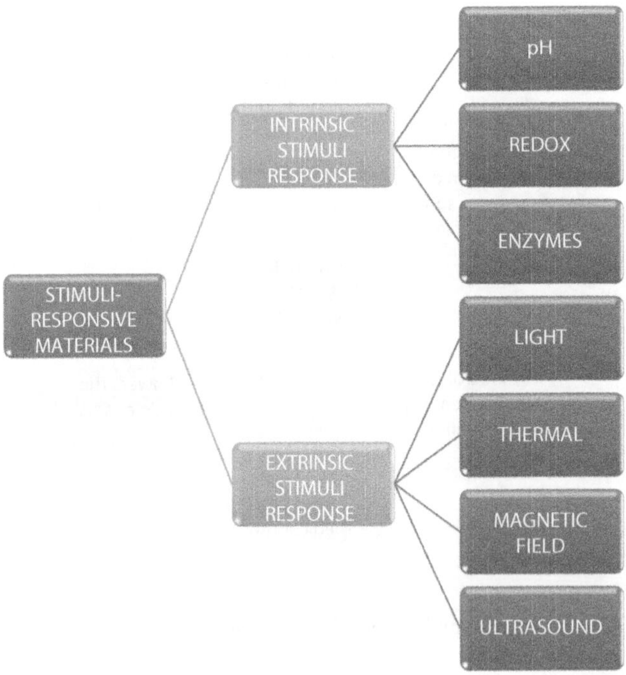

Figure 5.2 Types of stimuli-responsive material.

5.2.1 Physical Stimuli-Response

Physical stimuli, such as temperature changes, electrical fields, and mechanical stress, have been shown to affect biofilm formation and activity. The use of nanocarriers that respond to these physical stimuli could provide a novel approach to controlling biofilms. For example, thermo-responsive nanocarriers could deliver antimicrobials to the biofilm at elevated temperatures, which can disrupt the biofilm's structure [15].

5.2.1.1 Light Responsive Systems

Light is a non-harmful, excellent external biocompatible stimulus. Drugs for the treatment of biofilm microorganisms can be released under regulated conditions using light. Because the light source has characteristics that are unaffected by pH, temperature, and ionic strength. The usage of light-responsive smart materials is widespread in a number of industries, including biosensing, cancer therapy, and tissue repair [16]. Since the longer wavelength offers deeper tissue penetration and very little harm to living things, near-infrared (NIR) light is especially useful for treating biofilm infections. Near infra-red (NIR) light can also function as a thermal trigger because it can be converted into heat by photothermal agents. Following the light-induced excitation of the photosensitizer present in the photo-responsive material, the optical signal is changed into a physicochemical signal via the photochemical reaction process, resulting in the homologous physicochemical and physical properties being acquired by the photosensitive material [17].

The research of Li and his colleagues reported the thermosensitive phospholipid, distearoyl phosphatidylcholine (DSPC), and quaternized cholesterol were combined to generate NIR-responsive liposomes carrying antibiotics and NIR photothermal agent (cypate). By using electrostatic contact, the liposomes could enter biofilms and administer antibiotics since they were stable at physiologically normal temperatures. Due to the NIR radiation, the heat emitted by the photothermal effect of cypate turns liposomes into a liquid phase from their original gel phase, favoring the permeability of microbial cell membranes and the release of antibiotics. The results of the study confirmed that synergistic therapy efficiently eradicated pathogenic biofilms both *in vitro* and *in vivo* [18].

5.2.1.2 Photodynamic Therapy

Recently, Von Tappeinerin defined "photodynamic therapy" in 1904 to refer to the phenomenon of atmospheric oxygen and light aiding in wound healing. In photodynamic therapy, the photosensitizer, photon, and oxygens interaction are essential where the photosensitizer absorbs the light of specific wavelengths and turns into an excited state. Photodynamic therapy (PDT) requires interaction between the photon, ambient oxygen, and photosensitizer. The photosensitizer absorbs a certain wavelength of light, which excites it into the singlet state. After that, it passes through the Triplet State Intersystem Crossing, it either transfers the energy to molecular oxygen to create singlet oxygen or interacts with biomolecules to produce other reactive oxygen species such as superoxide, hydrogen peroxide, etc. It is generally known that singlet oxygen and too much ROS inside cells can cause cell toxicity [19].

Recent research by Chatterjee [20] and colleagues showed that combining antibiotics with manufactured nanoparticles (NPs) with band gap energy of 2.4 eV for regulated superoxide production could greatly boost the antibacterial impact. In 2020 Yan and his colleagues [21] developed a photosensitizer-modified biodegradable Zeolitic imidazolate scaffold-8 (ZIF-8) nanocomposite for enhanced antibacterial applications via PDT and Zn2+ ion against *Staphylococcus aureus* and methicillin-resistant *S. aureus* (MRSA). Chlorin e6 (Ce6), a PDT agent, and Zn2+ ions, a toxin, make up the nanocomposite. It's interesting to note that the nanocomposite has a rough surface that significantly affects how it interacts with bacterial cells. The nanocomposite's surface area is increased by the rough surface, increasing the likelihood that germs will be eliminated by the contact-killing process. By using poly(acrylic acid sodium salt) (PAAS) as a flexible template and adding cationic Zn2+ that was bound to the negative carboxylate group in the PAAS polymer chain, ZIF-8 nanocomposite was created. Next, a ZIF-8 nanocrystal was produced on the PAAS nanosphere as a result of a surface interaction with 2-methylimidazole (2-MIMs) and Zn2+. Finally, Ce6 conjugated with 3-aminopropyl triethoxysilane (APTES) was integrated into the nanocomposite to produce drug-loaded nanoparticles (NPs). The ZIF-8 nanocomposites had a diameter of 50–90 nm, making them ideal for use as nanodrug delivery systems (NDDSs).

They demonstrated the ZIF-8 nanocomposites' concentration-dependent antibacterial activity.

At high concentrations (200 mg/mL), the poisonous effect of Zn2+ ion prevented the influence of light contact from being seen due to its ability to stop the growth of *S. aureus* and MRSA. However, in the presence of 650 nm

LED light, the NPs exerted full suppression of *S. aureus* and MRSA at lower doses (22.2 and 7.4 g/mL). Importantly, when compared to free Ce6, immobilization of Ce6 to ZIF-8 NPs showed a considerable improvement in ROS production capacity and photostability.

In a recent study, Zhdanova and his team [22] demonstrated that pluronic F-127 polymer micelles are excellent nanosystems for the solubilization of cationic porphyrins, such as meso-arylporphyrins that contain pyridyl. In the absence and presence of light, respectively, Pluronic F-127 nanomicelles decreased the photosensitizer dose by two and four times. When compared to the killing impact on *S. aureus*, the suppression of *Escherichia coli* bacteria by Pluronic F-127 nanomicelles of pyridyl-containing meso-arylporphyrins was somewhat lower.

Hou and his colleague's [23] study revealed that polymeric micelles combined with silver nanoparticles Ag NPs can achieve a synergistic antibacterial effect against drug-resistant *S. aureus*. In this research, they prepared photosensitizer protoporphyrin IX (PpIX) and poly(poly aspartic acid)-block-poly(e-caprolactone) copolymer micelles fused with silver nanoparticles (Ag NPs) showed effective eradication of subcutaneous infections caused by drug-resistant *S. aureus*.

Boukherroub and his team [24] created a carbon dot nanodrug delivery system by combining ampicillin drugs with PDT to intensify the antibacterial effect on exposure to visible light. They utilized the amine-functionalized carbon dots (CD-NH2) as the ampicillin drug carrier vesicle. The results showed the carbon dots were seen to be readily absorbed by the cells and primarily located in the cytoplasm. The minimum inhibitory concentration MIC of *E. coli* treated with CD-NH2 was reduced by 44% compared to the ampicillin drug alone. Carbon dot with ampicillin drug (CD-NH2) also increased the stability of the drug ampicillin and its effective antibacterial activity against *E. coli* even after preserving it for two weeks. In comparison to AMP or CDs-NH2 (PDT agent), CDs-AMP in the presence of visible light had increased bactericidal activity, illustrating the synergistic impact.

5.2.2 Chemodynamic Therapy

Chemodynamic therapy is a further uprising prominent strategy utilized for effective antibacterial application. In this method, the chemo-dynamic therapeutic agents change the endogenous hydrogen peroxide (H_2O_2) into extremely reactive hydroxyl radical (OH) through the Fenton reaction. Here the hydroxyl radical oxidizes double bonds in lipids and thiol groups in the proteins at the surface membranes eliminating the cell wall [25].

Wu and his team [26] offered a combination strategy for rapid of bacterial destruction and fast wound healing by combining PTT, PDT, and CDT in 2018. In this study, they reported hybrid modified Mesoporous silica nanoparticles (MSNs) CuS nanoparticles and hydrogel consisting of 3-(trimethoxysilyl)propyl methacrylate for the treatment of *S. aureus* and *E. coli*. This nanomaterial demonstrated controlled photothermal and photodynamic capabilities after being exposed to an 808 nm laser, and it was 99.80% and 99.94% effective against *S. aureus* and *E. coli* within 10 minutes of exposure.

The same team [26] also created a nano-catalytic antibacterial system made of gold nanoparticles (Au NPs) and polydopamine-assisted hydroxyapatite nanorods to treat skin wounds using a combination of PTT and CDT. The nanosystem demonstrated improved antibacterial efficiency of 96.8% and 95.2 percent against *E. coli* and *S. aureus*, respectively, under exposure to 808 nm light at a temperature of 45 °C with no harm to healthy tissues. PDA@Au-HAp NPs also encouraged the expression of associated genes in NIH3T3 cells and aided in the production of collagen and granulation tissue to speed up the healing of wounds. Overall, this synergistic antibacterial mechanism has been used to efficiently destroy germs and support the regeneration of shin tissue.

5.3 Thermal Responsive Systems

The thermo-responsive system is a widely researched method for drug administration that responds to stimuli for a number of disorders. The medicine that is put into thermo- responsive nanocarriers can be released when the temperature of their environment changes. These nanocarriers only have one component made of a temperature-sensitive material. These systems are capable to deliver their loaded medicine at locations of locally heated tissues, which are typically present at the sites of inflammation, injury, infection, and malignancy [27]. These thermo-responsive materials are classified into two types (i) water-insoluble lower critical solution temperature (LCST), and (ii) water-insoluble below the upper critical solution temperature (UCST). These thermo-responsive systems are linked to the temperature decrease phenomenon where the porosity of the nanocarrier system may improve due to a local cooling effect, allowing the medicines to freely diffuse at the target site. Poly (N-isopropyl acrylamide) is one of the commonly used components for the preparation of thermo-responsive polymeric nanocarrier due to its ability to transition between different temperatures [28]. The difficult procedure of modifying the transition temperature of PEG analogs and the limited functionality of PNIPAM provide difficulties for thermo-responsive materials.

5.3.1 Photothermal Release

Photothermal (PTT) materials when illuminated by the light of a particular wavelength, absorb the light and turn it into heat, which leads to the thermal ablation of unnecessary cells at the target site by causing protein denaturation, cellular cavitation, rupture, etc. Photothermal (PTT) materials emerge as an attractive method for their promising local heating strategy due to their high inherent specificity, and minimal intrusive load [12]. Because of their huge absorbance cross sections, variable optical characteristics, and very effective light-to-heat conversion, plasmonically active metal nanoparticles including Au NPs, Ag NPs, and Cu NPs provide exceptionally potent photothermal contrast agents [29]. Metallic NPs and their surroundings are heated as a result of electron-photon collisions that release the energy accumulated by the electrons during photo-excitation at this resonance which can be either used for drug delivery for the destruction of encapsulating material or thermal heat deposition at a local site.

Radt and his team [30], implanted lysozyme crystal and gold nanoparticles Au NPs into the polyelectrolyte shell prepared by layer-by-layer deposition of positively charged poly9allylamine hydrochloride), and negatively charged poly(sodium 4-styrene sulfonate). The capsules were introduced to a suspension of Micrococcus lysodeikticus after being exposed to a 10 ns pulsed laser (1064 nm) at 50 mJ cm^2 for 5 min at a frequency of 10 Hz. The rapid digestion of bacteria was observed after the release of lysozymes from the capsules through photothermal radiation.

Amoli-Diva *et al.* [31] had done work on implanting Ag NPs in a poly(butyl methacrylate-co-acrylamide-co-methacrylic acid) hydrogel for the controlled release of ofloxacin antibiotic with laser irradiation with a 405nm laser for 15 s at regular intervals. A higher amount of ofloxacin (greater than 80% of the total weight) was released from the laser-irradiated sample after 70 minutes from the start of the experiment compared to the non-irradiated samples.

In subsequent work, Wang Liu [32] and his team created an effective antibacterial hybrid based on gold nanorods coated with polydopamine (PDA) for combined chemo-photothermal therapy. The studies showed the high efficiency of the polydopamine (PDA) coating with high silver ion loading capacity and chitosan functionalization. The hybrid was tested against gram-negative methicillin-resistant *S. aureus* (MRSA) and gram-negative *E. coli* in *in vitro* conditions where the samples were irradiated at 808nm-laser at 0.5Wcm^2 for 7 min. Additionally, photothermal therapy was implemented in order to create a chemo-photothermal combinational bactericide system and enhance the antibacterial effect. The hybrid had shown great results against MRSA 99.6% and *E. coli* 96.8% in *in-vitro* conditions. In addition, using the mouse subcutaneous abscess model the *in vivo* biocompatibility and *in vivo* antibacterial efficacy was examined. The mice showed no evidence of inflammation and abscess when the hybrid system was intravenously injected and irradiated in the near-infrared region (NIR) at 808nm-laser at 0.8 Wcm^2 for 6 min. The scientists conclude by proposing that this antimicrobial depot will be able to freely circulate in the bloodstream, collect in acidic infectious locations where they become positively charged, and specifically target negatively-charged bacteria-cell interfaces. They will produce heat near the target bacterium when the external light is activated.

Borzenkov and his team [33] developed a poly (vinyl alcohol) film entailing five-branched gold nanorods which exhibited enhanced photothermal efficiency when irradiated at three near-infrared (NIR) laser wavelengths such as 730, 680, and 1064 nm. *E. coli* that had developed on the film surface was demonstrated to be completely destroyed by the local temperature increase caused by NIR.

Wang *et al.* [34] developed an anti-biofilm phototherapy platform with high thermal support for antimicrobial capacity by depositing a photothermal multifunctional composite coating, specifically Ti-PDA/BP/ZnO, on a titanium (Ti) substrate. The constructed coating not only showed great photothermal conversion ability but also improvised the susceptibility of bacteria to Zn2+ while boosting the degradation of *E. coli* and *S. aureus* biofilms when exposed under 808 nm NIR irradiation. Therefore, the NIR-triggered ZnO-PTT synergistic antibacterial strategy provided a favorable idea for *in situ* anti-biofilm adhesion of medical implant coatings, opening up new opportunities for the prevention and treatment of clinical implant infections.

5.3.2 Magnetothermal Release

Superparamagnetic iron oxide nanoparticles (SPIONs) have been used to demonstrate the magneto-thermally induced release of antibiotics from the drug delivery systems. SPIONs comprises a core made up of iron oxides which can be used to target specific site using magnets. SPIONs are preferred for their biocompatibility, safety, and surface-active molecules which makes them more applicable in various activities like drug delivery, biosensing, medical screening, and regenerative medicine [35]. SPION drug delivery systems are made up of magnetite and maghemite nanoparticles either with organic or inorganic coating. These magnetic drug-carrying nanostructures navigate to their target tissues using an external magnetic field with increased drug release efficiency and reduced toxicity or overdosage at a specific site. Magnetic delivery systems, such as magnetic capsules, magneto liposomes, and magneto dendrimers, prolong the blood-circulation period of medications and decrease their clearance [36]. The alternating magnetic field causes a phase lag between the applied field and the NP magnetization during magnetothermal heating, generating a hysteresis loop. Changes in spin state and the induced rotation that causes friction between the particle and the surrounding fluid, respectively, are what cause the hysteresis' surplus energy to be thermally released. These processes are referred to as Néel and Brownian relaxations [37]. SPIONs with the help of an external magnetic field are magnetized to their saturation magnetization and acquire a high magnetic susceptibility and upon removing the magnetic field, no magnetic interactions are seen [38].

Mohapatra and associates demonstrated the on-demand release of vancomycin using chitosan-based microbeads that were loaded with magnetite SPIONs [38]. The cross-linking of vancomycin and chitosan to poly(ethylene glycol) dimethacrylate in this system results in the incorporation of SPIONs during the microbead production process (PEGDMA). The SPIONs loaded microbeads were exposed to the altering magnetic field of 25mT at 109.9 kHz for half an hour with the generation of magnetothermal heat fuelling the ester bonds cleavage linking with the vancomycin to the microbeads. The results showed a higher release of vancomycin from the microbeads compared to the vancomycin released from the non-stimuli-responsive controlled release in the levels of minimum inhibitory concentration (MIC).

Sirivisoot and his team also worked on the release of ciprofloxacin encapsulated within polycaprolactone-iron-oxide NPs after exposing it to the magnetic field for one day in a *S. aureus* broth inhibition assay for the use of SPIONs in the antimicrobial field [39]. They stated that the magnetothermal response of SPIONs varies from one iron oxide form to another. In fact, co-loading ciprofloxacin with either maghemite or hematite nanoparticles in polycaprolactone microspheres revealed variances in the drug's release and bactericidal action against *S. aureus* following exposure to an alternating magnetic field. The maghemite-loaded (γ-Fe2O3) nanoparticles showed high efficacy than the hematite-loaded (α-Fe2O3) nanospheres.

5.4 Ultrasound-Responsive Systems

For many years, ultrasound has been used in applications for diagnostic imaging. Acoustic waves play an important role as a trigger for medication delivery, as well as making cell membranes more permeable, in addition to their non-invasive nature. The two principal physical mechanisms that underlie the effects of ultrasound on biological systems are heat and pressure. By using extended, high-energy pulsed pulses, temperature-responsive nanocarriers are able to control drug delivery. With shorter pulsed impulses, however, pressure-mediated release can be obtained without hyperthermia. The most extensively studied ultrasound-responsive nanocarriers include micelles, perfluorocarbon-filled lipid nanobubbles, and liposomes containing PEG implanted in their lipid bilayer. PNPs have been used in a number of novel ultrasonic imaging applications, either as carriers for imaging contrast agents or as contrast agents [40].

Ultrasound-triggered drug delivery systems have received noteworthy attention in recent years due to their controlled therapeutic delivery and minimized adverse effects on healthy tissues. The non-invasiveness, lack of ionizing radiation requirement, ease of exposure, depth of tissue penetration, etc. of such a system are some of its major benefits. Ultrasound waves use the radiation phenomenon for the release of therapeutics such as genetic materials, proteins, etc by thermal or physical pressure without overheating [41]. Cavitation is the occurrence and/or movement of gas-filled bubbles in an environment exposed to ultrasound where the gas bubbles of any size expand at low pressure and compress at high pressure as the pressure wave travels through the medium. Cavitation plays a major role in two factors i) the release of medicine by disrupting the carrier vesicle structure, ii) the increase in the permeability of drugs in the cell membrane and capillaries [42].

In a study by Dong and his team [43], *two Staphylococcus epidermidis* infected catheters were implanted into an *in vivo* rabbit model before microbubbles were injected subcutaneously to promote ultrasound cavitation. The vancomycin drug-loaded system was irradiated with ultrasound 300 kHz, 0.5 W/cm^2, 50% duty cycle, and the activity was evaluated with the control group. The vancomycin-loaded drug system has shown increased activity than the control group due to the ultrasound-triggered system which proves that ultrasound cavitation can be a potential agent in the treatment of infections caused by microorganisms that can attach to both abiotic and biotic surfaces.

In another study, Dong Ma [44] and his colleagues reported that the focused ultrasound and gentamicin-loaded liposomes have shown increased activity against *Ralstonia insidiosa* biofilms by 72% comparatively than the ultrasound-treated or gentamicin-treated or gentamicin along with the high-intensity ultrasound-focused system. These results revealed that non-focused ultrasound can promote the adhesion of liposomes to the biofilm, as shown by the same authors, while focused ultrasound can kill bacteria within synthetic alginate biofilms when used in conjunction with an antibiotic strategy.

5.5 Magnet Responsive Systems

Magnetic fields with both static and alternating magnetic fields are the second most transduction method for the external activation of drug delivery vehicles. The most widely used transducing material in these magnetic drug delivery vehicles is superparamagnetic iron oxide nanoparticles (SPIONs) with 5-20 nm in size single crystal nanoparticles. These SPIONs have been exhibited to have easy synthesis, immune capability, and unambiguous drug release and particle size suitable for various applications such as MRI scanning, drug delivery vehicle, hyperthermia, and tracking of cells [37]. SPIONs are preferred for their nontoxic nature if degraded, unlike other nanoparticles in the transduction of extrinsic signals. The SPIONs can act as drug delivery vehicles for targeted drug delivery as well as a factor for drug release (static or alternating magnetic fields) enhancing the on-demand drug delivery and specific-site delivery [14].

Zink and his colleagues [45] fabricated a multi-stimulus responsive magnetic nanocarrier for the delivery of large and low molecular weight drugs against pathogenic biofilms. This nanocarrier was made of mesoporous silica nanoparticles (MSNs) compressed with β-cyclodextrin (β-CD)-modified polyethyleneimine (PEICD) in their large pores (MSNLP) and adamantane (ADA) decorated MSNs with a magnetic core containing cucurbit[6]uril (CB[6]). The formation of co-assemblies (H+G) between the host MSNs and guest MSNs occurs due to the host-guest interaction between β-cyclodextrin (β-CD) and adamantane (ADA). The study showed that in the heat produced by the alternating magnetic field, the nanocarriers released the large (peptide melittin) and low (ofloxacin) molecular weight drugs efficiently compared to the drugs (OFL and MEL) alone or unattached MSNs. The co-assembled nanocarrier showed potential biofilm eradication and no toxicity and inflammation in the host tissues. As a result, co-delivering and multi-stimuli-responsive nanocarriers could overcome the challenges associated with treating infections that are caused by protective biofilms.

Geilich and his colleagues [46] substantiated the capability of the iron oxide polymersome nanoparticles (IONPs) in biofilm eradication which is encapsulated methicillin with methoxy poly(ethylene glycol)-b-poly(d,l-lactic acid) in aqueous core and hydrophobic SPIONs inside the membrane layer. The biofilm-forming microbe *Staphylococcus epidermis* was incubated with IONPs containing different concentrations of SPIONs and methicillin for 24 hours along with a magnetic field to draw the IONPs through the biofilms. This application of magnetic field along with IONPs showed increased biofilm degradation efficiency through penetration of IONPs in the biofilm. The study revealed the potential capability of IONPs loaded with SPIONs and methicillin can destroy the biofilm with a magnetic field process. Importantly, utilizing IONPs containing 40 g mL1 SPIONs and 20 g mL1 methicillin, which is substantially less than the MBC value (500 g mL1) of free methicillin for *S. epidermidis* growing within a biofilm, resulted in the total eradication of the biofilms.

Wu and his colleagues [47] manufactured dual-stimuli targeting and microwave-responsive nanocarriers comprising both magnetic targeting and microwave caloric therapy for distinctive synergistic therapy for the treatment of MRSA-induced osteomyelitis. The mesoporous iron oxide nanoparticles (IOPs) were altered with carbon nanotubes (CNTs) of efficient microwave absorption potential which was further encapsulated with gentamicin and 1-tetradecanol (phase-changing material for controlling gentamicin release) for synergistic chemotherapy. The fused Fe_3O_4/CNT/Gnt nanocarrier showed an effective capacity for trapping the MRSA selectively due to their strong contact between nanocomposites and the amino groups in the bacterial surface. The study revealed that upon microwave stimulation, Fe_3O_4/CNT/Gnt had shown increased biopotential efficacy with a raise in the local temperature against the infection caused by MRSA in the tibial plateau effectively by constricting the spread of MRSA via blood to other organs and preventing other severe infections. This study used microwaves as a novel responsive trigger and showed promising outcomes for treating deep tissue infections. It also offered a promising substitute for visible light or NIR light-based phototherapy in clinical settings.

In another study, Wu and his co-workers [48] constructed a Prussian blue (PB) metal-organic framework (MOF) as a microwave-responsive material with a perspective for rapid treatment of osteomyelitis. Prussian blue (PB) is a typical mesoporous metal-organic framework with exceptional capabilities to release iron ions and transform MV energy into heat. The insertion of Na+ hastened the release of Fe(II) and Fe(III) ions from PB and made the iron ion's spin state irreversible which results in the increased iron ion's permeability into the bacterial membranes and causing cell death by producing a Fenton reaction. In order to cure deep tissue infections, external magnetic fields or microwaves with strong penetrating power have a lot of promise to substitute light. In the case of an external magnetic field stimulation, it may direct the nanocarriers to a specific region and concentrate them there, particularly *in vivo*. Magneto-responsive nanocarriers also have the ability to segregate and catch germs, which lessens the risk of bacterial transmission and the negative impacts of bacterial waste.

5.6 Electrical Responsive Systems

Electrical stimulation has strong spatial and temporal controllability, quick and reverse induction, and non-invasiveness. Conductive materials such as conductive polymers, conductive ions, or other types of conductive materials are necessary for electric response antibacterial systems. As smart actuators for soft robotics, drug administration, sensors, and artificial muscles, electro-responsive materials are becoming more and more significant. Electric fields can be utilized to provide on-demand drug delivery through a variety of various actuation techniques. An applied electric field is most typically utilized to perform reversible redox chemistry, which can transiently change the chemistry of a delivery vehicle to promote drug release. Different varieties of delivery vehicles respond to electrical field-induced changes such as conductive polymer polypyrrole due to redox reaction allowing for reversible, on-demand modifications in the release of encapsulated medicines [49]. Redox-stimulated polymer degradation provides on-demand release potential that is comparable to that of self-immolating polymers, however, in this case, burst release control is offered rather than on/off control [50]. But these approaches have been practically constrained in a therapeutic context due to the restricted tissue penetration of electric fields often necessitating the use of implanted electronics and the requirement to restrict the applied voltage to prevent unintended local cell/tissue damage [51]. Additionally, redox chemistry induced by the body or nonspecific fouling can impair the *in vivo* performance of conductive polymers resulting in inflammation or fibrosis and the latter, inactivity, especially over a prolonged period.

Electrically triggered release of drugs from the electro-responsive hydrogel matrix is based on the fact that most hydrogels are protein-repellent and can at least deal with the fouling problems brought on by other electroactive release vehicles. Hydrogels doped with conductive nanoparticles enhance the size and/or mobility of the electroactive reactions by improvising the stability of the material design [52].

Another strategy in electro-responsive vesicles is the electrically activated microchips method for triggerable release of customizable drug mixtures with great spatial and temporal precision, and/or including a level of automation that is difficult to achieve with traditional pulsatile delivery systems. But one of the drawbacks of these vesicles is the need for surgical implantation for their functioning which limits their use in precise temperature of controlled drug mixtures in successful therapy [53]. Electro-responsive vesicles also provide a distinctive ability to integrate electrochemical sensors with controlled drug-release vehicles to enable closed-loop drug release without the need for external monitoring or interference.

To eliminate pathogenic bacteria, an electric-driven luminescent system based on electrochemiluminescence (ECL) can be employed with cationic oligo (phenylene vinylene) (OPV). The negative membrane of dangerous bacteria will adsorb to the cationic OPV through electrostatic and/or hydrophobic contacts when energy is transferred from ECL to OPV, and the resulting ROS will kill them all at once. A frequently used conductive substance is graphene. A graphene oxide-quaternized coated with CS nanohybrid electrode can eliminate *E. coli* in contaminated water at least 99% of the time by using a capacitive deionization disinfection procedure. Additionally, numerous additional conductive polymers, including PANI, polythiophene, polypyrrole, and their different derivatives, have been extensively used in biomedical applications [54].

Wang and his colleagues for the first time have synthesized an electric-driven luminous system for the destruction of infectious bacterial communities where they used electrochemiluminescence (ECL) for exciting the photosensitizer instead of a physical light source for producing reactive oxygen species (ROS) [18]. The precise research mechanism depended on the luminol and photosensitizer's flawless spectrum overlap and energy transfer, which could change the ambient oxygen molecules into ROS.

Specifically, 5-amino-2,3-dihydro-1,4-naphthalenedione was employed as an ECL reagent; when hydrogen peroxide (H_2O_2) was present in an alkaline solution, the wavelength of the emission was between 400 and 550 nm. The maximal emission wavelength of the cationic PS oligomer (p-phenylene vinylene) (OPV) was 550 nm, with a rather broad absorption range of 350–550 nm. The energy produced by electrochemiluminescence (ECL) was absorbed by the oligo (p-phenylene vinylene) (OPV) which excites the surrounding oxygen molecules to form more reactive oxygen species (ROS) for sterilization. The luminol hydrogen peroxide oligo(p-phenylene vinylene) (L-H_2O_2-OPV) hydrogel had shown increased efficacy against *C. albicans* around 80% when compared with the control (L-H_2O_2-OPV) hydrogel showed less efficacy of 25%. With this approach, the issue of light's restricted penetration depth is resolved, opening up the prospect of sterilizing in deeper tissue treatments and electric fields can also control the release of drugs.

Baolin Guo and his colleagues [55] designed multi-stimuli responsive injectable hydrogels as drug delivery systems that responds to the electrical field and pH against bacteria. The hydrogel is a combination of chitosan-graft-polyaniline (CP) copolymer and a cross-linker of oxidized dextran (OD). The release rate of CP/OD hydrogels loaded with amoxicillin and ibuprofen drugs increased with the increase in voltage in an electric driven release. The hydrogel's antibacterial properties were enhanced by the presence of chitosan and polyaniline, both of which have inherent antibacterial properties. The study result shows that the multi-stimuli responsive injectable hydrogels can act as smart drug delivery vehicles for the controlled release of drugs.

5.7 Conclusion

Nanomaterials enrolled in biomedical applications are mostly modified for tissue recognition with various ligands. These fabricated nanocarrier's response to external or internal stimuli is emerging as a potential strategy in clinical settings. This review focused on the existing research works on the physical or external stimuli-responsive systems like light, heat, ultrasound, and magnetic fields in the treatment of biofilm eradication through specific target site-controlled drug release. In conclusion, nanocarriers offer an innovative approach to addressing the issue of biofilm formation and activity. By using nanoparticles to target and disrupt biofilms, researchers can increase the effectiveness of traditional antimicrobial agents while minimizing the development of antibiotic resistance. Remote triggering offers predictable, on-demand doses with lower systemic toxicity and higher treatment effectiveness. The stimuli-responsive nanoparticles application has a significant long-term remedy for the two primary problems with antimicrobial therapy i.e. resistance and severe side effects. The spatial and temporal release of antibiotics from drug carriers at an infection site is one of the key advantages of these stimuli-responsive nanoparticles. The controlled and accurate distribution of the antibacterial agents improves the therapeutic efficacy primarily on abscesses and infected wounds, with a decrease in systemic side effects, making this technique appear superior to the current one. Further, the penetration capability of the antibiotics can be increased efficiently by the inorganic nanoparticles through external stimulus in biofilm treatment. Another advantage of this approach is their manufacturing, inorganic nanoparticles are unquestionably preferable since they can provide stimuli-responsive characteristics in a more straightforward and repeatable method. However, these methods also have some drawbacks even though they have efficient antibacterial activity since this approach is in its early stage and needs more development. Like light has limited tissue penetration which reduces its capability in the treatment of peripheral infections even though it could be treated with near-infrared radiation (NIR) due to its great tissue penetration ability. Critically, the toxicity produced by inorganic materials remains a major threat in clinical settings. Additionally, a problem with magnetic-based systems is the requirement for the joint creation of a cost-effective magnetic field generator that is appropriate for use against a variety of infection types. Furthermore, the use of physical stimuli-responsive nanocarriers provides a novel approach to controlling biofilm activity. Future research in this area will undoubtedly lead to more effective treatments for biofilms in medicine and industry.

References

1. Canaparo, R., Foglietta, F., Giuntini, F., Pepa, C.D., Dosio, F., Serpe, L., Recent developments in antibacterial therapy: Focus on stimuli-responsive drug-delivery systems and therapeutic nanoparticles. *Molecules*, 24, 10, 2019, MDPI AG, https://doi.org/10.3390/molecules24101991.
2. Moorcroft, S.C.T., Jayne, D.G., Evans, S.D., Ong, Z.Y., Stimuli-Responsive Release of Antimicrobials Using Hybrid Inorganic Nanoparticle-Associated Drug-Delivery Systems. *Macromol. Biosci.*, 18, 12, 2018, Wiley-VCH Verlag, https://doi.org/10.1002/mabi.201800207.
3. Rajivgandhi, G.N., Ramachandran, G., Maruthupandy, M., Manoharan, N., Alharbi, N.S., Kadaikunnan, S., Khaled, J.M., Almanaa, T.N., Li, W.J., Anti-oxidant, anti-bacterial and anti-biofilm activity of biosynthesized silver nanoparticles using Gracilaria corticata against biofilm producing K. pneumoniae. *Colloids Surf., A, 600*, 2020, https://doi.org/10.1016/j.colsurfa.2020.124830.
4. Hu, Y., Ruan, X., Lv, X., Xu, Y., Wang, W., Cai, Y., Ding, M., Dong, H., Shao, J., Yang, D., Dong, X., Biofilm microenvironment-responsive nanoparticles for the treatment of bacterial infection. *Nano Today*, 46, 2022, Elsevier B.V., https://doi.org/10.1016/j.nantod.2022.101602.
5. Akbarian, M., Chen, S.H., Kianpour, M., Farjadian, F., Tayebi, L., Uversky, V.N., A review on biofilms and the currently available antibiofilm approaches:

Matrix-destabilizing hydrolases and anti-bacterial peptides as promising candidates for the food industries. *Int. J. Biol. Macromol.*, 219, 1163–1179, 2022, Elsevier B.V., https://doi.org/10.1016/j.ijbiomac.2022.08.192.

6. Wang, S., Gao, Y., Jin, Q., Ji, J., Emerging antibacterial nanomedicine for enhanced antibiotic therapy. *Biomaterials Sci.*, *8*, 24, 6825–6839, 2020, https://doi.org/10.1039/D0BM00974A.
7. Colilla, M. and Vallet-Regí, M., Targeted stimuli-responsive mesoporous silica nanoparticles for bacterial infection treatment. *Int. J. Mol. Sci.*, 21, 22, 1–32, 2020, MDPI AG, https://doi.org/10.3390/ijms21228605.
8. Maruthupandy, M., Rajivgandhi, G.N., Quero, F., Li, W.J., Antiquorum sensing and anti-biofilm activity of nickel oxide nanoparticles against Pseudomonas aeruginosa. *J. Environ. Chem. Eng.*, *8*, 6, 2020, https://doi.org/10.1016/j.jece.2020.104533.
9. Abbamondi, G.R. and Tommonaro, G., Research Progress and Hopeful Strategies of Application of Quorum Sensing in Food, Agriculture and Nanomedicine. *Microorganisms*, *10*, 6, 2022, https://doi.org/10.3390/microorganisms10061192.
10. Blackman, L.D., Qu, Y., Cass, P., Locock, K.E.S., Approaches for the inhibition and elimination of microbial biofilms using macromolecular agents. *Chem. Soc. Rev.*, 50, 3, 1587–1616, 2021, Royal Society of Chemistry, https://doi.org/10.1039/d0cs00986e.
11. Liu G., Lovell J.F., Zhang L., Zhang Y. Stimulus-Responsive Nanomedicines for Disease Diagnosis and Treatment. *Int. J. Mol. Sci.*, 21, 1–43, 2020.
12. Wang, X., Shan, M., Zhang, S., Chen, X., Liu, W., Chen, J., Liu, X., Stimuli-Responsive Antibacterial Materials: Molecular Structures, Design Principles, and Biomedical Applications. *Adv. Sci.*, 9, 13, 2022, John Wiley and Sons Inc., https://doi.org/10.1002/advs.202104843.
13. Sikder, A., Chaudhuri, A., Mondal, S., Singh, N.D.P., Recent Advances on Stimuli-Responsive Combination Therapy against Multidrug-Resistant Bacteria and Biofilm. *ACS Appl. Bio Mater.*, 4, 6, 4667–4683, 2021, American Chemical Society, https://doi.org/10.1021/acsabm.1c00150.
14. Said, S.S., Campbell, S., Hoare, T., Externally Addressable Smart Drug Delivery Vehicles: Current Technologies and Future Directions. *Chem. Mater.*, 31, 14, 4971–4989), 2019b, American Chemical Society, https://doi.org/10.1021/acs.chemmater.9b01798.
15. Devnarain, N., Osman, N., Fasiku, V.O., Makhathini, S., Salih, M., Ibrahim, U.H., Govender, T., Intrinsic stimuli-responsive nanocarriers for smart drug delivery of antibacterial agents—An in-depth review of the last two decades. *Wiley Interdiscip. Rev. Nanomed. Nanobiotechnol.*, 13, 1, 2021, Wiley-Blackwell, https://doi.org/10.1002/wnan.1664.
16. Cao, S., Xia, Y., Shao, J., Guo, B., Dong, Y., Pijpers, I.A.B., Zhong, Z., Meng, F., Abdelmohsen, L.K.E.A., Williams, D.S., van Hest, J.C.M., Biodegradable Polymersomes with Structure Inherent Fluorescence and Targeting Capacity

for Enhanced Photo-Dynamic Therapy. *Angew. Chem. Int. Ed.*, *60*, 32, 17629–17637, 2021, https://doi.org/10.1002/anie.202105103.

17. Weng, Z., Xu, Y., Gao, J., Wang, X., Research progress of stimuli-responsive ZnO-based nanomaterials in biomedical applications. *Biomater. Sci.*, 11, 1, 76–95), 2022, Royal Society of Chemistry, https://doi.org/10.1039/d2bm01460b.
18. Dai, X., Zhao, Y., Yu, Y., Chen, X., Wei, X., Zhang, X., Li, C., Allin-one NIR-activated nanoplatforms for enhanced bacterial biofilm eradication. *Nanoscale*, *10*, 39, 18520–18530, 2018, https://doi.org/10.1039/C8NR04748K.
19. Garapati, C., Boddu, S.H., Jacob, S., Ranch, K.M., Patel, C., Babu, R.J., Tiwari, A.K., Yasin, H., Photodynamic therapy: A special emphasis on nanocarrier-mediated delivery of photosensitizers in antimicrobial therapy. *Arabian J. Chem.*, 16, 4, 2023, Elsevier B.V., https://doi.org/10.1016/j.arabjc.2023.104583.
20. Courtney C.M., Goodman S.M., Nagy T.A., Levy M., Bhusal P., Madinger N.E., Detweiler C.S., Nagpal P., Chatterjee A., Potentiating antibiotics in drug-resistant clinical isolates via stimuli-activated superoxide generation. *Sci. Adv.*, 4, 1–10, 2017. https://www.science.org.
21. Li, J., Gopal, A., Karaosmanoglu, S., Lin, J., Munshi, T., Zhang, W., Chen, X., Yan, L., Photosensitizer doped zeolitic imidazolate framework-8 nanocomposites for combined antibacterial therapy to overcome methicillin-resistant Staphylococcus aureus (MRSA). *Colloids Surf., B*, *190*, 2020, https://doi.org/10.1016/j.colsurfb.2020.110900.
22. Zhdanova, K.A., Savelyeva, I.O., Ignatova, A.A., Gradova, M.A., Gradov, O.V., Lobanov, A.V., Feofanov, A.V., Mironov, A.F., Bragina, N.A., Synthesis and photodynamic antimicrobial activity of amphiphilic mesoarylporphyrins with pyridyl moieties. *Dyes Pigm.*, *181*, 108561, 2020, https://doi.org/10.1016/j.dyepig.2020.108561.
23. Hou, X., Yang, L., Liu, J., Zhang, Y., Chu, L., Ren, C., Huang, F., Liu, J., Silver-decorated, light-activatable polymeric antimicrobials for combined chemo-photodynamic therapy of drug-resistant bacterial infection. *Biomater. Sci.*, *8*, 22, 6350–6361, 2020, https://doi.org/10.1039/D0BM01084G.
24. Jijie, R., Barras, A., Bouckaert, J., Dumitrascu, N., Szunerits, S., Boukherroub, R., Enhanced antibacterial activity of carbon dots functionalized with ampicillin combined with visible light triggered photodynamic effects. *Colloids Surf., B*, *170*, 347–354, 2018, https://doi.org/10.1016/j.colsurfb.2018.06.040.
25. Kobayashi, Y., Hayashi, M., Yoshino, F., Tamura, M., Yoshida, A., Ibi, H., Lee, M.-C., Ochiai, K., Ogiso, B., Bactericidal effect of hydroxyl radicals generated from a low concentration hydrogen peroxide with ultrasound in endodontic treatment. *J. Clin. Biochem. Nutr.*, *54*, 3, 161–165, 2014, https://doi.org/10.3164/jcbn.13-86.
26. Xu, X., Liu, X., Tan, L., Cui, Z., Yang, X., Zhu, S., Li, Z., Yuan, X., Zheng, Y., Yeung, K.W.K., Chu, P.K., Wu, S., Controlled-temperature photothermal and oxidative bacteria killing and acceleration of wound healing by

polydopamine-assisted Au-hydroxyapatite nanorods. *Acta Biomater.*, *77*, 352–364, 2018, https://doi.org/10.1016/j.actbio.2018.07.030.

27. Majumder, J. and Minko, T., Multifunctional and stimuli-responsive nanocarriers for targeted therapeutic delivery. *Expert Opin. Drug Deliv.*, 18, 2, 205–227, 2021, Taylor and Francis Ltd, https://doi.org/10.1080/17425247.2021.1828339.
28. Santhamoorthy, M., Thi Tuong, V.P., Vanaraj, R., Chaitany, J.R., Kokila, T., and Seong-Cheol, K. Thermo-Sensitive Poly (N-isopropylacrylamide-co-polyacrylamide) Hydrogel for pH-Responsive Therapeutic Delivery, *Polymers,* 14, 1–14, 2022.
29. Jaque, D., Martínez Maestro, L., Del Rosal, B., Haro-Gonzalez, P., Benayas, A., Plaza, J.L., Martín Rodríguez, E., García Solé, J., Nanoparticles for photothermal therapies. *Nanoscale*, *6*, 16, 9494–9530, 2014, https://doi.org/10.1039/c4nr00708e.
30. Radt, B., Smith, T.A., Caruso, F., Optically addressable nanostructured capsules. *Adv. Mater.*, *16*, 23–24, 2184–2189, 2004, https://doi.org/10.1002/adma.200400920.
31. Amoli-Diva, M., Sadighi-Bonabi, R., Pourghazi, K., Laser-assisted triggered-drug release from silver nanoparticles-grafted dual-responsive polymer. *Mater. Sci. Eng., C*, *76*, 536–542, 2017, https://doi.org/10.1016/j.msec.2017.03.089.
32. Liu, M., He, D., Yang, T., Liu, W., Mao, L., Zhu, Y., Wu, J., Luo, G., Deng, J., An efficient antimicrobial depot for infectious site-targeted chemophotothermal therapy. *J. Nanobiotechnol.*, *16*, 1, 2018, https://doi.org/10.1186/s12951-018-0348-z.
33. Borzenkov, M., Moros, M., Tortiglione, C., Bertoldi, S., Contessi, N., Faré, S., Taglietti, A., D'Agostino, A., Pallavicini, P., Collini, M., Chirico, G., Fabrication of photothermally active poly(vinyl alcohol) films with gold nanostars for antibacterial applications. *Beilstein J. Nanotechnol.*, *9*, 1, 2040–2048, 2018, https://doi.org/10.3762/bjnano.9.193.
34. Fang, J., Wan, Y., Sun, Y., Sun, X., Qi, M., Cheng, S., Li, C., Zhou, Y., Xu, L., Dong, B., Wang, L., Near-infrared-activated nanohybrid coating with black phosphorus/zinc oxide for efficient biofilm eradication against implant-associated infections. *Chem. Eng. J.*, *435*, 134935, 2022, https://doi.org/10.1016/j.cej.2022.134935.
35. Vasilakaki, M., Binns, C., Trohidou, K.N., Susceptibility losses in heating of magnetic core/shell nanoparticles for hyperthermia: A Monte Carlo study of shape and size effects. *Nanoscale*, *7*, 17, 7753–7762, 2015, https://doi.org/10.1039/c4nr07576e.
36. Nawaz, M., Sliman, Y., Ercan, I., Lima-Tenório, M.K., Tenório-Neto, E.T., Kaewsaneha, C., Elaissari, A., Magnetic and pH-responsive magnetic nanocarriers, in: *Stimuli Responsive Polymeric Nanocarriers for Drug Delivery Applications: Volume 2: Advanced Nanocarriers for Therapeutics*, 2, 37–85,

Elsevier, Woodhead Publishing, United Kingdom, 2018, https://doi.org/10.1016/B978-0-08-101995-5.00002-7.

37. Laurent, S., Saei, A.A., Behzadi, S., Panahifar, A., Mahmoudi, M., Superparamagnetic iron oxide nanoparticles for delivery of therapeutic agents: Opportunities and challenges. *Expert Opin. Drug Deliv.*, 11, 9, 1449–1470, 2014, Informa Healthcare, https://doi.org/10.1517/17425247.2014.924501.
38. Harris, M., Ahmed, H., Barr, B., LeVine, D., Pace, L., Mohapatra, A., Morshed, B., Bumgardner, J.D., Jennings, J.A., Magnetic stimuli-responsive chitosan-based drug delivery biocomposite for multiple triggered release. *Int. J. Biol. Macromol.*, *104*, 1407–1414, 2017, https://doi.org/10.1016/j.ijbiomac.2017.03.141.
39. Sirivisoot, S. and Harrison, B.S., Magnetically stimulated ciprofloxacin release from polymeric microspheres entrapping iron oxide nanoparticles. *Int. J. Nanomed.*, *10*, 4447–4458, 2015, https://doi.org/10.2147/IJN.S82830.
40. Crucho, C., II, Stimuli-responsive polymeric nanoparticles for nanomedicine. *ChemMedChem*, 10, 1, 24–38, 2015, John Wiley and Sons Ltd., https://doi.org/10.1002/cmdc.201402290.
41. Chen, K.J., Liang, H.F., Chen, H.L., Wang, Y., Cheng, P.Y., Liu, H.L., Xia, Y., Sung, H.W., A thermoresponsive bubble-generating liposomal system for triggering localized extracellular drug delivery. *ACS Nano*, *7*, 1, 438–446, 2013b, https://doi.org/10.1021/nn304474j.
42. Percival, S.L., Suleman, L., Vuotto, C., Donelli, G., Healthcare-Associated infections, medical devices and biofilms: Risk, tolerance and control. *J. Med. Microbiol.*, 64, 4, 323–334, 2015, Microbiology Society, https://doi.org/10.1099/jmm.0.000032.
43. Dong, Y., Li, J., Li, P., Yu, J., Ultrasound microbubbles enhance the activity of vancomycin against staphylococcus epidermidis biofilms *in Vivo*. *J. Ultrasound Med.*, *37*, 6, 1379–1387, 2018, https://doi.org/10.1002/jum.14475.
44. Ma, D. and Wu, J., Biofilm mitigation by drug (gentamicin)-loaded liposomes promoted by pulsed ultrasound. *J. Acoust. Soc. Am.*, *140*, 6, EL534–EL538, 2016, https://doi.org/10.1121/1.4972336.
45. Yu, Q., Deng, T., Lin, F.C., Zhang, B., Zink, J., II, Supramolecular Assemblies of Heterogeneous Mesoporous Silica Nanoparticles to Co-deliver Antimicrobial Peptides and Antibiotics for Synergistic Eradication of Pathogenic Biofilms. *ACS Nano*, *14*, 5, 5926–5937, 2020, https://doi.org/10.1021/acsnano.0c01336.
46. Geilich, B.M., Gelfat, I., Sridhar, S., van de Ven, A.L., Webster, T.J., Superparamagnetic iron oxide-encapsulating polymersome nanocarriers for biofilm eradication. *Biomaterials*, *119*, 78–85, 2017, https://doi.org/10.1016/j.biomaterials.2016.12.011.
47. Qiao, Y., Liu, X., Li, B., Han, Y., Zheng, Y., Yeung, K.W.K., Li, C., Cui, Z., Liang, Y., Li, Z., Zhu, S., Wang, X., Wu, S., Treatment of MRSA-infected osteomyelitis using bacterial capturing, magnetically targeted composites

with microwave-assisted bacterial killing. *Nat. Commun.*, *11*, 1, 2020, https://doi.org/10.1038/s41467-020-18268-0.

48. Wei, S., Qiao, Y., Wu, Z., Liu, X., Li, Y., Cui, Z., Li, C., Zheng, Y., Liang, Y., Li, Z., Zhu, S., Wang, H., Wang, X., Che, R., Wu, S., Na+ inserted metal-organic framework for rapid therapy of bacteria-infected osteomyelitis through microwave strengthened Fenton reaction and thermal effects. *Nano Today*, *37*, 2021, https://doi.org/10.1016/j.nantod.2021.101090.
49. Ge, J., Neofytou, E., Cahill, T.J., Beygui, R.E., Zare, R.N., Drug release from electric-field-responsive nanoparticles. *ACS Nano*, *6*, 1, 227–233, 2012, https://doi.org/10.1021/nn203430m.
50. Hardy, J.G., Mouser, D.J., Arroyo-Currás, N., Geissler, S., Chow, J.K., Nguy, L., Kim, J.M., Schmidt, C.E., Biodegradable electroactive polymers for electrochemically-triggered drug delivery. *J. Mater. Chem. B*, *2*, 39, 6809–6822, 2014, https://doi.org/10.1039/c4tb00355a.
51. Staples, M., Microchips and controlled-release drug reservoirs. *Wiley Interdiscip. Rev. Nanomed. Nanobiotechnol.*, *2*, 4, 400–417, 2010, https://doi.org/10.1002/wnan.93.
52. Sun, Y. and Davis, E., Nanoplatforms for targeted stimuli-responsive drug delivery: A review of platform materials and stimuli-responsive release and targeting mechanisms. *Nanomaterials*, 11, 3, 1–103), 2021, MDPI AG, https://doi.org/10.3390/nano11030746.
53. Xiao, Y., Xu, M., Lv, N., Cheng, C., Huang, P., Li, J., Hu, Y., Sun, M., Dual stimuli-responsive metal-organic framework-based nanosystem for synergistic photothermal/pharmacological antibacterial therapy. *Acta Biomater.*, *122*, 291–305, 2021, https://doi.org/10.1016/j.actbio.2020.12.045.
54. Zhang, A., Jung, K., Li, A., Liu, J., Boyer, C., Recent advances in stimuli-responsive polymer systems for remotely controlled drug release. *Prog. Polym. Sci.*, 99, 2019, Elsevier Ltd, https://doi.org/10.1016/j.progpolymsci.2019.101164.
55. Qu, J., Zhao, X., Ma, P.X., Guo, B., Injectable antibacterial conductive hydrogels with dual response to an electric field and pH for localized "smart" drug release. *Acta Biomater.*, *72*, 55–69, 2018, https://doi.org/10.1016/j.actbio.2018.03.018.

6

Current Advances in the Use of Functionalized Nanoparticles for the Diagnosis and Treatment of Microbial Infections in Aquaculture

Kannan Rangesh[1], Muthusamy Anand[1]*, Subbiah Padmapriya[1] and Muthuchamy Maruthupandy[2]

[1]*Department of Marine and Coastal studies, Madurai Kamaraj University, Madurai, Tamil Nadu, India*
[2]*Department of Health Sciences, Dong-A University, Busan, South Korea*

Abstract

Aquaculture is regarded as the production of vital nourishment systems and plays a substantial role in commercial, economic, and food security, and its continuous progress is an essential component of the global nutritional safety strategy. Diverse fish diseases are now being reported in aquaculture, causing economic losses in the seafood trade. Several techniques are used to control the fish disease, but they are expensive. One of the strategies proposed by recent research finding is to use the nanoparticles in aquaculture diseases. Nanotechnology has grown to be a significant science and technology research due to its exceptional role in cutting-edge applications. Nanoparticles have been used in a broad range of medical applications, including diagnostics, immunomodulation, immunostimulation, effective drug delivery, and tissue engineering. As a peculiar and sensitive application for diagnosing fungal, bacterial, and viral diseases, nanoparticles have piqued the interest of aquaculture researchers. We concentrated on the antibacterial consequence of nanoparticles, with a specific emphasis on the difficulties of antibiotic-resistant microbes in aquaculture and solutions using the bioencapsulation technique in this chapter. A novel nanoparticle was used to encapsulate antigen in alive biodelivery module, like Artemia and organic delivery system, such as nanoliposome and alginate. In fish, both synthetic and natural polymers

**Corresponding authors*: anandm21@yahoo.com

Sathish-Kumar Kamaraj, Arun Thirumurugan, Muthuchamy Maruthupandy, Mercedes Guadalupe López Pérez and Shanmuga Sundar Dhanabalan (eds.) Nanoscience and Nanotechnology for Smart Prevention, Diagnostics and Therapeutics: Fundamentals to Applications, (117–142) © 2024 Scrivener Publishing LLC

revealed to progress antigen intake and specific immunity production with alginate microparticles, as well as variety of graphene-based nanovehicles, including nanoliposomes, chitosan nanoparticles, and poly lactic-co-glycolic acid. A recent development in fish disease and treatment signifies the exploration of nanoparticles related vaccinations to protect from a variety of bacterial and viral infections. Thus, one of the top priorities for the fisheries and aquaculture industries is the treatment and prevention of bacterial diseases. Hence, nanotechnology is very useful in targeted drug delivery with compact nanostructures that have yet to be identified in fish disease and treatment.

***Keywords*:** Aquaculture, fish disease, nanoparticles, antimicrobial activity, bioencapsulation

6.1 Introduction

According to FAO (2022), the capture fishery and fish culture segments are progressively known due to their vital support to global food security in the 21st century. Global production of fisheries and other aquatic animals was valued at 178 million tonnes (mt) in 2020 (Figure 6.1) [1]. In 2020, worldwide marine captures were recorded as 78.8 mt, and it is a decline of 6.8% from the top most of 84.5 mt in the year 2018, when comparatively excessive anchoveta fish catches were claimed in Chile and Peru. Capture fisheries provided 90 mt (51%) and aquaculture production provided 88 mt (49%). From the total production, 63% (112 mt) was used to harvest in marine waters especially 70% from capture fishery and 30% from aquaculture. Remaining 66 mt (37%) was used from inland water

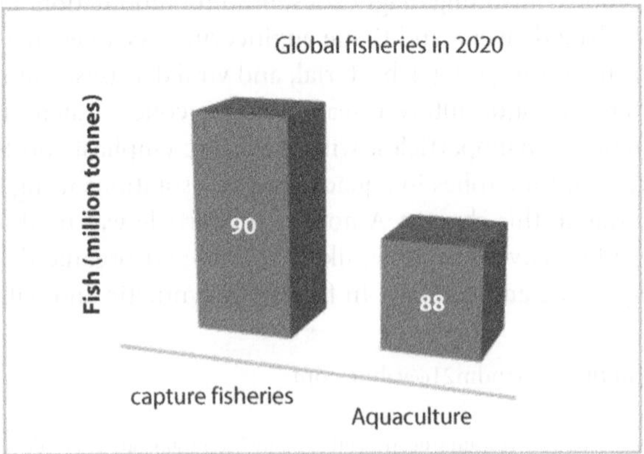

Figure 6.1 Global capture fisheries and aquaculture production in 2020.

harvesting (Aquaculture is 83%; Capture fishery is 17%). In total, production's marketing worth in worldwide was 406 billion (USD), encompassing 2,65,000 million USD for aquaculture and USD 1,41,000 million for capture fisheries. Moreover of aquatic fauna, 36 million tonnes of wet weight seaweed were produced in 2020, about 97% was get from seaweed culture, mainly mariculture. Global consumption of marine based foods improved up to 3.0% of yearly average rate from the year 1961 to 2019. Aquaculture contributed for 56% of the food production from aquatic fauna available for human food. Past few decades, per capita consumption of aquaculture based foods was prompted by augmented supplies and varying consumer favourites as well as technology advancements and growth in income.

Aquatic foods continue to be among the furthermost marketed food merchandises all over the world, with 225 states recording aquaculture and fisheries product trading activity in 2020. Global exports of fishery and aquatic products without seaweed totally harvested about live weight of 60 million tonnes and its worth is USD 1,51,000 million. Efforts are undertaken and improved in aquaculture especially practices of seaweed cultivation [2]. In 2019, fishery stocks of the global mariculture were 64.6% of the trawled fishing showing the biologically sustainable levels. The world's estimated fishing vessels is 4.1 million in numbers. In 2020, 5.85 million fishermen were reported as full-time, part-time, regular/occasional and unspecified personnel in aquaculture and capture fisheries, and in which women constituted 21%. The employment distribution across sectors showed that 35% worked in aquaculture while 65% were engaged in capture fisheries [1].

6.1.1 Utilization and Processing of Fisheries and Aquaculture Production

Aquatic foods are important for a healthy and balanced diet [3]. Aquatic foods, even in small amounts, could constitute a significant nutritious impact with essential nutritions, which are lacking in plant based foods. Aquatic foods are providing vitamins-A, B, and D, phosphorus, essential amino acids, high quality proteins and minerals including calcium, iron, zinc, magnesium, iodine, potassium, and selenium. They serve as a chief source of omega-3 fatty acids in the diet which is heart-healthy (Figure 6.2). Aquatic food products have widely different levels and amounts of nutrients based on the species. The main noticeable difference has always been fat content: fatty species include salmon, sardines, and tuna, whereas cod and catfish are slender. Docosahexaenoic acid (DHA) and eicosapentaenoic

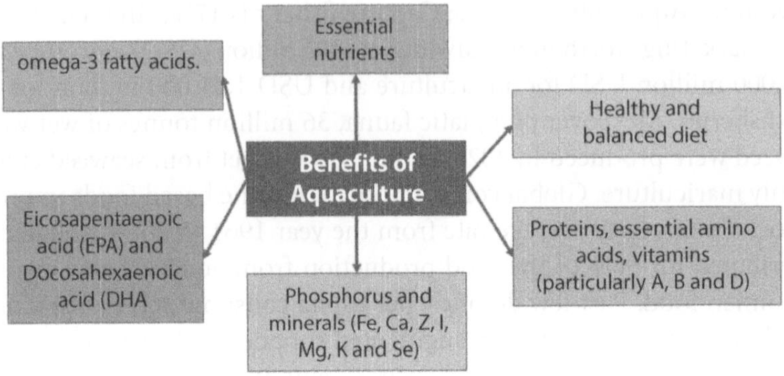

Figure 6.2 Utilization and processing of fisheries.

acid (EPA) are the two omega-3 fatty acids usually present in aquatic species. The human body cannot produce omega-3 fatty acids so they must be sourced through food [4].

Fisheries and aquaculture yields are used for broad range of products with diverse characteristics and flavor based on the fish variety, preserving method and final product. Significant advancements in fish product processing, its refrigeration, freezing, storage, and transport have allowed for longer shelf-life with distribution across far distances and international borders, and a greater diversity of products. Due to high perishability of aquatic products, special attention is crucial during harvesting and throughout the entire supply chain. Improper treatment after harvesting can lead to rapid spoilage and potential health hazards due to microbial growth, chemical changes, break down by enzymes, and cross-contamination, posing risks to food safety. Implementing appropriate handling, processing, preservation, packaging, and storage techniques is vital to prolong shelf life, ensure food safety, preserve quality and nutritional properties, and prevent unnecessary loss and waste [5].

Preservation and processing methods have crucial roles in enabling the supply and marketing of aquatic products both domestically and internationally. These techniques involve chilling/freezing (temperature reduction), heat treatment process (canning/boiling process/smoking), removal of water content (drying/salting process/smoking), and alteration of the storage setup (modified atmosphere packing/vacuum packing process/refrigeration). Nutritional properties of fishery food have varieties based on the way it is processed and packed. Heating process (sterilization/pasteurization/cooking/hot smoking) can reduce the quantity of thermolabile nutrition, including vitamins. Nevertheless, the concentration of some nutrients increase during heating process to remove water. Numerous countries are currently experiencing noteworthy advancements in food processing and packaging technologies. These developments have led to enhanced efficiency, effectiveness, and profitable utilization of raw materials. Moreover, there have been innovative strides in diversifying food products for both human consumption and the production of fishmeal, fish oil, and other valuable purposes [6].

In the fisheries and aquaculture field, approximately 35% of the global production may lost or wasted annually. The total fish loss and waste at a global level are estimated to range between 30% and 35% [7].

6.1.2 Aquaculture Biosecurity

The strengthening of aquaculture and worldwide trade in fishery products should have to deal with infectious diseases. SDG goal 14.1—aimed in preventing and considerable reduction of marine pollution. SDG Target 14.2—sustainably manage and to keep protected the marine and coastal ecosystems to elude significant adverse effects [8].

Considering the pollution in the fisheries sector deal with abandoned, lost, or otherwise discarded fishing gear (ALDFG) is an increasingly influential reason because of its harmful consequences to environment and economics along with navigational associated safety risks. ALDFG is continued to catch aquatic animals in an uncontrolled manner that is "ghost fishing"—considered as detrimental effects on fishery stocks along with possible consequences on endangered organisms and marine benthic

Figure 6.3 Controlling ghost fisheries by avoid ALDFG for SDG target 14.1 and 14.2 (image courtesy: https://textilesforsdgs.org/sdgs/goals/14-life-below-water/).

environments. ALDFG became international problem of marine plastics/litters, considered by international organizations, their activities and agreements focusing on reducing marine debris with numerous national and local level implementations across countries (Figure 6.3). Total aquaculture production (excluding algae) and capture fisheries is expected to expand further and touch 202 mt in the year 2030. In general, it is estimated that by 2030, the quantity of aquatic food available for human consumption will rise by 24 million tonnes in comparison to 2020, reaching a total of 182 million tonnes [9].

6.2 Fishery Disease Outbreaks

The first outbreak was found in 1970s, during which, disease caused on both cultured fish and wild fish hence fisherman and fishery farmers suffered due to significant loss in their income. Farmed and wild fish worldwide are affected, with natural infection having been reported for some 76 finfish species (e.g. barbs, breams, catfish, gourami, tigerfish, tilapias, mullet, eel, pike, and snakehead) [10].

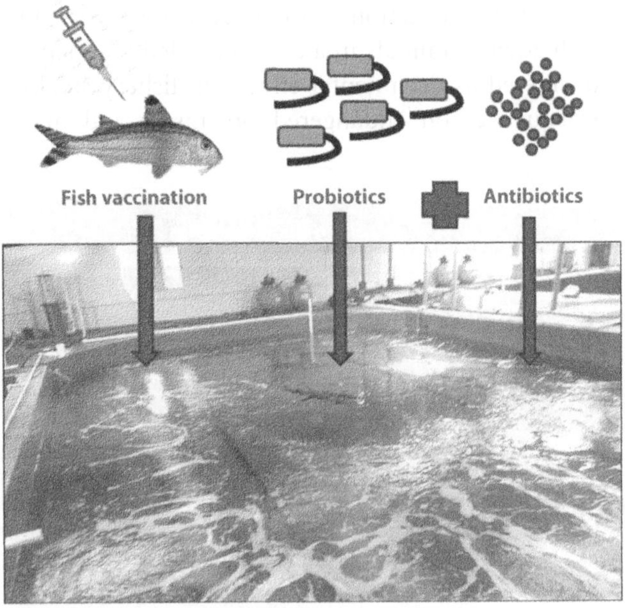

Figure 6.4 Control of fish disease outbreaks using the probiotics, vaccination, and fish vaccines.

The first report of epizootic ulcerative syndrome (EUS) were from Japan in 1971, which recognized *A. piscicida* for the cause of the disease "ulcerative mycosis." Following findings from Australia in the year 1972, the same organism was implicated of being the cause of red spot disease. It was initially identified as "ulcerative menhaden" disease in the United States in 1978, and it was referred to as EUS in South and Southeast Asia in 1986. Following an outbreak in 2006, the initial reports of EUS in southern Africa were officially documented. EUS now affects 24 countries on four continents (Africa, Asia, Australia, and North America) [10].

Most countries, particularly those in Asia (e.g., Bangladesh, Cambodia, India, Nepal, Philippines, Thailand, and Vietnam), and also Australia and Japan, have reported EUS among wild fish populations [11]. Fish health against disease outbreaks can be improved with fish vaccination, antibiotics and probiotics in aquaculture practices (Figure 6.4).

6.2.1 Fish Vaccination

Numerous factors contribute to a significant decline in aquaculture production. Foremost among these factors is the occurrence of diseases, which poses a severe constraint, leading to detrimental effects on the livelihood of farmers, job losses, food insecurity, and reduced incomes. According to studies, diseases, that are much more severe in developing countries, account for nearly half of all production losses. This is due to the fact that 90% of the aquaculture farm is located in the industrialized regions. Disease related revenue losses reached $6 billion per year. In Chile, for example, contagious salmon anaemia only costs around $2 billion dollars and has resulted in the layoff of 20,000 workers. Diseases have cost China, one of the world's leading aquaculture producers, 15% of total fish production [12]. To counteract losses due to diseases in aquaculture, it should be addressed each health constraint in a scientific and recommended manner, as well as in a locally adaptable manner. Challenges in aquaculture because of climate change, limited water sources, and the growth drive the requirement for epidemiological approaches in keeping aquatic animal health safe [13]. As "prevention is better than treatment," it is advisable to focus on preventing the occurrence of disease rather than treating it [14]. The usages and practices of improved management/husbandry practices, movement limitations, genetically resistant stock maintenance, dietary supplements, probiotics prebiotics, nonspecific immunostimulants, vaccine, medicinal plant products, antimicrobial compounds, water disinfection biological control, water disinfection, and control of movement are the best approaches in control of infectious diseases of fish [15]. Due to drug resistance and residue issues, antibiotic utilization is strictly controlled and regulated [16]. For decades, vaccine production has been increased for infectious disease control in response to reduce an antibiotic usage in fish. Vaccines got broader acceptance for the fact there is no risk of drug resistance development in vaccinated animals and protection of minority unvaccinated animals because of herd immunity [17]. A solo approach to prevent as well as maintenance of aquaculture health is not only considered successful practices. Effectiveness lies in adopting a combination of diverse strategies. It is essential to establish an exchange platform between aquaculture farmers and responsible entities. In addition to implementing all these approaches, disease surveillance and employing sensitive and specific diagnostic tests are invaluable for ensuring the well-being of fish populations [18].

Fish vaccination was initiated by vaccinating against *Aeromonas salmonicida* infection in 1942 in Cutthroat [19]. Currently, used vaccines are the type of conventional vaccines because no advanced level information on the fish immunology. Vaccines now available are injectable vaccines and oil adjuvant [19]. The salmon genome has been completely sequenced [20], for itself and other several fish species as well. These led to the findings of novel vaccine development strategies in near future [21]. Vaccines for combating intracellular bacteria and viral pathogens will be the big challenges for the upcoming years. DNA vaccine will play a role in such cases [20]. Modern vaccines are groped as DNA vaccines, recombinant vector, genetically modified, synthetic peptide, killed, attenuated and subunit vaccines. Whole organism vaccines exhibited better advantage than other vaccines. However, most of the vaccines do not completely prevent disease [22].

6.2.2 The Use of Antibiotics in Aquaculture

Currently, it is widely accepted that antibiotics used to protect fish from disease should be kept at a minimum [23]. Antibiotics used in large level can result in resistance development. The usage of antibiotics for aquaculture practices has no primary treatment option. Despite the fact that the use of antibacterial agents in animal species for food, including fish, is regulated, especially in Europe and the United States, a large variety of medically and veterinary inhibitory compounds are used in fish farming [24]. Antimicrobial administration in aquaculture differs from administration in land animals. This variation particularly due to adding drugs to the water with or without feedstuffs results in eco/environmental disturbance of the microbiota [25]. Using antibiotics commonly used in human medicine in aquaculture fisheries is a risky practise. Aquaculturists have been recommended to employ alternative prevention methods instead of relying on antibiotics. However, in certain situations where the use of antibiotics is necessary, they should strictly administer only approved drugs for fish [26]. FDA-USA developed a list of antibiotics for usage in aquaculture which was undergone for review and is classified as new animal drugs of minimum regulatory priority. Such methods encompass various compounds like carbon dioxide gas, acetic acid, hydrogen, NaCl, and even garlic employed towards controlling helminth and sea lice copepods. Onion is used to treat external crustacean parasites, and ice is utilized to reduce the biological impact and fish metabolic rate level during transportation [27].

6.2.3 The Usage of Probiotics in Aquaculture for Disease Control

Biological disease control in fish farming aquaculture is one among the best strategy in infectious disease control [28]. Probiotics is a bacterial culture of the bacterial strains which are non-pathogenic to fish [29]. Following administration to fish, these microorganisms undergo multiplication within the fish's gut, aiding the normal microflora and maintaining microbial balance in the host [30]. Many microbes have been evaluated as potential probiotics in fish farming and aquaculture. Some of these include *Bacillus subtilis*, *Lactobacillus sakei*, *L. acidophilus*, and *Shewanella putrefaciens*. These probiotics can be used in fish species and other cultured animals to stop diseases and enhanced weight gain. They can be incorporated into the feed or added directly to the water. Encapsulation is the other administration strategy. Encapsulation helps by improving nutritional values and proper delivery of the microbe to the host without waste of live organisms [31].

6.2.4 Administration Strategies of Probiotics

Probiotics could be introduced into the water tank or incorporated directly into feed pellets, making it one of the most crucial and widely used methods of administration. Another widely adopted approach is microencapsulation, where the probiotic organism's cells are encased in a colloidal matrix, utilizing substances like alginate, chitosan, carboxymethyl cellulose, or pectin to provide both physical and chemical protection to the microorganisms [32]. Immobilized probiotics is another technology used to make the wall of immobilization formed by the cell arrest and released in alginate based gel matrix around the core substance. Probiotic immobilization is a relatively new technology that is widely used within the dairy and pharma domains.

6.3 Nanotechnology in Aquaculture

Nanobiotechnology has extensive potential to bring about innovative developments to aquaculture systems in order to minimize costs, increase efficiency, and reduce our environmental impact, as a vital demand affecting our ability to sustain the 7 billion plus people of the world [33, 34]. The worldwide nanotechnology in the food sector is forecasted to rise at a rate of more than 24% per year between 2019 and 2023, reaching $112.48 billion, and the growing applications in nutraceuticals primarily responsible for this market boost [35].

Nanotechnology involves a wide array of disciplines, such as physics, chemistry, biotechnology, and engineering [36]. It operates within the 1 to 100 nm range and exhibits diverse dimensions in nanomaterials, including zero-dimensional structures (nanoparticles/nanoclusters/quantum dots), one-dimensional structures (carbon nanotubes/multi-walled nanotubes), two-dimensional structures (graphene layers/ultrathin films), and three-dimensional structures (nanostructured materials). Different forms and shapes of nanoparticles are used, such as nanotubes, dendrimers, nanocapsules, nanospheres, etc. [37, 38].

6.3.1 Advantages of Nanotechnology in Aquaculture

Nanomaterials offer several benefits, such as tissue-specific targeting, reduced dosage and toxicity, enhanced bioavailability, improved drug efficacy, and decreased secondary adverse effects [39]. The physicochemical properties of nanomaterials enable a broad spectrum of applications, including food preservation, water treatment, and healthcare [40]. Nanotechnology finds numerous applications in aquaculture and holds the potential to revolutionize the industry significantly. Among its current applications are pathogen detection and control, water treatment, pond sterilization, and efficient nutrient and drug delivery. DNA based nano vaccines, for example, have been utilized to improve the fish immune response. On the other hand, iron nanoparticles, can be used to boost fish growth [41].

6.3.2 Seafood Processing Using Nanotechnology

The main objectives of the food industry is to extend the shelf life of products to retain both freshness and food quality. Nowadays, nanotechnology can develop many advancements in food technology and processing divisions, aimed at increasing product shelf life by avoiding gas flow across packaging, and also providing information about potentially spoilt ingredients [42]. Nanostructures, such as nanoemulsion, nanofibers, as well as nanoparticles can be employed to slow down the degradation of quality, preserving the product's color and flavor [43].

Moreover, ice infused with nanoparticles presents a wide range of possibilities for food packaging applications. For instance, researchers have successfully incorporated silver nanoparticles (AgNP) derived from midrib part of banana into nano-ice, which has demonstrated the ability to decrease the microbial load on the surface of flat head grey mullet (*Mugil cephalus*), effectively preventing *Acinetobacter*'s growth. In the domain of seafood preservation, made from organic silver nanoparticles are being prioritized over chemical synthesis [44].

Various types of chitosan nanoparticles having antimicrobial properties have been investigated. Example includes nanocomposite films produced with chitosan nanoparticles and gelatin, adding oregano essential oil, result in enhanced antimicrobial activity, including against the four most common test food pathogens (*Escherichia coli, Listeria monocytogenes Salmonella enteritidis,* and *S. aureus*). Chitosan is considered as a potent antimicrobial substance because of the poly cationic form. Furthermore, chitosan nanoparticles present extraordinary physiochemical properties, including bioactivity [45].

6.3.3 Cerium Oxide (CeO_2) as Potential Nanoparticle for Fish Disease

Controlling microbes such as viruses, bacteria, and fungi would involve early detection, and pathogen elimination could be accomplished using nanomaterials, which operate on the same scale as a particle infecting a virus or disease [46]. The antibacterial activity of five different nanoparticles such as Al_2O_3, Fe_3O_4, CeO_2, ZrO_2, and MgO against fish bacterial pathogens such as *Aeromonas hydrophila, Bacillus subtilis, serratia* sp, *Vibrio harveyi,* and *Vibrio parahaemolyticus* were studied using time kill assay method, minimum bactericidal concentration, inhibitory concentration and well diffusion method. Previously chemicals treatments were carried out using copper sulphate, formalin, malachite green, methyl blue, and potassium permanganate to remedy the bacterial fish diseases. Most biological resources, including mangroves, seaweeds, seagrasses, and sponges, as well as AgNP, demonstrated both antibacterial and antifungal activity. However, using metal nanoparticles against fish pathogens is a new technology. But recently, the nanoparticles have the efficient antibacterial activities. The maximum antibacterial activity recorded in CeO_2 using the time kill assay method. The CeO_2 nanoparticles recorded a maximum inhibition rate at the concentration of 20 μg mL^{-1} against *Bacillus subtilis* and 30 μg mL^{-1} against *Vibrio harveyi* than the other nanoparticles. Compared to CeO_2, the ZnO nanoparticles also showed antibacterial activity against various bacterial pathogens includes *E. coli*, *Staphylococcus aureus*, and *Bacillus Subtilis* [47]. It was due to the ROS (reactive oxygen species) mechanism, which can produce significant oxidative stress and altered the cell wall system into equally permeable levels in bacteria [47].

6.3.4 Silver Nanoparticle for Fish Bacterial Disease

The administration of antimicrobial drugs in aquatic medicine poses a significant environmental concern due to the fast dispersion of antibiotics in water. The utilization of metallic silver for antibacterial agent is recognized, and many types of silver compounds are used day-to-day, for example, silver dressings, silver nitrate, silver zeolite, and silver nanoparticles are utilized for a variety of antimicrobial purposes. Since the bacterial resistance is developed, innovative antimicrobial agents are needed. One of the way is to use AgNP as antimicrobial drugs. AgNP have proved to be among the effective metallic nanoparticles as they have a good antimicrobial efficacy against some bacteria, viruses and fungi. The assessment process of the effects of various silver compositions on a variety of pathogenic bacteria is critical due to the risk of silver toxic effects, which can pose agyrosis and argyria [48]. Silver nanoparticles have more preventing effects than probably because of the different cell wall membrane structures between Gram-positive and Gram-negative bacteria [49]. The same mechanism of the silver nanoparticles' inhibitory effects on microorganisms is unknown. Several studies have indicated that the antimicrobial activity of silver ions relies on their positive charge, as it facilitates electrostatic attraction between the negatively charged cell membrane of microorganisms and the positively charged nanoparticles [50]. The antimicrobial impact of silver nanoparticles on Gram-negative bacteria was concentration dependent, and this was closely associated with the formation of "pits" in the cell wall of bacteria, which results in permeable bacterial membranes and consequent bacterial cell death [49]. Soltani *et al.* found that AgNP was able to diminish the growth of *A. hydrophila, Lactococcus garveiae, Streptococcus iniae,* and *Yersinia ruckeri,* after 30 to 90 minutes of exposure with AgNP 5 μg/mL. Hence, it is possible to use such a nanomaterial to control the disease outbreaks against bacterial fish pathogens. This is particularly applicable in the regions where rivers are commonly used as the water source for fisheries [51].

6.3.5 Use of Gold Nanoparticles as Efficient Diagnosis of Fish Disease

AuNPs (gold nanoparticles) received a lot of attention on account of its exceptional optical properties generated by surface plasmon resonance with its high scattering and absorption. Furthermore, Au NPs are easily synthesized and functionalized with a variety of biomolecules, such as pathogen-specific nucleotide sequences or antibodies. Recently, AuNPs have been applied in hybridization assays to identify unamplified nucleic acids. Single- and double-stranded oligonucleotides can be adsorbed on Au NPs in colloidal solution under certain conditions. The hybridization phenomenon is diagnosed visually within a minute after adding AuNPs, when the color indication from red to blue in case of positive reaction or remains red in case of negative reaction. Nanoparticle-based techniques like these hold significant promise as potent methods for diagnosing fish pathogens [52].

When Au NPs are in solution, the presence of citrate ions (negative) adsorbed on their exteriors inhibit particle gathering due to the stronger van der Waals attraction between the AuNPs. If adding NaCl, the attraction will be stronger than counteraction, which can lead to an aggregation of Au NPs and solution's color changes from red to blue. Using this property, AuNPs-based DNA detection can be identified either by labeling or label-free. In the labeled technique, Gold nanoparticles (Au NPs) are primarily functionalized with thiolated single-stranded DNA (ssDNA) and subsequently permitted to form hybrid complexes with a sequence having complementary target gene. However, the Au NPs must be pre-modified, this strategy is laborious, complicated, and time-consuming [52].

6.4 Immunomodulation and Immunostimulation

Immunostimulants are employed as dietary supplements in farmed fish to enhance feed efficiency, weight gain, and disease resistance, thereby promoting fish health and overall performance [53]. Immunomodulatory drugs can alter the immune system either by increasing immune response (immunostimulators) or decreasing (immunosuppressives) serum antibody production. While immunosuppressants generally have a dose-dependent effect of globally impairing the host immune response, immunomodulators may exhibit more selective action by targeting specific portions of the immune system, posing a lower risk of immune-related complications.

6.4.1 Chitosan Nanoparticles for Immunomodulation in Fish

Chitosan and its derivatives are widely used in aquaculture diseases. Furthermore, chitosan has been utilized for encapsulating various bioactive substances, such as metal ions, vitamins, inactivated bacterial strains for vaccines, proteins, and nucleic acids, in numerous fish farming treatments. Chitosan and its derivatives are becoming increasingly popular as dietary additives in the fish farming industry. Simultaneous to the immune response, chitosan also increasing antioxidant responses in fish. In loach fish, the levels of superoxide dismutase (SOD), phenoloxidase, and glutathione peroxidase (GPx) exhibited an increase after 12 weeks of chitosan supplementation [54]. Dietary supplementation of chitosan in Tilapia fish resulted in elevated activity of catalase and SOD after 56 days. The protective response of chitosan from oxidative stress was also reported in olive flounder (*Paralichthys olivaceus*) challenged with H_2O_2 [54].

6.4.2 Chitosan Nanoparticle as Dietary Supplementation

Supplementing the diet with chitosan nanoparticles combined with vitamin C and thymol appears to be high effective in immunity enhancement compared to using each additive individually. Dietary chitosan + vitamin C nanoparticles raised growth performance of Tilapia fish. Upon exposure to imidacloprid-polluted water, fish fed with chitosan–vitamin C nanoparticles exhibited notably improved immunity and antioxidant activity. This was evidenced by enhanced activities of lysozyme, glutathione reductase, catalase (CAT), and immunoglobulins. Furthermore, the combination of chitosan and thymol demonstrated a synergistic effect on lymphocytes and monocyte leukocytes [55].

Fish secrete a complex, viscous substance known as mucus, which coats the epithelial cells. It acts as essential external barriers. Its functional characteristics rely on its capability to create a gel-like layer on the epithelial surface. Through osmoregulation process, mucus performs the host defense system. Chitosan-based nanoparticles to fish's mucoadhesive properties have recently began as an approach of mucosal drug release.

Chitosan nanoparticles have previously been used to deliver bacterial antigens, vitamin C, RNA, and plasmid DNA. The use of intraperitoneal (IP) immunomodulators in fish shows good, but it is time consuming and costly. The most frequently employed method, especially for younger fish, involves using modulators through immersion or bath application, leading to absorption through the skin, gills, and intestine (via drinking). In aquaculture, the primary concern regarding the effectiveness of immunomodulators is linked to their method of exposure to intraperitoneal and their absorption, which can significantly vary and depend on the dosage, potentially leading to a decrease in the drug's bioavailability. The interaction between a mucoadhesive system, Poly lactic-co-glycolic acid with chitosan-coated nanoparticles mucosal adhesive system (PLGA-NPMA), and fish's external mucus suggests that using mucoadhesive nanocarriers could serve as an alternative for drug and immunomodulator administration in immersion systems. The nanosystem can adhere to the fish's mucosal surface with minimal residual effect in the water [54].

6.4.3 Selenium Nanoparticles for Immunomodulation in Fish

Selenium (Se) is a vital element for all animals. Commonly found Se is inorganic selenite. And organic form is selenomethionine. Selenoproteins have an important role in human health. In humans and animals, more than 25 selenoproteins were identified, including glutathione peroxidase (GPx), which plays pivotal roles in immune and antioxidant responses as well as other metabolism regulation. Se NPs' capacity to modulate fish's immune system is one of their promises. Se NPs were discovered to induce T-helper cell activity and expression of cytokine in the fight against tumours. They have also shown to enhance both the systemic and mucosal immune responses. Selenium nanoparticles (Se NPs) have been suggested as potential adjuvants for vaccines due to their ability to trigger a strong Th1 cytokine response. Se NPs are also considered a promising form of selenium formulation for regulating antioxidant functions. Chitosan is used as a stabiliser in the production of Se NPs. Se NPs' protection against *A. hydrophilla* infection, resulting in preventing the disease. Se NPs could increase phagocytic respiratory burst, lysozyme activity, and proliferation of lymphocytes. Excessive production of reactive oxygen species (ROS) can lead to oxidative damage of biomolecules such as lipids, proteins, and DNA. This oxidative damage has been associated with aging and various diseases, including cardiovascular, cancer, respiratory, neurodegenerative, and digestive disorders. Thus, Se NP highly reduces the ROS by regulating antioxidant function [56].

6.4.4 Nanoparticles for Infectious Fish Disease

Infectious diseases could be the main reason for economic losses in the world aquaculture. Senegalese sole (*Solea senegalensis*) is the species considered for aquaculture industries in southern Europe, their production is lowered by the bacterial diseases photobacteriosis, it is a septicaemia caused by subsp. *piscicida* of *Photobacterium damselae.* Ulvan is a marine derived polysaccharide from green seaweed of the *Ulva* species that possess immunomodulating, antioxidant, anticoagulant, anticancer, antiviral and anti-hyperlipidemic activities. Studies have reported the potential immunomodulatory activity of ulvan from *U. ohnoi*, highlighting its role as a bioactive compound and its possible application as a vaccine adjuvant against *P. damselae* in Senegalese sole [57]. As drug delivery vector for fish vaccination and protection against diseases, the utilization of chitosan within the aquaculture sector has great potential due to its biocompatibility, bioadhesion, biodegradability and immunomodulatory properties. Ulvan boost the antibacterial activity of chitosan nanoparticles against *P. damselae* [57].

6.4.5 Nanomaterials as Efficient Diagnosis of White Spot Disease

Nanoparticle-based diagnostic techniques offer diverse bio-applications, contributing to the advancement of significant industrial and engineering systems through biotechnological tools. In aquaculture and fisheries, nanotechnology finds application in water purification to eliminate microbes, organic chemicals, and metals, demonstrating its potential in these fields. White spot disease (WSD) is indeed a lethal and infective agent in shrimps and shellfishes by White spot syndrome virus (WSSV) that kills all shrimp in a short period of time. The crucial step is to notice it immediately by regular screening to control the outbreak of the disease. A first clinical symptom of WSSV is the white spots in the shellfish's carapace caused by calcium salt deposition by the cuticular epidermis. Other clinical signs include the animal's body color changing to pale or reddish, lethargy, and reduced feed intake. Stressor exposure increases the risk of WSSV because stressors impair the shrimp immune system [58].

Traditional WSSV identification methods include conventional approaches such as direct observation, histology, and electron microscopy. Other than these, DNA based probes, PCR, real-time PCR, Multiplex reverse transcription-polymerase chain reaction (mRT-PCR), single-step immunochromatographic methods are used for detection. These

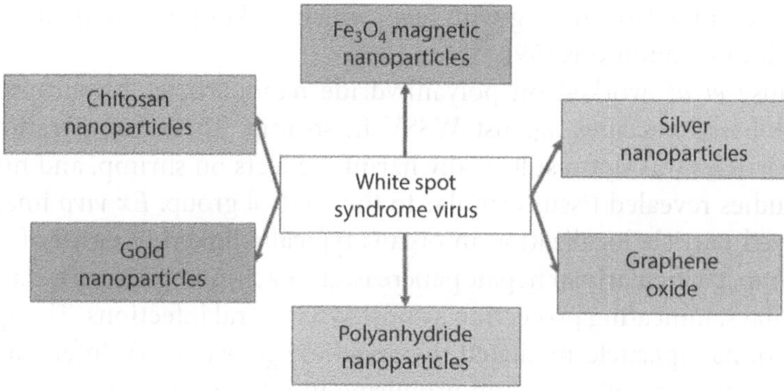

Figure 6.5 Nanomaterials as efficient diagnosis of white spot disease.

nano-technological approaches comprise nanoarrays, protein arrays, nanopores, nanoparticles and nano-vaccine as a platform for nano based sensors (Figure 6.5). A simple and rapid immunochromatographic test-strip kit was used for WSSV detection by colloidal gold nanoparticles used as the detector antibody developed by Thailand researchers, and they were conjugated with W29 monoclonal antibody, which targeted the VP28 structural protein. Thiruppathiraja *et al.* used gold nanoparticles in associated with a secondary antibody conjugated with alkaline phosphatase which raised against WSSV. They successfully increased the sensitivity of the immuno-dot blot test was performed by 80-fold over the conventional approach, which can detect at the resolution of 1 ng/mL of purified WSSV visually [59].

6.4.6 Vaccine Delivery for WSSV Control Using Nanoparticles

Vaccine delivery systems is the advanced tools using liposomes, recombinant vaccine, plasmids, DNA vaccines through oral or intramuscular injection, polymer and carbon-based nano delivery systems etc. The application of the ionic gelation technique in developing CS/TPP nanoparticles for delivering pVP28DNA in shrimp through the oral route showed promising results. This technique utilizes the polyelectrolytes to form cross-linked bonds and increase the contact surface for effective delivery. Additionally, nano-vaccines based polyanhydride nanoparticles was effectively utilized for encapsulating and releasing vaccine antigens. Hence, there is a potential to improve the immunogenicity by modulating the vaccine formulation using nanotechnology for mass immunization of shrimp via concentration or milling with feed, which makes possible their use in field relevant conditions [59].

Phanse *et al.* worked on polyanhydride nanoparticles to encapsulate dsRNA-based vaccines against WSSV in shrimp. The administration of nanoparticles was not result in any harmful effects on shrimp, and histology studies revealed tissues similar to the control group. *Ex vivo* imaging displayed particle localization in organs typically linked to entry of virus in shrimp, particularly in hepatopancreas, stomach, and gills, with stability for 28 days, indicating protection against WSSV viral infections. The application of nanoparticle-mediated therapeutics against WSSV infection has been shown to be effective and promising in inhibiting viral disease. The significance of silver nanoparticles in enhancing the survival of WSSV-infected *Penaeus vannamei* shrimps was demonstrated in the experiment. Here, the commercial non-toxic polyvinylpyrrolidone (PVP) coated silver nanoparticles was utilized to improve the immune response of WSSV-infected shrimps in a single dose of silver nanoparticles treatment (111 nmol/shrimp) [60].

In another study, the impact of dietary chitosan nanoparticles on WSSV challenged crayfish *Procambarus clarkia* was discussed. The crayfish feed with 10 mg/g of chitosan nanoparticles resulted in high survival rates (65%) in crayfish due to the improved immunological parameters. In another study by Juarez-Moreno *et al.*, silver nanoparticles in white shrimp via the intramuscular injection led to a high survival rate (80%) after 4 days without any observed toxicity or alteration in metabolic rate. Taken altogether, these results show the protective efficiency of nanoparticles against WSSV infection can be utilized as an effective modality for the treatment of viral diseases in aquaculture [61].

6.5 Nanoparticles for Bioencapsulation

Nanotechnology based on biodegradable materials appears to be a fascinating way to apply in aquaculture. To progress in fish aquaculture nutrition, we can consider nanoparticles as a novel tool for studying the role of certain micronutrients during fish development. New technologies, that is controlled release systems and encapsulation can be used readily in scaling-up operations to improve the delivery of bioactive compounds and ultimately raised the production and high profits in aquaculture. Bioencapsulation or bioenrichment refers to enhancing the nutritional content of live food organisms by introducing various types of nutrients into their diet or incorporating them within their bodies. This can be achieved through experimental enrichment diets like unicellular algae, emulsions, liposomes, and microencapsulated diets. In this context, copepods and *Artemia* sp have been employed as carriers to deliver compounds with diverse nutritional and therapeutic benefits to larval stages of aquatic animals, including fish (Figure 6.6). This is the enrichment process, also known as bioencapsulation, is effective to allow *Artemia* nauplii to utilize as larval feed at certain stages of marine finfish and shellfish rearing. Selection of the live food should be based on its size. Commonly used carriers are micro algae, rotifers, *Copopods*, Moina, *Artemia*, and *Daphnia*. One such way to enrich *Artemia* is to use liposomes with free amino acids (FAA). Liposomes looks spherical vesicles (size range, 10 nm to 20 μm) where an liquid volume is trapped by a lipid membrane in the form of phospholipids. The efficacy of nanoparticles incorporation into microdiets to ensure a better delivery of vitamin C in marine fish larvae is subject of experiments currently in progress [62].

Artemia nauplii can also be utilized to inoculate the digestive tracts of target species with probiotic bacteria via bioencapsulation and feeding. Probionts are live microbial feed supplements that provide benefits to the host animal through enhancing its intestinal balance. Bioencapsulation based lactic acid bacteria were subjected into turbot larvae, resulting in noteworthy increases in survival. Bacteria with varying properties were incorporated into *Artemia* nauplii to induce an oral challenge in turbot

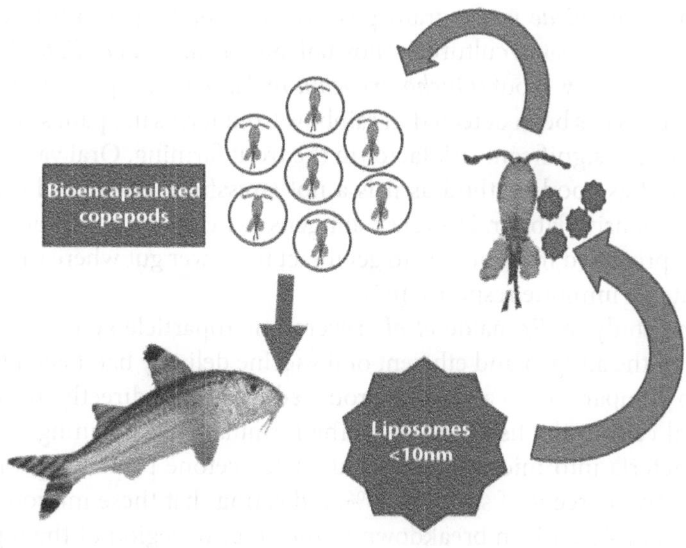

Figure 6.6 Bioencapsulation of liposome to fish via *Copepod.*

fish larvae, they were exposed to a pathogenic *Vibrio anguillarum*. This method has also been used to immunize sea bass larvae, juvenile carp, and fish fry [63].

6.5.1 Nanoencapsulation Improves Seafood Product

Poly unsaturated fatty acids (PUFAs) are beneficial to human health by save from cardiovascular diseases, cancer/autoimmune disorders, brain function, etc. Therefore, it is crucial to have a sufficient intake of omega PUFA. Fish and other marine organisms are among the primary sources of PUFA in the human diet. Fish oils offer numerous dietary advantages, but their application in food formulations is restricted due to their pronounced odors and rapid degradation. To overcome this issue, nano-liposomes are employed to encapsulate fish oil at the nanoscale, and the encapsulated fish oil is then utilized to fortify yogurt. The use of nanoliposomes reduced acidity, syneresis, and peroxide value significantly. Yogurt enriched with nanoencapsulated fish oil exhibited elevated levels of DHA and EPA compared to yogurt fortified with free fish oil. The addition of nano-encapsulated fish oil to yogurt also enhances sensory characteristics in comparison to yogurt fortified with free fish oil [64].

6.5.2 Alginate-Encapsulated Vaccine as Effective Oral Booster for *Lactococcus* Disease

Lactococcus garvieae is the gram-positive cocci pathogen to fish, considered a serious threat in cultured yellowtail (*Seriola quinqueradiata*) fishes in Japan and rainbow trout (*Onchorrrynchus mykiss*) in Europe and Australia. *Lactococcosis* has been detected in rainbow trout farms in Spain since 1991, emerging as a significant risk factor in the trout farming. Oral vaccines are considered as good method as it is a not stressful method and does not require extensive labour. However, antigens to be used in oral vaccination must be protected in some way to get intact the lower gut where they could stimulate an immune response [65].

In the study by Romalde *et al.*, several microparticle systems for protection of the antigen and efficient oral vaccine delivery had been tested in trout in comparison to a vaccine produced by adding directly inactivated bacterial cells to the fish food [65]. The formulation containing encapsulated bacteria into microspheres of alginate-acetone provided protection with relative percent of survival 50%, indicating that these microparticles appear to avoid antigen breakdown in the anterior region of the digestive tract due to low pH and proteases. Efficacy of various oral vaccine formulations against fish *lactococcosis*, established an appropriate vaccination module to prevent *L. garvieae* attack. Recently, the alginate microparticle is replaced by alginate nanoparticles used for encapsulated radiovaccine in rainbow trout fish against the protozoan pathogen *Ichthyophytirius multifiliis*. The creatinine elevation observed in fish treated with encapsulated radiovaccine suggests that the maintenance of an activated immune system is energetically demanding and may compete with other energy-intensive processes, such as growth. This suggests that alginate nanoparticles may be a safe and efficient tool for oral delivery of a radiovaccine against *I. multifiliis* [66].

6.6 Conclusion

Nanomedicine has thus emerged as the most significant fields of nanotechnological study since it is further disputed that modern fish vaccination and other ways can considerably benefit from employing nanotechnology. It is currently focused on developing new methods of disease prevention, diagnosis, and treatment. The utilization of nanoparticle-based vaccines on a variety of bacterial and viral pathogens is a new arena of research in fish medicine. Thus, one of the top priorities for the fisheries and aquaculture industries is the treatment and prevention of bacterial diseases. Hence, nanotechnology is very useful in targeted drug delivery with compact nanostructures that have yet to be explored in fish medicine.

Acknowledgment

This book chapter work was partially supported by Madurai Kamaraj University's National Centre for Coastal Research Project (MoES-NCCR-MKU Project) Sanction Order: MoES/NCCR/ESM/GIA-MKU/2/2022 Dt. 10.08.2022.

References

1. FAO, *The State of World Fisheries and Aquaculture 2022. Towards Blue Transformation*, FAO, Rome, 2022, https://doi.org/10.4060/cc0461en.
2. Jaikumar, M., Dineshram, R., Imchen, T., Mandal, S., Rangesh, K., Seaweed Farming Potential in India- An Assessment and Review, in: *Global*

Blue Economy: Analysis, Developments, and Challenges, 1st Edition, M.N. Islam, and S.M. Bartell (Eds.), p. 530, CRC Press, Florida, 2022, https://doi.org/10.1201/9781003184287.
3. Ahern, M., Thilsted, S., Oenema, S., Kühnhold, H., The role of aquatic foods in sustainable healthy diets, UN Nutrition Discussion Paper, 58, 2021, https://www.unnutrition.org/wp-content/uploads/FINAL-UN-Nutrition-Aquatic-foods-Paper_EN_.pdf.
4. Willett, W., Rockström, J., Loken, B., Springmann, M., Lang, T., Vermeulen, S., Garnett, T. *et al.*, Food in the Anthropocene: The EAT–Lancet Commission on healthy diets from sustainable food systems. *Lancet*, 393, 10170, 447–492, 2019.
5. Getu, A., Misganaw, K., Bazezew, M., Post-harvesting and major related problems of fish production. *Fish. Aquac. J.*, 6, 4, 1–6, 2015.
6. Coles, R., McDowell, D., Kirwan, M.J. (Eds.), *Food packaging technology*, vol. 5, CRC Press, Florida, 2003.
7. FAO, *Global food losses and food waste – Extent, causes and prevention*, FAO, 2011b, www.fao.org/3/mb060e/mb060e.pdf.
8. FAO, *FAOLEX Database*, FAO, Rome, 2022c, 9 February 2022, www.fao.org/faolex/en.
9. GGGI (Global Ghost Gear Initiative), *Protecting our oceans and the life within them*, GGGI, Washington, 2021, 2 November 2021, www.ghostgear.org.
10. FAO, *Report of the international emergency disease investigation task force on a serious finfish disease in Southern Africa*, FAO, Rome, 2009, 18–26 May 2007.
11. Kamilya, D. and Baruah, A., Epizootic ulcerative syndrome (EUS) in fish: history and current status of understanding. *Rev. Fish Biol. Fish.*, 24, 1, 369–380, 2014.
12. Leung, T.L. and Bates, A.E., More rapid and severe disease outbreaks for aquaculture at the tropics: implications for food security. *J. Appl. Ecol.*, 50, 215–222, 2013.
13. Peeler, E.J. and Taylor, N.G., The application of epidemiology in aquatic animal health-opportunities and challenges. *Vet. Res.*, 42, 1, 1–15, 2011.
14. Romero, J., Feijoó, C.G., Navarrete, P., Antibiotics in aquaculture–use, abuse and alternatives. *Health Environ. Aquac.*, 159, 159–198, 2012.
15. Kumar, V., Roy, S., Meena, D.K., Sarkar, U.K., Application of probiotics in shrimp aquaculture: importance, mechanisms of action, and methods of administration. *Rev. Fish. Sci. Aquac.*, 24, 4, 342–368, 2016.
16. Harikrishnan, R., Balasundaram, C., Heo, M.S., Fish health aspects in grouper aquaculture. *Aquacul.*, 320, 1-2, 1–21, 2011.
17. Gudding, R. and Van Muiswinkel, W.B., A history of fish vaccination: science-based disease prevention in aquaculture. *Fish Shellfish Immunol.*, 35, 6, 1683–1688, 2013.

18. Assefa, A. and Abunna, F., Maintenance of fish health in aquaculture: review of epidemiological approaches for prevention and control of infectious disease of fish. *Vet. Med. Int.*, 1–10, 2018.
19. Gudding, R., Vaccination as a preventive measure, in: *Fish Vaccination*, pp. 12–21, 2014.
20. Lien, S., Koop, B.F., Sandve, S.R., Miller, J.R., Kent, M.P., Nome, T. *et al.*, The Atlantic salmon genome provides insights into rediploidization. *Nature*, 533, 7602, 200–205, 2016.
21. Earle, G. and Hintz, W., New approaches for controlling *saprolegnia parasitica*, the causal agent of a devastating fish disease. *Trop. Life Sci. Res.*, 25, 2, 101, 2014.
22. Dadar, M., Dhama, K., Vakharia, V.N., Hoseinifar, S.H., Karthik, K., Tiwari, R. *et al.*, Advances in aquaculture vaccines against fish pathogens: global status and current trends. *Rev. Fish. Sci. Aquac.*, 25, 3, 184–217, 2017.
23. Aly, S.M. and Albutti, A., Antimicrobials use in aquaculture and their public health impact. *J. Aquac. Res. Dev.*, 5, 4, 1, 2014.
24. Austin, B., Antibiotics and disinfectants, in: *Diagnosis and Control of Diseases of Fish and Shellfish*, pp. 263–278, 2017.
25. Plumb, J.A. and Hanson, L.A., *Health maintenance and principal microbial diseases of cultured fishes*, John Wiley & Sons, Florida, 2010.
26. Rodgers, C.J. and Furones, M.D., Antimicrobial agents in aquaculture: practice, needs and issues. *Options Mediterr.*, 86, 41–59, 2009.
27. Bentzon-Tilia, M., Sonnenschein, E.C., Gram, L., Monitoring and managing microbes in aquaculture–Towards a sustainable industry. *Microb. Biotechnol.*, 9, 5, 576–584, 2016.
28. Maqsood, S., Singh, P., Samoon, M.H., Munir, K., Emerging role of immunostimulants in combating the disease outbreak in aquaculture. *Int. Aquat. Res.*, 3, 3, 147–163, 2011.
29. Sharifuzzaman, S.M. and Austin, B., Probiotics for disease control in aquaculture, in: *Diagnosis and control of diseases of fish and shellfish*, pp. 189–222, 2017.
30. Mastan, S.A., Use of immunostimulants in aquaculture disease management. *Int. J. Fish. Aquat.*, 2, 4, 277–280, 2015.
31. De, C., Meena, D.K., Behera, B.K., Das, P., Das Mohapatra, P.K., Sharma, A.P., Probiotics in fish and shellfish culture: immunomodulatory and ecophysiological responses. *Fish Physiol. Biochem.*, 40, 3, 921–971, 2014.
32. Martínez Cruz, P., Ibáñez, A.L., Monroy Hermosillo, O.A., Ramírez Saad, H.C., Use of probiotics in aquaculture, *ISRN Microbiology 2012,* 1–13, 2012.
33. Fajardo, C., Martinez-Rodriguez, G., Blasco, J., Mancera, J.M., Thomas, B., De Donato, M., Nanotechnology in aquaculture: Applications, perspectives and regulatory challenges. *Aquac. Fish*, 7, 2, 185–200, 2022.
34. Chena, H. and Yadab, R., Nanotechnologies in agriculture: New tools for sustainable development. *Trends Food Sci. Technol.*, 22, 11, 585–594, 2011, https://doi.org/10.1016/j.tifs.2011.09.004.

35. Technavio, *Global food nanotechnology market 2019-2023: Growing applications in nutraceuticals to boost the market*, Technavio, Toronto, Canada, 2019.
36. Chandra, P., *Nano-biosensors for personalized and onsite biomedical diagnosis*, 1st, p. 637, The Institution of Engineering and Technology, Stevenage, England, 2016.
37. Torchilin, V.P., *Nanoparticulates as drug carriers*, Imperial College Press, London, 2006.
38. Donbrow, M., *Microcapsules and nanoparticles in medicine and pharmacy*, CRC Press, Florida, 1991.
39. Shah, B.R. and Mraz, J., Advances in nanotechnology for sustainable aquaculture and fisheries. *Rev. Aquac.*, 12, 2, 925–942, 2020, https://doi.org/10.1111/raq.12356.
40. Ogunkalu, O.A., Utilization of nanotechnology in aquaculture and seafood sectors. *EJFST*, 19, 26–33, 2019.
41. Mohammadi, N. and Tukmechi, A., The effects of iron nanoparticles in combination with Lactobacillus casei on growth parameters and probiotic counts in rainbow trout (*Oncorhynchus mykiss*) intestine. *J. Vet. Res.*, 70, 47–53, 2015, https://doi.org/10.22059/JVR.2015.52968.
42. Nickols-Richardson, S.M. and Piehowski, K.E., Nanotechnology in nutritional sciences. *Minerva Biotecnol.*, 20, 3, 117, 2008.
43. Chellaram, C., Murugaboopathi, G., John, A.A., Sivakumar, R., Ganesan, S., Krithika, S., Krithika, S., Priya, G., Significance of nanotechnology in food industry. *APCBEE Proc.*, 8, 109–113, 2014, https://doi.org/10.1016/j.apcbee.2014.03.010.
44. Huang, S., Wang, L., Liu, L., Hou, Y., Li, L., Nanotechnology in agriculture, livestock, and aquaculture in China. A Review. *Agron. Sustain. Dev.*, 35, 2, 369–400, 2015, https://doi.org/10.1007/s13593-014-0274-x.
45. Yang, H.C., Wang, W.H., Huang, K.S., Hon, M.H., Preparation and application of nanochitosan to finishing treatment with anti-microbial and anti-shrinking properties. *Carbohydr. Polym.*, 79, 1, 176–179, 2010.
46. Nasr-Eldahan, S., Nabil-Adam, A., Shreadah, M.A., Maher, A.M., El-Sayed Ali, T., A review article on nanotechnology in aquaculture sustainability as a novel tool in fish disease control. *Aquac. Int.*, 29, 4, 1459–1480, 2021.
47. Ravikumar, S., Gokulakrishnan, R., Raj, J.A., Nanoparticles as a source for the treatment of fish diseases. *Asian Pac. J. Trop. Dis.*, 2, S703–S706, 2012.
48. Lansdown, A.B., Silver I: its antibacterial properties and mechanism of action. *J. Wound Care*, 11, 4, 125–130, 2002.
49. Pazos-Ortiz, E., Roque-Ruiz, J.H., Hinojos-Márquez, E.A., López-Esparza, J., Donohué-Cornejo, A., Cuevas-González, J.C. *et al.*, Dose-dependent antimicrobial activity of silver nanoparticles on *polycaprolactone fibers* against gram-positive and gram-negative bacteria. *J. Nanomater.*, 1–9, 2017.
50. Tang, S. and Zheng, J., Antibacterial activity of silver nanoparticles: structural effects. *Adv. Healthc. Mater*, 7, 13, 1701503, 2018.

51. Soltani, M., Ghodratnema, M., Ahari, H., Ebrahimzadeh, M.H., Atei, M., Dastmalchi, F.R.N.J., Rahmania, J., The inhibitory effect of silver nanoparticles on the bacterial fish pathogens, *Streptococcus iniae, Lactococcus garvieae, Yersinia ruckeri Aeromonas hydrophila. Iran. J. Vet. Res.*, 3, 2, 137–142, 2009.
52. Saleh, M., Soliman, H., El-Matbouli, M., Gold nanoparticles as a potential tool for diagnosis of fish diseases, in: *Veterinary infection biology: Molecular diagnostics and high-throughput strategies*, pp. 245–252, Humana Press, New York, NY, 2015.
53. Vallejos-Vidal, E., Reyes-López, F., Teles, M., MacKenzie, S., The response of fish to immunostimulant diets. *Fish Shellfish Immunol.*, 56, 34–69, 2016.
54. Charlie-Silva, I., de Melo, N.F.S., Gomes, J.M., Fraceto, L.F., de Melo, D.C., de Oliveira Silva, J. *et al.*, Nanoparticle mucoadhesive system as a new tool for fish immune system modulation. *Fish Shellfish Immunol.*, 80, 651–654, 2018.
55. Wu, Y., Rashidpour, A., Almajano, M.P., Metón, I., Chitosan-based drug delivery system: Applications in fish biotechnology. *Polymers*, 12, 5, 1177, 1–24, 2020.
56. Xia, I.F., Kong, H.K., Wu, M.M., Lu, Y., Wong, K.H., Kwok, K.W., Selenium Nanoparticles (SeNPs) Immunomodulation Is More Than Redox Improvement: Serum Proteomics and Transcriptomic Analyses. *Antioxidants*, 11, 5, 964, 2022.
57. Ponce, M., Zuasti, E., Anguís, V., Fernández-Díaz, C., Effects of the sulfated polysaccharide ulvan from Ulva ohnoi on the modulation of the immune response in Senegalese sole (*Solea senegalensis*). *Fish Shellfish Immunol.*, 100, 27–40, 2020.
58. Govindaraju, K., Dilip Itroutwar, P., Veeramani, V., Ashok Kumar, T., Tamilselvan, S., Application of nanotechnology in diagnosis and disease management of white spot syndrome virus (WSSV) in aquaculture. *J. Clust. Sci.*, 31, 6, 1163–1171, 2020.
59. Thiruppathiraja, C., Kumar, S., Murugan, V., Adaikkappan, P., Sankaran, K., Alagar, M., An enhanced immuno-dot blot assay for the detection of white spot syndrome virus in shrimp using antibody conjugated gold nanoparticles probe. *Aquaculture*, 318, 3-4, 262–267, 2011.
60. Phanse, Y., Puttamreddy, S., Loy, D., Ross, K., Ramirez, J.V., Narasimhan, B., Bartholomay, L., Driving Biomaterial Innovation and the Race to Translation, p. 15, Society for Biomaterials, Charlotte, 2015.
61. Juarez-Moreno, K., Mejía-Ruiz, C.H., Díaz, F., Reyna-Verdugo, H., Re, A.D., Vazquez-Felix, E.F. *et al.*, Effect of silver nanoparticles on the metabolic rate, hematological response, and survival of juvenile white shrimp *Litopenaeus vannamei. Chemosphere*, 169, 716–724, 2017.
62. Imelda, J., Bioencapsulation of live feeds, pp. 1–6, CMFRI, Cochin, 2003.
63. Gomez-Gil, B., Herrera-Vega, M.A., Abreu-Grobois, F.A., Roque, A., Bioencapsulation of two different Vibrio species in nauplii of the brine shrimp (*Artemia franciscana*). *Appl. Environ. Microbiol.*, 64, 6, 2318–2322, 1998.

64. Ghorbanzade, T., Jafari, S.M., Akhavan, S., Hadavi, R., Nano-encapsulation of fish oil in nano-liposomes and its application in fortification of yogurt. *Food Chem.*, 216, 146–152, 2017.
65. Romalde, J.L., Luzardo-Alvarez, A., Ravelo, C., Toranzo, A.E., Blanco-Mendez, J., Oral immunization using alginate microparticles as a useful strategy for booster vaccination against fish lactococcosis. *Aquaculture*, 236, 1-4, 119–129, 2004.
66. Heidarieh, M., Moodi, S., Katuli, K.K., Unger, H., Biochemical Effects of encapsulated radiovaccine via alginate nanoparticles as useful strategy for booster in immunized rainbow trout against *ichthyophytirius multifiliis*. *Acta Sci. Vet.*, 43, 1330, 2015.

7

Nanotechnological Strategy for the Diagnosis of Infectious Diseases: Recent Developments and Opportunities

Vennila Thirumalaiswamy[1], C.V. Vaishali[2], Sathyavathi Sundararaju[3], Chockalingam Muthiah Ramakritinan[4*], Muneeswaran Thillaichidambaram[5†] and Franck Quero[5‡]

[1]*Department of Chemistry, Sri Sairam Engineering College, Chennai, Tamil Nadu, India*
[2]*School of Life Sciences, JSS Academy of Higher Education & Research, Mysore, Karnataka, India*
[3]*Department of Translational Medicine, Cancer Immunology Laboratory, Sidra Medicine, Doha, Qatar*
[4]*Department of Marine Biotechnology, Academy of Maritime Education and Training (AMET) Deemed to be University, Kanathur, Chennai, Tamil Nadu, India*
[5]*Laboratorio de Nanocelulosa y Biomateriales, Departamento de Ingeniería Química, Biotecnología y Materiales, Facultad de Ciencias Físicas y Matemáticas, Universidad de Chile, Avenida Beauchef, Santiago, Chile*

Abstract

Recently, the earth has witnessed a novel coronavirus disease outbreak and for the prevention of disease, a lockdown has been implemented. Infectious diseases cause economic burdens to almost all developing and developed countries and are also responsible for major deaths globally. To minimize transmission, end the pandemic, and effectively treat infectious diseases, rapid, cost-effective, and accurate screening is essential. Recent advancement in nanotechnology has paved the way to improve the sensitivity of disease diagnosis. Nanomaterials pose several unique properties, which makes them an ideal candidate for disease diagnosis

**Corresponding author*: ramakritinan@gmail.com
†*Corresponding author*: tmuneeswaran@hotmail.com
‡*Corresponding author*: fquero@ing.uchile.cl

Sathish-Kumar Kamaraj, Arun Thirumurugan, Muthuchamy Maruthupandy, Mercedes Guadalupe López Pérez and Shanmuga Sundar Dhanabalan (eds.) Nanoscience and Nanotechnology for Smart Prevention, Diagnostics and Therapeutics: Fundamentals to Applications, (143–182) © 2024 Scrivener Publishing LLC

and treatment. In addition, it is possible to develop optical and electrochemical nano biosensors in point-of-care devices with hassle-free detection of infectious diseases. Further, nanotechnology could improve the sensitivity of conventional methods. This chapter highlights the importance of nanotechnology in infectious disease diagnosis and various nanomaterials used for the diagnosis and their limitations. Further, it emphasizes the diagnosis of several major viral, bacterial, and vector-borne infectious diseases based on nanoplatforms. In addition, it also deals with the safety concerns that arise over commercial production and their clinical use, especially in disease diagnosis.

Keywords: Smart devices, disease diagnosis, infectious diseases, COVID-19, nanomaterials

7.1 Introduction

In the recent pandemic situation, lots of people suffering from emerging diseases caused by bacteria, viruses and vectors. Pathogens are the major cause of infectious diseases, which leads to even death and their prevention is challenging because of the diagnosing issues. The treatment of fatal diseases needs early detection of the disease-causing etiological agent. Fast and sensitive detection methods are very much needed to overcome such a pandemic situation. A variety of techniques is used to detect infectious diseases. The conventional culturing technique is a gold standard, which is laborious and time-consuming. Other techniques include PCR-based methods, ELISA, and immunoassays are expensive and also needs skilled labor, and highly equipped laboratory. Hence, rapid, simple, sensitive, and cost effective diagnosis is need of the hour for the early detection of pathogens in developing/growing countries where resources are limited. Biosensor-based methods are more sensitive than regular diagnostic methods, which could serve as an alternative to conventional methods. In brief, a biosensor can measure a signal generated from biological and chemical reactions, which is proportional to the concentration of an analyte [1].

The physicochemical transduction, such as electrochemical, optical, thermal, and piezoelectric biosensors, or the type of biorecognition element, could be used to categorize biosensors [2]. With high surface-to-volume ratios of nanomaterials, it is possible to change the detection limit and duration, which may enhance the interactions between the sensor and the analyte. The key characteristic of a biosensor is its ability to specifically interact with an analyte molecule, leading to changes in the generation of ions, colored moieties, electrons, gases, heat, mass, or light (Figure 7.1),

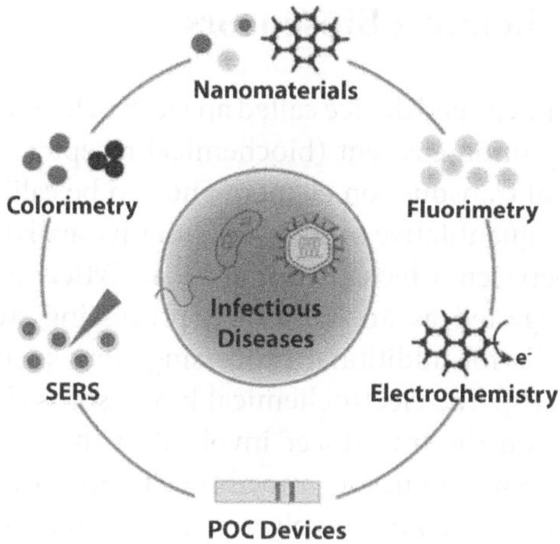

Figure 7.1 Nanosensors for diagnostics of infectious diseases (permission from [4]).

which can then be boosted and turned into outcomes that are simple to interpret [3].

7.2 Optical Biosensors

Optical Biosensors (OB) are one of the most significant biosensors, which have high specificity, a small footprint, and a low cost of production. OB gather information by measuring the photons rather than electrons and the evaluation is based on absorbance, reflectance/fluorescence emissions from the target [5]. The produced signal can be converted into measurable value (e.g., current, temperature alteration) by the transducer [6]. OB can be either label-free or labelled. In label-free OB, the signal is produced directly from the interaction between the transducer and the analyte, and in labelled OB, a label is needed to produce a signal which further measured by colorimetric, luminescent, or fluorescent spectrophotometers [7]. Optical biosensors widely employed in virus detection because it is safe to use, cost-effective, and devoid of nucleic acid amplification [8]. Optical biosensors can be easily miniaturized and present potential for chip integration also based on Surface Plasmon Resonance (SPR), Localized Surface Plasmon Resonance based (LSPR), optical waveguided, optical resonator, photonic crystal (PC), and optical fiber [9].

7.3 Electrochemical Biosensors

A self-contained integrated device called an electrochemical biosensor uses a biological recognition element (biochemical receptor) that is kept near an electrochemical transduction element and can be calibrated repeatedly to provide precise quantitative or semi-quantitative analytical information. To distinguish between a biosensor and bioanalytical processes, reagent addition was suggested as an additional processing step. Bioanalytical system, which calls for additional processing steps including the inclusion of reagents [10]. The electrochemical biosensor is classified into four types, depending on the transducer involved in the measurements such as amperometric, potentiometric, impedimetric, and conductometric biosensors. Here, various methods involved in the diagnosis of viral, bacterial, and vector-borne diseases based on the optical and electrochemical biosensors are discussed.

7.4 Detection of Viral Diseases

Viruses are present almost in all environments and most of them are regarded as pathogens and their early identification is the most important during outbreaks or pandemics. Nanomaterials, including metallic nanoparticles, polymeric nanoparticles, silica nanoparticles, carbon nanotubes, and quantum dots, are used for the detection of viruses [3] among them AuNPs widely used because of its potential optical and electrochemical properties [11]. Plasmons, known as surface plasmon resonance are caused to travel away from the atomic core when a free electron on AuNPs exerts light (SPR). Applications for biosensors use the quenching effects of SPR in Au NPs to interact with a new substance and serve as an indication [12]. Metal-enhanced fluorescence (MEF) occurs when proximal fluorophores interact with metallic surfaces to significantly increase the intensity of the fluorescent signal [13]. Because they may display robust surface plasmon resonance in the visible wavelength region, Gold (Au) and Silver (Ag) nanoparticles are typically employed for MEF. AgNPs show greater signal amplification than Au NPs.

In recent years, one of the most common and fatal infection is COVID-19 caused by the SARS-CoV2 virus, which is a challenging outbreak due to the high reproduction and transmission rate [14]. SARS-CoV2 is a beta coronavirus, which has a single-positive strand RNA genome. CoV2 genomes encode four structural proteins: Spike (S), Envelope (E), Matrix (M), and nucleocapsid (N) [15]. Structural and nonstructural proteins of the

CoV2 virus are considered an antiviral target, including phosphorylated nucleocapsid, envelope, membrane glycoproteins, spike, helicase, papain-like protease, and 3-chymotrypsin-like protease [16]. Nucleic acid–based detection is appropriate in the early stages of illness or even before symptoms appear. The extracted nucleic acid is often employed to detect viruses in clinical samples like saliva or nasopharyngeal secretions [17]. Otherwise, blood samples from infected patients are used for antibody (IgG/IgM) screening [18]. During infection, the amount of IgM level increases in the blood whether it is symptomatic or asymptomatic [19]. A primary or acute infection is demonstrated by a positive IgM with a negative IgG or a positive IgM with a positive IgG, whereas a secondary or advanced stage of infection is demonstrated by a positive IgG with a negative IgM [20]. Even though antibody screening is feasible, the quantity of IgM levels gradually decreases after 2 weeks [18].

The well-established POC diagnostics for detecting SARS-CoV2 infections are nucleic acid and antibody (Ab) testing [21]. Lateral flow assay is used for the detection of viruses, especially for the SARS-CoV2. The analytical membrane is coated with nanoparticles, particularly Au NP which is conjugated with antihuman antibodies. By using capillary action, the patient's sample drowns into the modified membrane, and the binding of antigen to the AuNP changes the color and acts as an indicator. The assay protocol needs approximately 15 minutes to detect and show advantages over the molecular detection method, which has the above-said pitfall. In addition, lateral flow assay needs a low volume (10-20 μl of a serum) sample with better stability, and minimal training is enough to carry out the test procedure without hassle [19]. The AuNP-LF (Gold Nanoparticle-Lateral Flow) assay was helpful for rapid diagnosis and onsite detection with sensitivity and specificity of 100 and 93.3% and consistent with RT-PCR results. The study by [22] postulated that among 60 covid positive patients, 13 shows a negative for IgM antibody screening and the IgG titer was higher than the IgM titer. Hence, a combined kit for analyzing IgM and IgG is a better option for diagnosing SARS-CoV2. Li *et al.* [23] developed a fluorescence immunochromatography-based lateral flow immune assay technique that uses IgM-IgG combined antibody test to diagnose SARS-CoV2 with respective sensitivity and specificity of 88.66% and 90.63%. In the combined IgM and IgG assays, the sensitivity and specificity showed 89.22% and 96.86%, respectively. Sometimes, the antibody assay produces false-positive results due to the interfering substances and cross-reactivity with other viruses. So, it has to be fully evaluated to avoid cross-reactivity [24].

Gold nanoparticles (AuNPs), coated with thiol-modified antisense oligonucleotides (ASOs) specific for N-gene (nucleocapsid phosphoprotein) of SARS-CoV2 used for colorimetric detection. The change in plasmon resonance indicates that the ASO-capped Au NPs agglomerated in the presence of the target RNA sequence of SARS-CoV2 and the addition of RNase breaks the RNA strand, which forms visually detectable precipitate [25]. Qiu *et al.* [26] developed a dual functional plasmonic biosensor for detecting SARS-CoV2 by combining the plasmonic photothermal (PPT) effect and LSPR with gold nanoislands (Au NIs). This novel technique utilizes the heat from the photothermal reaction for the hybridization of RNA-dependent RNA polymerase sequence and the complementary DNA. Hybridization of two strands will occur just below the melting temperature and the unhybridized strands eventually decrease. The increasing temperature could remove all the noncomplementary strands and thus enabling label-free detection also enhancing the stability, sensitivity, and reliability of biosensors. Moreover, significant heating energy will be relaxed when the absorption of light and exhibition of the optical cross section can be relaxed nonradioactive in exhibited state of the plasmonic nanoparticles [27].

Electrochemical methods that do not require sandwich assays or extra signaling molecules, such as electrochemical impedance, cyclic voltammetry, and amperometry, can allow label-free detection [29]. The detailed fabrication process and the detection principle of electrochemical and optical biosensors for the detection of respiratory disease-causing viruses were shown in Figure 7.2. The electrochemical immunosensor stands out among a wide range of potential substitutes for its capacity to carry out reasonably quick, sensitive, and economical detection and, more crucially, its potential for downsizing for quick, on-site measurement. The label-free electrochemical technique has advantages over labeled electrochemical immunosensor assay due to the lack of involvement of enzyme (Sandwich type) or antibody coating [30]. Tripathy and Singh [31] developed a label-free electrochemical immunosensor using Au Nps as a transducer attached with a smartphone-based readout for POC testing. The target sequence attached to the signaling electrode is highly stable in harsh chemical conditions. The viral RNA or cDNA can be detected after the hybridization takes place in the complementary thiolated probe.

An in-house built biosensor device (eCovSens) was fabricated to detect the spike protein domain I of the SARS-CoV2 virus based on the electrical conductivity changes when antigen binds with antibody immobilized on the surface of carbon electrode within 10 to 30 s, which needs 10 to 20 μl of a sample. The working range of eCovSens was reported to be ~10 fM^{-1} μM and the limit of detection ~90 fM and FTO/Au NP potentiostat showed high sensitivity with a detection range from 1fM – 1 μM [32].

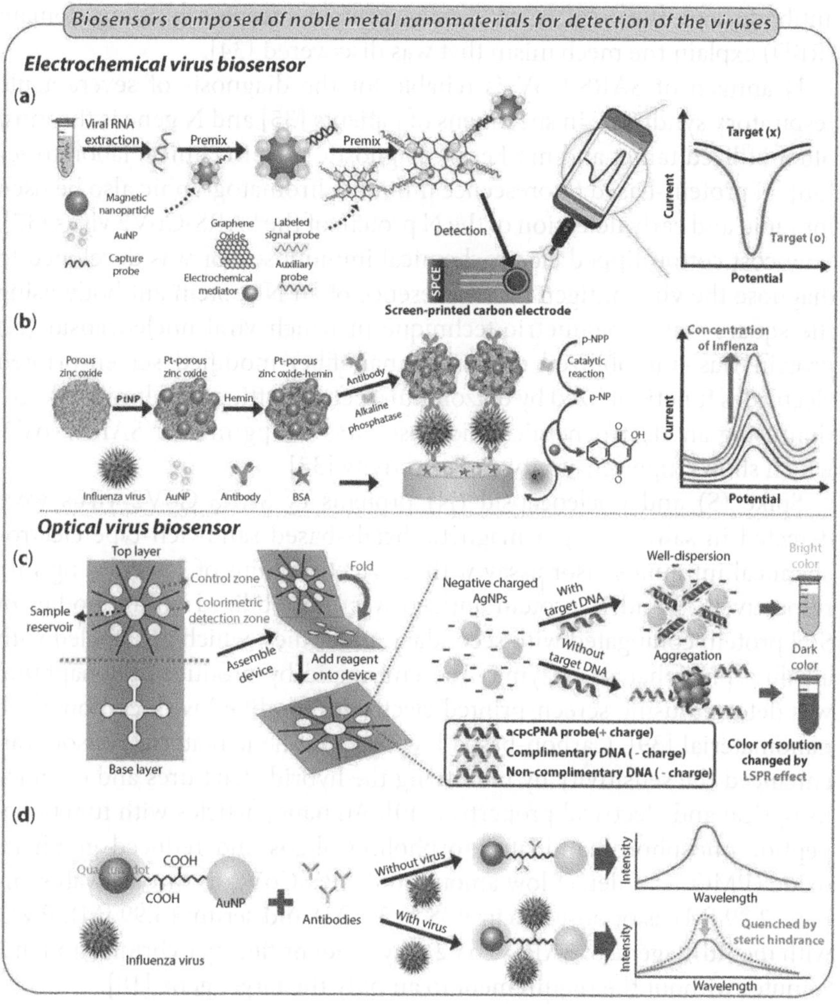

Figure 7.2 Fabrication process and detection principle of respiratory disease-causing viruses using electrochemical and optical biosensors (permission obtained from [28]).

The total impedance can change as a result of the binding events because they can change the double layer's charge distribution, shape, and size, which in turn affects the adsorbed layer's conductivity and dielectric constant [33]. A novel impedance sensing platform was developed for the detection of SARS-CoV2 spike protein (SCSP) subsequently tested with anti-SARS-CoV2 monoclonal antibody CR3022 which could able to differentiate spike protein from negative samples along with consistency with ELISA test results. The kinetics of interaction between the anti-SARS-CoV2 antibody and the SARS-CoV2 spike protein receptor binding domain (RBD) explain the mechanism that was discovered [34].

N antigen of SARS-CoV is reliable for the diagnosis of severe acute respiratory syndrome in specimens of patients [35] and N gene is the most often utilized target and marker in diagnostic tests in clinical laboratories [36]. N protein-based fluorescence immunochromatographic also be used for rapid and early detection of the N protein of the SARS-CoV2 virus [37]. Low-cost cotton tipped electrochemical immunosensor was developed to diagnose the viral antigen in the presence of an N protein antibody using the square wave voltametric technique in which viral nucleocapsid (N) protein was immobilized on carbon nanofiber-modified screen-printed electrodes functionalized by diazonium electro grafting. The limit of detection using an electrochemical biosensor was 0.8 pg/mL for SARS-CoV2, which shows high sensitivity and selectivity [38].

Spike (S) and nucleocapsid (N) proteins of SARS-CoV2 virus were detected in saliva using a magnetic beads-based sandwich-type electrochemical immunosensor assay with a detection limit of 19 and 8 ng/mL, respectively. S and N protein antigen was immobilized on antibodies of S/N protein conjugated with secondary antibodies, which are labeled with alkaline phosphatase enzyme. The enzymatic by-product of 1-naphthol was detected using screen-printed electrodes modified with carbon black nanomaterial [39]. Carbon-based low-cost graphene material sensors can enhanced the sensitivity by modifying the hybrid structures and enhance its optical and electrical properties [40]. Au nanoparticles with functional peptide phosphorodiamidate morpholino oligos and reduced graphene oxide (PMO) can detect low amount of SARS-CoV2 in throat swab samples (2.29 fM) as opposed to in PBS (0.37 fM) and serum (3.99 fM). PMO with the RdRp gene of SARS-CoV2 may be identified by hybridization in 2 minutes without the requirement to amplify the target gene [18].

7.4.1 Influenza Virus

Influenza cause respiratory tract infection; divided into four types namely Influenza A, B, C, and D [41] based on the Nucleoprotein (NP) and Matrix protein (M) Further sub-divided based on combinations of their surface glycoproteins hemagglutinin (HA-18 subtypes (H1 to H18)) and neuraminidase (NA-11 subtypes (N1 to N11)) [42]. Influenza epidemics cause serious illness and even death because of its severe range of symptoms and higher transmission rate. The mutagenic/antigenic drift makes diagnostics difficult. Rapid detection of Influenza using point-of-care diagnostics widely used and different kinds of biosensors are available for the detection of Influenza utilizing Nanotechnology and recently reported elsewhere [43].

Optical biosensor based on nanostructure has potential attractive characteristics such as rapidity, low cost, and high sensitivity. [44] proposed peptide-immobilized nanopillar structures with label-free detection of the Influenza A virus in the range of 10^3 to 10^5 PFU through simple measurements of reflectance. It used polymerized dopamine, a sialic acid-mimic peptide that can specifically bind to hemagglutinin on the surface of the influenza virus was immobilized onto the nanopillars. Carbon-based quantum dots (CNQDs) has excellent optical properties, low-cost preparation, and good biocompatibility [45]. Graphitic carbon nitrides QDs (gCNQDs) are nontoxic nanomaterials made of carbon that have exceptional optical properties that rival those of conventional heavy metal-based QDs and emerging as competitive materials as nanosensors [46–49]. A magnetic/plasmonic-assisted fluoro-immunoassay system was developed using fluorescent graphitic carbon nitride quantum dots (gCNQDs and magnetic-derivatized plasmonic molybdenum trioxide quantum dots (MP-MoO_3 QDs) as magnetic probes and plasmonic/magnetic agents to detect influenza A virus. Formation of immunocomplex between Influenza A virus and Influenza specific antibody conjugated QDs enhances the fluorescent intensity which is proportional to the concentration of the virus. While the detection ranged from 45 to 25, 000 PFU/mL in clinical samples, influenza virus (H3N2) was detected at 0.25 pg/mL in DI water and 0.9 pg/mL in human serum [45].

Fast Evaluation of Viral Emerging Risks (FEVER) pipeline which detected influenza RNA as low as 50 and 1 nM based on a waveguide-based optical biosensor and a flow cytometry, respectively [50]. The same approach used for the detection of SARS-Cov2 to design high-coverage MB probes [51]. Conductive polymers have high electrical conductivity, orientation, charge distribution, high capacitance, high porosity, and low-cost preparation making them a perfect tool for analytical investigations [52]. An instantaneous translation of stimuli into response-enabled conductive polymers has been reported as a quick detection assay for intermolecular interaction of cellular activity and associated biorecognition elements, such as antibodies, enzymes, and proteins [53, 54]. Rapid system that uses conductive polymer [poly(aniline-co-pyrrole)] based nanoparticles encapsulated and functionalized with peptides can be used to quantify influenza A virus (H1N1) [42, 52, 55].

Raman optical spectroscopy uses a scattering spectrum to enable the label-free detection of bacteria across a wide range of bacterial cells. Particularly, aptasensor systems based on surface-enhanced Raman scattering can detect infections with high sensitivity [56]. To identify and trap viruses, Raman reporter molecules modified aptamers are often mounted on a nanostructure metal-dielectric substrate. Due to the interaction between aptamers and viruses during laser irradiation, photon energy varies, allowing for the quantification of targets [57]. In aptamer-based biosensors (aptasensor), aptamers act as a recognition element that specifically binds to the target material with high specificity and affinity without interfering with other materials. Then, the output from the interaction of both the aptamer and target will be developed by the transducer [58]. The detailed fabrication of aptasensors and their detection mechanism was shown in Figure 7.3. When A/H1N1 viruses are added to the substrate, A/H1N1 target recognition causes the aptamers to alter in conformation, and the aptamer and A/H1N1 coupling results in a steady decline in Raman signal intensities with the addition of more A/H1N1 viruses [56]. Rapid SERS-based aptamer detect the Influenza virus with the limit of detection of 10^4 virus particles per sample or 10^{-4} HAU per sample [59]. RHA0385 aptamer used for the detection of H1, H3, and H5 IV subtypes. SERS-based techniques least sensitive than ELISA or PCR. The highly sensitive SERS-based aptasensor has poor reproducibility [56, 58].

Magnetic particle spectroscopy (MPS) system combined with magnetic nanoparticles (MNP) can detect Influenza A virus in which multiple epitopes of H1N1 nucleoprotein can bind with MNPs functionalized with IgG antibodies specific for the Influenza A virus. Cross-linking H1N1 protein and IgG antibodies produce differentiation in the magnetic response/harmonic changes detected by external oscillating magnetic fields.

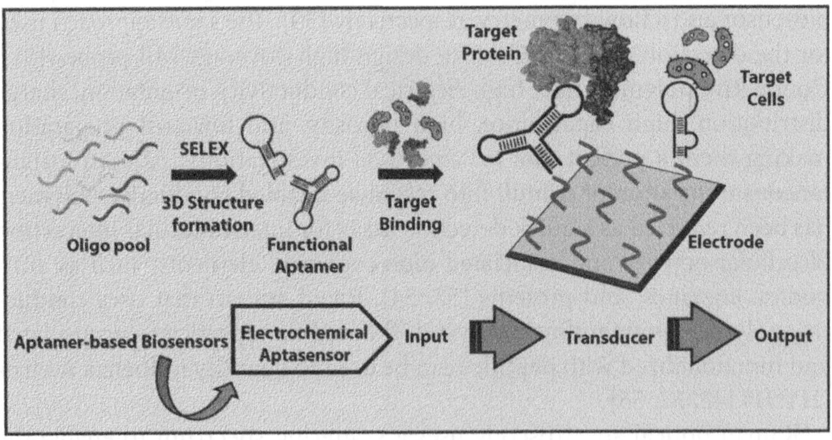

Figure 7.3 Schematic representation of fabrication of aptasensors for the detection of various target molecules (permission from [60]).

The harmonic ratio of R35 metrics was reported with a concentration as low as 44 nM (442 p mole) of H1N1 nucleoprotein. The detection of H1N1 by the MPS system is rapid, easy to use, wash-free, sensitive, and cost-effective [61]. Magnetic bead-based enzyme-linked immunoassay (MELISA) can able to identify the nucleoprotein (NuP) as low as 3×10^3 PFU mL^{-1} and 4×10^4 PFU mL^{-1} of the H1N1 and H3N2 virus with the combination of immunomagnetic beads and biotinylated silica NPs as capturing and detection elements. In contrast, a classical ELISA test based on the same antibody sandwich showed a detection limit of 1.2×10^7 PFU mL^{-1} for H1N1 [62].

Label-free detection of Influenza viruses based on an electroactive optical waveguide was demonstrated to detect the HA antigen. The electroactive waveguide surface was functionalized with antibodies specific for the target antigen, which is captured by secondary antibodies incorporated with methylene blue dye to act as a redox active probe for antibody labeling. The potential of the redox probe provides a self-referenced detection protocol for direct quantification of the virus analyte in relatively small times [63].

7.4.2 Chikungunya and Zika

Chikungunya (CHIK) is a disease caused by the Chikungunya virus (CHIKV) transmitted by mosquitoes, namely *Aedes agypti* and *Aedes albopictis* [64]. It is an RNA virus belonging to the *Alphavirus* genus and *Togaviridae* family [65]. The *Flavivirus* genus is related to the family *Flaviviridae,* which contains single-stranded RNA. *Flavivirus* is transmitted to people through the bites of infected ticks or mosquitoes [66]. The most frequent *Flavivirus* infections from mosquitoes caused by the dengue virus (DENV), Zika virus (ZIKV), and Japanese encephalitis virus (JEV) [67]. There is an urgent need for improved diagnostics tools needed for the early diagnosis of viral infections, such as DENV, ZIKV, and CHIKV.

Electrochemical impedance spectroscopy (EIS) is one of the promising techniques for detecting viruses/pathogens among other electrochemical diagnostics. It is based on the detection of changes in resistance in electrodes before and after the addition of biological materials [68]. Nonstructural protein 3 (NS3) was used to detect CHIKV using a self-assembled monolayer of 11-mercaptoundecanoic acid and 6-mercaptohexanol by modifying a gold surface electrode with LOD and LOQ were 8 and 17 ng/ml, respectively, which is highly specific to CHIKV-nsP3 [69]. Early detection of infection may increase the survival rate of the patients but the onset of infection contains only a small number of pathogens. Developing sensitive techniques for the detection even in small quantities is prerequisite. Dual-mode immunosensors is used to overcome the pitfall. A fluorescent and electrochemical-based dual-mode diagnosing system used to detect recombinant CHIKV E1 protein (CHIK VP) which contains fluorescent and an electrochemical sensing probe serving dual responsive sensing to make more reliable results from real samples [70]. The detection signal can be amplified in both processes of fluorescence and electrochemical differential pulse voltammetry (DPV).

Although nanotechnology platform shows higher sensitivity and multiplex system detecting multiple pathogens is limited. The high surface area of the sensing platform enhances the immobilization eventually increasing the signal strength and the sensitivity. Nanocubes structures have high surface area added advantage of sensing potential. Gold shells coated metal nanocubes (Fe3O4@Au) which amplify the electrochemical reaction are fabricated on electrochemical paper analytical devices (ePADs) to detect CHIK DNA (LOD of 0.1 nM) [64]. Label-free electrochemical biosensors for the detection of the Zika virus are limited and numerous bio-recognition components have been used on the surface of label-free electrochemical biosensors. The nonstructural protein 1 (NS1) served as a key molecule in the biosensors' specific detection of ZIKV which can selectively diagnose ZIKV without reacting with other flaviviruses [71]. However, studies suggested it may cross-react with other flavivirus like dengue. Hence, finding a specific NS1 protein for ZIKV is important during screening [68].

Antigen-specific monoclonal antibodies improve the detection performance of the system. SERS-based immunoassay using Au NPs (100 nm) and silica shell (4 nm) which prevents the aggregation of nanoparticles, coated with Nile Blue (NB) functionalized with ZIKV NS1 monoclonal antibodies which provide high specificity and less cross-reactivity with DENV NS1 antigen [72]. Electrochemical impedance spectroscopy-based microfluidic paper chip with a printed electrode was developed to detect ZIKV in which labeling of viral lysate with Pt metal enhances the selectivity and specificity by increasing the conductivity [12]. ZnO nanostructures immobilized with ZIKV mAbs on a Printed Circuit Board (PCB) based on electrochemical impedance spectroscopy showed specificity for of ZIKV NS1 antigen without cross-reaction with DENV, with a linear range from 0.1ng mL^{-1} to 100 ng mL^{-1} and the limit of detection is lower than 1.00 pg·mL^{-1} [68].

Simultaneous detection of two or more pathogens in a single diagnosis method has a high advantage as it is rapid and high sensitivity. SERS-based multiplexed assay can simultaneously detect NS1 antigen of dengue and Zika viruses with has 15-fold and 7-fold lower detection limits for Zika NS1 and dengue NS1, respectively with LOD values down to 0.72 ng·mL⁻ and 7.67 $ng \cdot mL^{-1}$ [73]. Similarly [70] developed electrochemical impedimetric immunosensor based on Au–PAni-N, SGQDs nanocomposite conjugated with corresponding DENV, ZIKV, and CHIKV antibody for their simultaneous detection. The carboxylic groups on GQDs are attached to the free amino group of antibodies through covalent bonding, which makes the electrode surface stable and specific. Au–PAni enhances the electron transfer, which improves the electrochemical response.

7.4.3 HIV/AIDS

Human immunodeficiency virus (HIV) is an RNA virus causing the most fatal acquired immunodeficiency syndrome (AIDS) in humans. There are two types of HIV, they are HIV-1 and HIV-2. Among these two, HIV 1 is a major concern because it is the most fatal and causes a pandemic [74]. Many methods are used to diagnose HIV and molecular assays are preferred because of high precision. However, they are time-consuming and expensive. Nowadays biosensor-based diagnostic test also available, which is simple, rapid, and cost-effective. The construction of viral DNA/RNA-based biosensors with nanomaterial poses highly sensitive and selective diagnosis techniques. When compared to other metal-based nanomaterials, AuNPs have received a lot of attention from both academia and industry as a result of their advantageous characteristics, which include a high surface area-to-volume ratio, biocompatibility, distinctive optical and electrical properties, and ease of surface modification. Additionally, biomolecules like proteins, antibodies, and aptamers can be successfully adsorbed onto the surfaces of AuNPs. The optical and electrical characteristics of AuNPs have been carefully crafted for various biosensing applications [75]. Sensitive electrochemical biosensor for the detection of HIV-related DNA based on TSDR and cruciform DNA crystal was developed [76]. A label-free electrochemical impedimetric sensor was developed using carbonized glass fiber–coal tar pitch (GF–CTP) electrodes with gold nanoparticles (AuNPs) for the detection of the HIV-1. The proposed ssDNA sensor showed a lower LOD with a broad detection range for HIV-1 DNA due to the highly conductive carbon fiber structure that provides a high active surface area. The calibration of the sensor was obtained using EIS between 0.1 pM and 10 nM (1×10^{-13} and 1×10^{-8} M) with 13 fM (1.3×10^{-14} M) detection limit [77].

Raman reporter-labeled AuNPs were applied in SERS to detect HIV-1 DNA markers. The SERS-based lateral flow assay monitors the characteristic of the Raman peak intensity of the DNA-conjugated AuNPs with a detection limit of 0.24 pg mL^{-1} [78]. Recently, metal-organic framework based electrochemical biosensor was developed from a complex of metal oxide containing CoO, NiCo2O4, and metallic Co/Ni oxide to detect HIV-1 gene exhibiting an excellent sensing performance with a low LOD of 16.7 fM to determine HIV-1 DNA. The nanocomposites bound with the DNA probe through a noncovalent bond, which enhances the sensing efficiency of the target HIV-1 gene [79]. Copper (II) zwitterionic carboxylate MOF interacts with two different carboxyfluorescein (FAM)-labeled probe ssDNA by electrostatic, π-stacking, and/or hydrogen-bonding interactions to form two P-DNA which extinguish the fluorescence of FAM by the photoinduced electron transfer process. This P-DNA acts as a fluorescent sensor that identifies HIV1 ds DNA and Sudan virus RNA with detection limits of 196 and 73 pM, respectively [80].

7.4.4 Hepatitis

Figure 7.4 depicts calorimetric sensors used for the detection of infectious diseases based on ELISA and immunoassays. For instance, the hepatitis B virus was detected using DNA microarray-based fluorescence enhancement strategy with Ag nanoparticle functioned with recognition probes (Cy3-probe) and hybrid probes (oligomer-A and oligomer-B) by increasing the number of fluorophores MEF effect, the fluorescent intensity enhanced extremely. Also showed a detection limit of 50 fM and a high specificity for distinguishing single-base mismatches [81]. Fluorescence resonance energy transfer (FRET) system with gold nanorods (AuNRs) and fluorescein (FAM) used to diagnose hepatitis B virus DNA sequences. Here, the emission spectrum of the donor is absorbed by the acceptor, which represents appreciable overlap. The decline of the fluorescence intensity of FAM was linearly related to the concentration of the complementary DNA in the range from 0.045 to 6.0 nmol L^{-1} [82].

Ag nanorice@Raman label@SiO_2 sandwich nanoparticles are used to build a SERS sensor, which is connected to a periodic Au triangle nanoarray by hepatitis B virus (HBV) DNA. A spatially increased EM field of the quasi-periodic array produced by such an architecture is anticipated to enable ultrasensitive SERS detection. When the Ag nano rice is linked to the periodic Au nanoarray rather than the planar film chip, the SERS signal is much improved. The resulting spatially expanded SPR field allows the SERS biosensor with a limit of detection of 50 aM toward hepatitis B viral DNA and the capacity to distinguish a single-base mutation of DNA to be used [83]. HCV surface antigen E2 Protein was detected with the aid of an electrochemical biosensor where antigen-specific ligand for host cell receptor (CD81) molecule act as the molecular recognition element in which short fragment of LEL (large extracellular loop) sequence of CD81 was immobilized on SPCE. Differential pulse voltammetry was used to evaluate the binding affinity of ligand molecules toward E2. The resulting HCV biosensor demonstrated selectivity toward E2 in the presence of interfering protein, conalbumin with the LOD of $2.1 \cdot 10^{-5}$ mg/mL both in 0.01 M PBS solution (pH 7.4) and in simulated blood plasma [84].

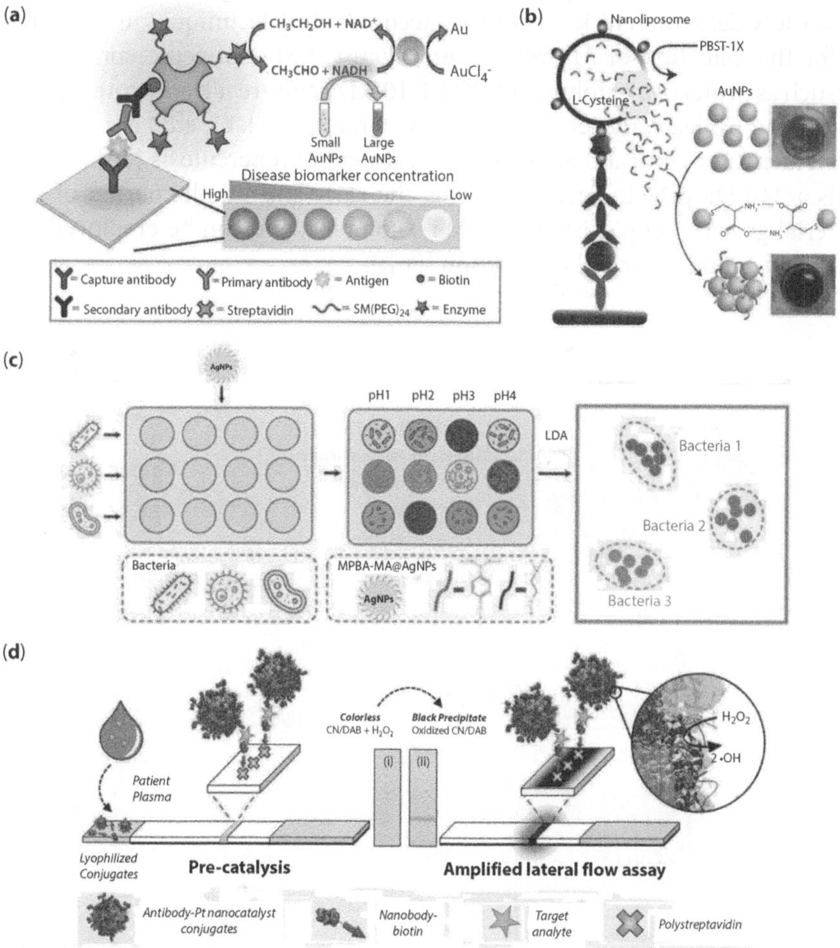

Figure 7.4 Colorimetric sensors for detection of various infectious diseases. (a) Alcohol dehydrogenase-based ELISA. (b) Liposome-amplified plasmonic immunoassay. (c) 4-Mercaptophenylboronic acid (MPBA)-mercaptoethylamine (MA) cofunctionalized AgNPs (MPBA-MA@AgNPs). (d) Pt nanocatalysts-based LFA with ultrabroad dynamic range (permission from [4]).

7.5 Detection of Bacterial Diseases

7.5.1 *Mycobacterium tuberculosis*

Tuberculosis (TB) is a contagious disease caused by the bacteria *Mycobacterium tuberculosis* (MTB) responsible for major health problems, especially multi-drug-resistant (MDR) TB. MDR TB is a major challenging and lethal infectious disease to treat and diagnose. Delayed diagnosis is a major concern for the spreading and mortality of a high number of people worldwide. Different kinds of nanotechnological techniques are available for the detection of *M. tuberculosis* (Figure 7.5). Antigenic components such as protective cytokines (MPT64, 10-kDa culture filtrate protein [CFP-10] and 6-kDa early secreted antigenic target [ESAT-6], etc.) are potential elements for clinical diagnosis and treatment of tuberculosis [85]. MPT64 is a 24-kDa protein secreted during the early and middle period of bacterial growth, only exists in the MTB complex, and can be employed as a specific antigen for MTB detection [86].

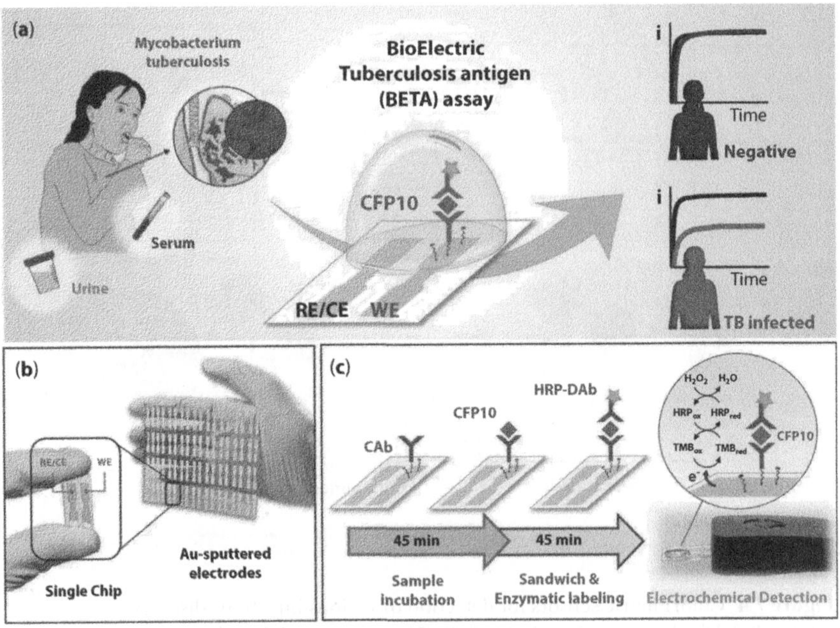

Figure 7.5 Workflow and function of novel bioelectronic TB antigen (BETA) assay representing the electrochemical sensors and the immunoreaction and transduction processes and detection of current (permission from [87]).

Seifert *et al.* [87] developed a novel bioelectronic tuberculosis antigen (BETA) assay to detect tuberculosis-specific antigen, CFP10 in unprocessed clinical samples. The estimated LOD for detection of Mtb CFP10 antigen in serum was 0.4 nM (44 pg in a 10 μL clinical sample droplet) and 0.7 nM in urine. The rapid system showed a high sensitivity and specificity toward the CFP10 antigen. This quantitative BETA assay has the potential to be used as a point-of-care diagnostic solution and treatment response monitoring tool at the lowest levels of the healthcare system due to its low cost of goods, straightforward workflow, small volume clinical sample requirement, and quick time to result (Figure 7.5).

Polyaniline (PAn) a conducting polymer has a wide variety of electrical and electrochemical properties, is simple to synthesize, is a supercapacitor, and has a highly controlled conductivity [88]. When PAn was doped with specific types of carbon nanomaterials, the designed nanohybrids demonstrated better redox activity and conductivity performances than pure PAn in neutral conditions. This led to the development of creative strategies for the generation and amplification of response signals in electrochemical biosensors to improve detection performances [89] and the inherent redox activity can be enhanced with carbon material [90, 91]. The limitation of using PAn is being active only in acidic pH, which is solved by the addition of some nanostructures like graphene material enhances the activity in the neutral pH [91].

Reusable Ni-rGO-PAn electrochemical-based immunosensor for early diagnosis of tuberculosis was developed using ESAT-6 biomarker via cyclic voltammetry [92]. Another study by Mohd *et al.* [93] also demonstrated a GP/PAn nanocomposite-based electrochemical immunosensor for the detection of CFP10 antigen. This system is based on the screen-printed gold electrode (SPGE), which has high reproducibility, low cost than metal-based electrodes, and easy disposability considered for better usage. Compared to single antigen biomarkers, CFP10-ESAT6 demonstrated greater sensitivity as an *M. tuberculosis* biomarker [94].

Sandwich electrochemical immunosensor based on Fe_3O_4/Au MNPs using the SPGE platform used to detect active TB in which antibody (Abs) was immobilized onto the graphene/polyaniline (GP/PANI)-modified gold screen-printed electrode with a limit of detection at 1.5 ng mL^{-1}. The incorporation of Au with a magnetic core enhances the chemical stability and biocompatibility [95]. When the electrode was modified with GP/PANI, the oxidation peak current was relocated to a lower positive potential, indicating that the created immunosensor has increased the electronic transport capacity. The improved electrochemical response in the oxidation processes demonstrated the electroconducting characteristics of the

GP/PANI nanocomposite in promoting the transfer of electrons between the electrode surface and chloride ion [96]. Electrochemical AuNPs-based probe hybridization for the detection of MTB showed higher sensitivity and specificity than in ELISA. Perumal *et al.* [97] designed an impedimetric-based DNA sensor using 3D graphene/Au nanocomposite. The proposed DNA biosensor displays the capability of detecting femtomolar DNA concentration and has high sensitivity and specificity.

Aptamer-based assays are one of the WHO-recommended "ASSURED (affordable, sensitive, specific, user-friendly, rapid and robust, equipment-free, and deliverable to the end user)" diagnostic techniques [98]. An ultrasensitive voltammetric aptasensor based on reduced graphene oxide doped with a metal-organic framework (MOF-rGO) for the detection of ESAT-6. Core-shell Au–Pt nanoparticles showed high conductivity and affinity with thiolate ESAT aptamer which facilitates the electron transfer, resulting in high sensing performance with the detection limit of (LOD) 3.3×10^{-5} ng·mL^{-1} [99]. Dual aptamers enhances the detection sensitivity and showed a higher signal compared to a single aptamer to capture the MPT64 antigen, based on Zr(IV)-based MOF using Au nanoparticles [99]. Au–Pt nanoparticles enhance the electron transfer and the immobilization of MPT64 antigen aptamer I (MAA I) on polyethyleneimine (PEI)-functionalized Fe-based metal-organic framework (P-MOF) which increases the electrical conductivity with a limit of detection (LOD) as low as 0.33 fg/mL. C_{60} enhances the electrical conductivity and mechanical strength when incorporated into CNTs but it has low dispersivity and stability, which can enhance by Graphene Oxide (GO) [100].

The quality of the used antibody and its capacity for sensitive and specific antigen recognition are crucial factors. Alterations in pH, temperature, and other external factors have the potential to affect sensitivity. Sandwich-type immunosensors provide increased sensitivity [101]. In electrochemical biosensors (EBS), the enzymes produces electroactive moieties that can be easily sensed by amperometric biosensor with lower LOD and sensitivity than other electrochemical platforms [102]. Aptamer-antibody sandwich-based amperometric sensing of *M. tuberculosis* MPT64 antigen system was developed where signal was determined using hydroquinone (HQ) as an intermediary for the transfer of electrons during the enzymatic reaction between the peroxidase and H_2O_2. In the measuring buffer and spiked-serum samples, the aptasensor could distinguish antigens as low as 1.11 ng mL^{-1} and 1.38 ng mL^{-1}, respectively with a diagnostic sensitivity and specificity of 88 and 100% respectively [103]. Similarly, amperometric dual aptasensor proposed for the simultaneous detection of CFP10 and

MPT64 antigen with respective detection limit of 1.68 ng mL^{-1} and 1.82 ng mL^{-1} [104].

The combination of composite materials preserves the features of each component with a synergistic impact, DNA biosensors based on a hybrid composite of CNTs and electroactive polymers have attracted a lot of interest [105, 106]. A straightforward-stacking contact between MWCNTs and the redox polymer can link them, which can facilitate electron transfer reactions. Additionally, a large surface area greatly raises the sensitivity of hybridization and the amount of DNA probe attachment [107]. Direct DNA biosensor was developed from MWCNTs and redox polymer constructed by covalently binding DNA with the carboxylic group of nanocomposite oligomer (Fc-ac-OMPA) showed a detection limit of 0.08 f.mol.L^{-1}. The potential DNA sensing can discriminate against SNP mutation of drug resistance [108]. This LOD was comparatively better than that of previous works, Fc-PPP/MWCNTs (detection range of 1 f.mol.L^{-1} to 10 p.mol.L^{-1}, LOD = 1.6 p.mol.L^{-1}) [105] and Fc-ac-OMPA (detection range of 1 f.mol.L^{-1} to 100 p.mol.L^{-1}, LOD = 0.2 f.mol.L^{-1}) [109], nanomaterials based on (MWCNTs/PPy/PAMAM) (detection range from 1 f.mol.L^{-1} to 100 n.mol.L^{-1}, LOD = 0.3 f.mol.L^{-1}) [110], Fe_3O_4@Ag/GQD nano triplex-based biosensor with LOD 0.33 ng/mL (detection range 0.005-500 μg/mL) [111].

7.5.2 *Salmonella Spp*

A biosensor based on ssDNA/rGO-CNT/GCE aptasensor to detect food-borne disease *S. enterica* through differential pulse voltammetry (DPV) technique. The enhanced electrochemical properties and chemical functionality is due to the synergistic effect of both graphene and CNTs for the stable interface. Conjugation of rGO with CNT prevented the aggregation of graphene sheets. In addition, aptasensor (ssDNA/rGO-CNT/GCE) could detect *S. typhimurium* with a LOD of 10^1 cfu mL^{-1} [112]. Label-free ssDNA aptamer specific to *Salmonella enterica* was immobilized on rGO-TiO_2 nanocomposite by noncovalent interactions. The inhibition of electron transfer occurs due to the interaction of aptamer with bacterial cells at the electrode interface which causes decreased DPV signal and it is proportional to the decreasing bacterial concentration. This aptasensor exhibited high sensitivity with a LOD of 10^1 cfu mL^{-1} and good selectivity for *Salmonella* [113]. A new nanozyme based on Pt Ps (Platinum functionalized polymer nanosphere) modified with streptavidin bound with biotin-antibody have good stability and sensitive for the colorimetric identification of *S. typhimurium* with LOD of 1000 CFU/mL with the naked eye [114].

7.5.3 *Clostridium Spp*

Kumar *et al.* [115] developed a label-free optical biosensor for the detection of *Clostridium botulinum* neurotoxin (BoNT serotype C) using a porous silicon Fabry–Pérot interferometer. The immunoassay uses specific antibodies to detect the target antigen followed by immunorecognition of HRP (horseradish peroxidase (HRP)) conjugated secondary antibodies for the amplification of optical response with a LOD of 4.8 pg mL^{-1}. An impedimetric immunosensor coated with Au nanoparticles/graphene-chitosan composite was fabricated to detect *botulinum* neurotoxin serotype A (BoNT/A). BoNT/A Ab was immobilized on Au nanoparticles/graphene-chitosan electrodes for signal amplification. The impedance changes during Ab-Ag interaction efficiently restricted the electron transfer of the redox probe which is used to detect BoNT/with a LOD of 0.11 pg mL^{-1} [116].

7.6 Vector-Borne Diseases

7.6.1 Malaria

Malaria is one of the major parasitic infections caused by Protozoan *Plasmodium sp.* transmitted by the bite of a female *Anopheles* mosquito [117]. The spreading of malarial infection is rapid due to mobility and tourism. Misdiagnosis with other diseases and incorrect prescription causes malaria drug resistance to pose a major challenge/difficulty of prevention and treatment [118]. Severe infection and mortality caused by *Plasmodium sp* increase the need for rapid and early diagnosis of malarial disease for effective treatment. Besides the conventional diagnostics method alternative rapid, simple, field-applicable, biomarker-based diagnostics methods are emerging worldwide. Different biomarkers are available for the detection of malaria includes, including *Plasmodium* liver-specific protein 2 (LISP2) [119], The sporozoite surface circumsporozoite proteins (CSP) [120], *P. falciparum* histidine-rich protein II (*Pf*HRP-II) [121], *P. falciparum* lactate dehydrogenase (*Pf*LDH) [122], *P. falciparum* aldolase (*Pf*ALD) [123], *P. falciparum* hypoxanthine guanine phosphoribosyl transferase (*Pf*HGPRT) [124] and *Plasmodium* glutamate dehydrogenase (*P*GluDH) [125], *P. vivax* merozoite surface protein 1 ($Pv\mathrm{MSP1}_{19}$) [126] and products such as hemozoin [127]. *P. falciparum*-specific histidine-rich protein 2 (PfHRP-2) is secreted into peripheral blood during parasite growth and development and is involved in heme detoxification [117].

Immunosensors are potential biosensor platform that uses antibody as recognition element which has a high affinity toward the target molecule, lowering the limit of detection and enhancing the specificity [128]. Dual electrochemical-based immunosensor developed by immobilizing *Pf*HRP2 antibody (Ab-*Pf*HRP2)/GE on DHP (di hexadecyl phosphate) for the detection of Ag-*Pf*HRP2, a specific biomarker for Malaria which showed a low detection limit than conventional ELISA [129]. Similarly, Gandarilla *et al.* [130] fabricated an electrochemical immunosensor to determine Ag-*Pv*-LDH in which Egg yolk IgY antibody (Ab-*Pv*LDH) was immobilized on the GE surface that showed a LOD of 455 ng mL^{-1}.

Microfluidic lab-on-a-chip sensors are tiny devices that combine several operations from different sensing platforms to be carried out simultaneously utilizing small amounts of material (Figure 7.6). This procedure uses less sample, reagents, and energy while also having interesting impacts on fluid properties at the micro-scale including lower viscosity, flow dynamics, and surface tension [131]. Microfluidic paper double-sided screen printed carbon electrode (MP-dsSPCE) was fabricated with magnetic beads and an immunomodified signal amplifier to detect *P. falciparum* LDH (*Pf*LDH)

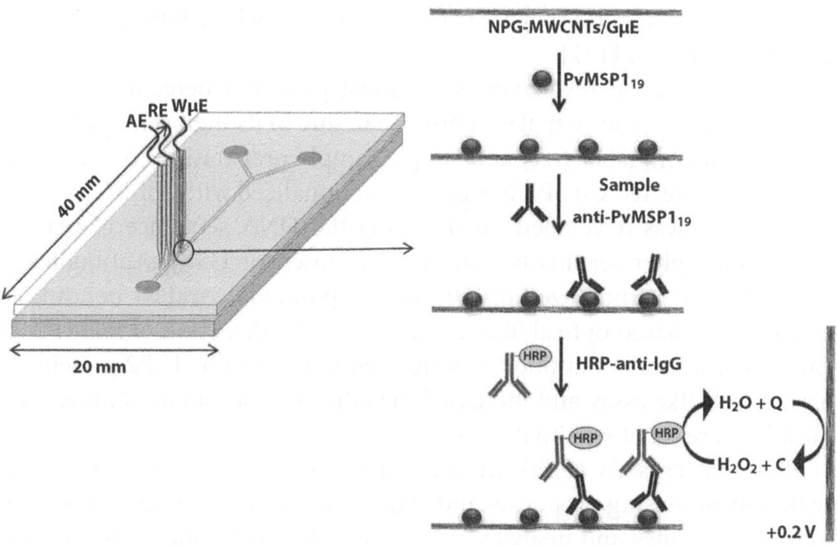

Figure 7.6 Microfluidic electrochemical immunosensor for the detection of anti-PvMSP119 in human serum samples.

in whole blood samples with a LOD of 200 ng mL^{-1} (corresponding to 300 parasites L^{-1}, which is near to the WHO criteria for clinical malaria diagnosis), parasitemia quantification in a wide range, and detected PfLDH in lysed whole blood in under 20 minutes (0.006–1.5% in cultured RBC) [132].

New bio-recognition components known as aptamers are made of either single-stranded nucleic acid (DNA or RNA) or a peptide with a molecular weight of less than 25 kDa [133]. "Systematic evolution of ligands by exponential enrichment" (SELEX), an *in vitro* method, allows the selection of aptamers unique to any target molecule [134]. An electrochemical *Pf*HRP2 aptamer (35 bp) biosensor tagged with methylene blue was developed to detect the *P. falciparum* histidine-rich protein II (*Pf*HRP2) and quantified by analyzing the change in electron transfer kinetics during the binding of *Pf*HRP2 by the aptamer-MB complex with a LOD of 3.73 nM [135]. Aptamer-based electrochemical impedance spectroscopic (EIS) sensor with good sensitivity and selectivity was developed by PfLDH-specific DNA aptamer functionalized on a gold electrode surface through Au-S linkages. The aptasensor showed a detection range from 100fM to 100nM with a limit of detection of 0.77 pM, and it has high specificity and reproducibility [136]. Electrochemical impedance-based spectroscopy analysis of *Pf*HRP-II was done based on the SELEX procedure which applies ssDNA sequence against HRP-II. The impedance value was gradually increased with the increasing HRP-II concentrations in the sample solution. The detection range was 1 pM to 500 pM with a limit of detection (LOD) of 3.15 pM [137].

Surface plasmon resonance is the most preferred detection platform for the identification of malarial protozoan due to its high sensitivity, low-cost production, label-free, and simple sample preparation. In expensive graphene protected Cu SPR biosensor functionalized with complementary DNA probe was developed for detecting the DNA sequence of malaria has 2-fold higher sensitivity than optical biosensor [118]. Antibodies of *Pf*HRP2 were immobilized on OF bio receptors via covalent bonding to develop SPR-based optical fiber biosensor for the detection of *Pf*HRP2. In this study, secondary antibodies were used to bind with HRP2 proteins in a sandwich-like assay and increased the affinity of secondary binders and 10-fold increase of sensitivity [138].

Photonic crystals (PCs) are dielectric and nanostructured materials with optical sensing properties and widely employed to detect various biological molecules and analytes [139]. There are many methods are used to evaluate the optical properties such as photonic bandgap, electric field intensity, reflectance, and transmittance of photonic crystals structure such as transfer matrix method (TMM), finite time domain method (FDTD), plane wave expansion (PWE) method [140]. Gold thin film coated photonic crystal fiber (PCF) based surface plasmon resonance (SPR) biosensor was developed to detect malaria by measuring variation of red blood cells (RBCs) [141].

7.6.2 Dengue

Dengue is one of the most prevalent mosquito borne viral diseases mainly spread through *Aedes aegypti*. It causes severe forms of dengue fever (DF) in some infected people such as dengue hemorrhagic fever (DHF), or dengue shock syndrome (DSS) which are life-threatening [142]. There are four dengue serotypes: DENV-1, 2, 3, and 4. Viral detection relies on, viral antigen or genomic RNA or viral-specific antibody detection. DENV consists of three structural proteins, a lipoprotein membrane, and seven nonstructural proteins. NS1 protein is prevalent among other serotypes and is significantly used to detect DENV [143]. Nonstructural protein (NS1) detection is also an efficient way of detecting the dengue virus and their detection in blood is very difficult than antibody isolation [144] due to their lower concentration (0.01–2 μg/ml) [145]. In whole blood, NS1 antigen of DENV was rapidly detected by LSPR-based lab-on-chip device fabricated with thermally annealing thin silver film deposited onto a silicon substrate within 30 minutes using a small volume of (10 μl) blood and the LOD was 0.06 μg/mL [146].

Functionalization of anti-DENV NS1 antibodies on metal electrodes has some advantages. For instance, single-walled carbon nanotube networks chemi-resistor transducer functionalized with anti-dengue NS1 monoclonal antibodies can detect as low of 0.09 ng/mL NS1 protein [147]. Similarly, Langmuir-Blodgett (LB) film deposition of molybdenum disulfide and gold nanoparticles (MoS2-AuNPs) composite onto indium tin oxide (ITO) coated glass substrates has high sensitivity toward anti-DENV NS1 [148]. The synergistic effect of metal nanoparticles, such as Au and two-dimensional material, enhances the charge transfer properties of the composite [149], and also, the increased electron transfer enhances the sensitivity and specificity due to the synergistic effect of the components of the composite electrode [150]. NS1 protein is secreted only from the fifth day of infection, which hinders the early detection of the disease that can be overcome by the detecting structural E protein which is secreted earlier during onset of the infection. Omar *et al.* [151] proposed a spin-coating approach for the detection of DENV E protein by immobilizing monoclonal antibody (IgM) on gold/Fe-MPA-NCC-CTAB ((3-mercapto propionic acid nanocellulose crystalline/hexadecyl trimethyl ammonium bromide/ EDCNHS (EDC-N-hydroxysuccinimide) thin film. Conjugation of specific antibodies with DENV E protein generated an SPR signal in the concentration ranges from 0.0001 to 10 nM with a sensitivity of 39.96 nM^{-1}.

DNA is a perfect target for the precise detection of viral infections and can increase the selectivity of the biosensor toward the dengue virus. DNA/ RNA biosensors are developed based on the hybridization with oligonucleotides of different types of DENV serotypes. Either by adding a hybridization solution to the surface of the electrode and recording the direct electrochemical response of hybrid duplex [152] or by the DNA probe which is complementary to the target DNA, immobilized on the sensor surface and signals from redox potential is recorded by voltammetric/ amperometric methods [153]. Amperometry is a key method applied biosensors where the reduction-oxidation of the ferri/ferrocyanide probe on the electrode surface, proportionate to the concentration of the target analyte in the sample [143]. An amperometric DNA biosensor was designed for the detection of DENV by deposition of Cu_2CdSnS_4 (CCTS) quaternary alloy nanostructure on an oxygen-etched silicon substrate (O_2/Si) via spin coating and annealed at 400°C and immobilized with dengue-specific DNA probes This amperometry biosensor recognizes ssDNA from 100 fM to 10 nM with LOD of 17 nM. DNA biosensor is rapid, inexpensive, highly sensitive, and it consumes low power [154].

An optical DNA biosensor was developed for the detection of DENV based on novel square-planar piperidine side chain-functionalized N,N′-bis-4-(hydroxysalicylidene)-phenylenediamine-nickel(II) with a limit of detection (LOD) at 0.2 aM. The salphen complex of target nucleic acid with a labeled probe changes the color rendered by the PSiNs (porous silica nanoparticles) could be quantified by the reflectance transduction [155]. Salphen complex, also called N, N0-bisphenylene (salisilidine imine), is a Schiff base ligand that has N_2O_2-donor atoms that donate a lone pair of electrons from their respective donor groups to the orbital d of the metal ion to form coordinate (dative covalent) bonds with the metal ion at the center of the ligand. The metal complex possesses a square-planar shape that allows it to intercalate within a DNA duplex through aromatic stacking contact. The metal ion is in the plane formed by the four metal-ligating N_2O_2-donor atoms [156].

7.7 Conclusion

Traditional diagnostic methods have faced limitations in sensitivity, specificity, and turnaround time, necessitating the development of more efficient and accurate techniques. The integration of nanotechnology with established diagnostic techniques like PCR, immunoassays, and biosensors amplifies the diagnostic performance, enabling early and precise identification of infectious diseases. The nanotechnology-based diagnostics has several advantages particularly the ability to detect multiple pathogens simultaneously, the potential for point-of-care testing, and the feasibility of remote monitoring which contribute to improved patient outcomes and efficient public health management. Nanotechnology has the potential to revolutionize the landscape of infectious disease diagnostics by enabling quicker and more accurate detection, leading to timely interventions and effective control of outbreaks. However, the implementation of nanotechnology in diagnostic settings also presents challenges. Cost-effectiveness, regulatory approvals, scalability, and ethical considerations are important factors that need to be addressed. Overcoming these challenges will require collaboration among scientists, clinicians, regulatory bodies, and policymakers to ensure the safe and effective integration of nanotechnology in healthcare.

Acknowledgment

T.M. and F.Q. acknowledge the financial support from FONDECYT (Chile) under the Postdoctoral Fellowship No. 3190072.

References

1. Bhalla, N., Jolly, P., Formisano, N., Estrela, P., Introduction to biosensors. *Essays Biochem.*, 60, 1, 1–8, 2016, 10.1042/EBC20150001.
2. Sawant, S.N., 13 - Development of Biosensors From Biopolymer Composites, in: *Biopolymer Composites in Electronics*, K.K. Sadasivuni, D. Ponnamma, J. Kim, J.J. Cabibihan, M.A. AlMaadeed, (Eds.), pp. 353–383, Elsevier, USA, 2017.
3. Campos, E.V.R., Pereira, A.E.S., de Oliveira, J.L., Carvalho, L.B., Guilger-Casagrande, M., de Lima, R., Fraceto, L.F., How can nanotechnology help to combat COVID-19? Opportunities and urgent need. *J. Nanobiotechnol.*, 18, 1, 125, 2020, 10.1186/s12951-020-00685-.
4. Deng, J., Zhao, S., Liu, Y., Liu, C., Sun, J., Nanosensors for Diagnosis of Infectious Diseases. *ACS Appl. Bio Mater.*, 4, 5, 3863–3879, 2021, 10.1021/acsabm.0c01247.
5. Eivazzadeh-Keihan, R., Pashazadeh-Panahi, P., Baradaran, B., de la Guardia, M., Hejazi, M., Sohrabi, H., Mokhtarzadeh, A., Maleki, A., Recent progress

in optical and electrochemical biosensors for sensing of Clostridium botulinum neurotoxin. *TrAC, Trends Anal. Chem.*, 103, 184–197, 2018, https:// doi.org/10.1016/j.trac.2018.03.019.
6. Herrera-Dominguez, M., Morales-Luna, G., Mahlknecht, J., Cheng, Q., Aguilar-Hernandez, I., Ornelas-Soto, N., Optical Biosensors and Their Applications for the Detection of Water Pollutants. *Biosensors (Basel)*, 13, 3, 2023, 10.3390/bios13030370.
7. Thamer, T., Noha, A., Zahrah, A.A., Sundos, A., Jhanjhi, N.Z., Nawaf, B.D., Mike, H.S., Optoelectronics and Optical Bio-Sensors, in: *Optoelectronics*, p. Ch. 2, IntechOpen, Rijeka, 2021.
8. Arranz, A. and Ripoll, J., Advances in optical imaging for pharmacological studies. *Front. Pharmacol.*, 6, 189, 2015, 10.3389/fphar.2015.00189.
9. Chen, C. and Wang, J., Optical biosensors: an exhaustive and comprehensive review. *Analyst*, 145, 5, 1605–1628, 2020, 10.1039/c9an01998g.
10. Thevenot, D.R., Toth, K., Durst, R.A., Wilson, G.S., Electrochemical biosensors: recommended definitions and classification. *Biosens. Bioelectron.*, 16, 1-2, 121–31, 2001. 10.1016/s0956-5663(01)00115-4.
11. Mokhtarzadeh, A., Eivazzadeh-Keihan, R., Pashazadeh, P., Hejazi, M., Gharaatifar, N., Hasanzadeh, M., Baradaran, B., de la Guardia, M., Nanomaterial-based biosensors for detection of pathogenic virus. *TrAC, Trends Anal. Chem.*, 97, 445–457, 2017, 10.1016/j.trac.2017.10.005.
12. Draz, M.S., Venkataramani, M., Lakshminarayanan, H., Saygili, E., Moazeni, M., Vasan, A., Li, Y., Sun, X., Hua, S., Yu, X.G., Shafiee, H., Nanoparticle-enhanced electrical detection of Zika virus on paper microchips. *Nanoscale*, 10, 25, 11841–11849, 2018, 10.1039/c8nr01646a.
13. Fothergill, S.M., Joyce, C., Xie, F., Metal enhanced fluorescence biosensing: from ultra-violet towards second near-infrared window. *Nanoscale*, 10, 45, 20914–20929, 2018, 10.1039/c8nr06156d.
14. Wang, C., Liu, M., Wang, Z., Li, S., Deng, Y., He, N., Point-of-care diagnostics for infectious diseases: From methods to devices. *Nano Today*, 37, 101092, 2021, 10.1016/j.nantod.2021.101092.
15. Seo, G., Lee, G., Kim, M.J., Baek, S.H., Choi, M., Ku, K.B., Lee, C.S., Jun, S., Park, D., Kim, H.G., Kim, S.J., Lee, J.O., Kim, B.T., Park, E.C., Kim, S., II, Rapid Detection of COVID-19 Causative Virus (SARS-CoV2) in Human Nasopharyngeal Swab Specimens Using Field-Effect Transistor-Based Biosensor. *ACS Nano*, 14, 4, 5135–5142, 2020, 10.1021/acsnano.0c02823.
16. Ulrich, H. and Pillat, M.M., CD147 as a Target for COVID-19 Treatment: Suggested Effects of Azithromycin and Stem Cell Engagement. *Stem Cell Rev. Rep.*, 16, 3, 434–440, 2020, 10.1007/s12015-020-09976-7.
17. Zhifeng, J., Feng, A., Li, T., Consistency analysis of COVID-19 nucleic acid tests and the changes of lung CT. *J. Clin. Virol.*, 127, 104359, 2020, 10.1016/j.jcv.2020.104359.
18. Li, J., Wu, D., Yu, Y., Li, T., Li, K., Xiao, M.M., Li, Y., Zhang, Z.Y., Zhang, G.J., Rapid and unamplified identification of COVID-19 with

morpholino-modified graphene field-effect transistor nanosensor. *Biosens. Bioelectron.*, 183, 113206, 2021, 10.1016/j.bios.2021.113206.

19. Huang, C., Wen, T., Shi, F.J., Zeng, X.Y., Jiao, Y.J., Rapid Detection of IgM Antibodies against the SARS-CoV2 Virus via Colloidal Gold Nanoparticle Based Lateral-Flow Assay. *ACS Omega*, 5, 21, 12550–12556, 2020, 10.1021/acsomega.0c01554.
20. Choi, J.R., Development of Point-of-Care Biosensors for COVID-19. *Front. Chem.*, 8, 517, 2020, 10.3389/fchem.2020.00517.
21. Sheridan, C., Fast, portable tests come online to curb coronavirus pandemic. *Nat. Biotechnol.*, 38, 5, 515–518, 2020, 10.1038/d41587-020-00010-2.
22. Du, Z., Zhu, F., Guo, F., Yang, B., Wang, T., Detection of antibodies against SARS-CoV2 in patients with COVID-19. *J. Med. Virol.*, 92, 10, 1735–1738, 2020, 10.1002/jmv.25820.
23. Li, Z., Yi, Y., Luo, X., Xiong, N., Liu, Y., Li, S., Sun, R., Wang, Y., Hu, B., Chen, W., Zhang, Y., Wang, J., Huang, B., Lin, Y., Yang, J., Cai, W., Wang, X., Cheng, J., Chen, Z., Sun, K., Pan, W., Zhan, Z., Chen, L., Ye, F., Development and clinical application of a rapid IgM-IgG combined antibody test for SARS-CoV-2 infection diagnosis. *J. Med. Virol.*, 92, 9, 1518–1524, 2020, 10.1002/jmv.25727.
24. Kang, K., Huang, L., Ouyang, C., Du, J., Yang, B., Chi, Y., He, S., Ying, L., Chen, G., Wang, J., Development, performance evaluation, and clinical application of a Rapid SARS-CoV2 IgM and IgG Test Kit based on automated fluorescence immunoassay. *J. Med. Virol.*, 93, 5, 2838–2847, 2021, 10.1002/jmv.26696.
25. Moitra, P., Alafeef, M., Dighe, K., Frieman, M.B., Pan, D., Selective Naked-Eye Detection of SARS-CoV2 Mediated by N Gene Targeted Antisense Oligonucleotide Capped Plasmonic Nanoparticles. *ACS Nano*, 14, 6, 7617–7627, 2020, 10.1021/acsnano.0c03822.
26. Qiu, G., Gai, Z., Tao, Y., Schmitt, J., Kullak-Ublick, G.A., Wang, J., Dual-Functional Plasmonic Photothermal Biosensors for Highly Accurate Severe Acute Respiratory Syndrome Coronavirus 2 Detection. *ACS Nano*, 14, 5, 5268–5277, 2020, 10.1021/acsnano.0c02439.
27. Jauffred, L., Samadi, A., Klingberg, H., Bendix, P.M., Oddershede, L.B., Plasmonic Heating of Nanostructures. *Chem. Rev.*, 119, 13, 8087–8130, 2019, 10.1021/acs.chemrev.8b00738.
28. Choi, H.K., Lee, M.J., Lee, S.N., Kim, T.H., Oh, B.K., Noble Metal Nanomaterial-Based Biosensors for Electrochemical and Optical Detection of Viruses Causing Respiratory Illnesses. *Front. Chem.*, 9, 672739, 2021, 10.3389/fchem.2021.672739.
29. Hasanzadeh, A., Alamdaran, M., Ahmadi, S., Nourizadeh, H., Bagherzadeh, M.A., Mofazzal Jahromi, M.A., Simon, P., Karimi, M., Hamblin, M.R., Nanotechnology against COVID-19: Immunization, diagnostic and therapeutic studies. *J. Controlled Release*, 336, 354–374, 2021, 10.1016/j.jconrel.2021.06.036.

30. Lu, D., Zhu, D.Z., Gan, H., Yao, Z., Fu, Q., Zhang, X., Prospects and challenges of using electrochemical immunosensors as an alternative detection method for SARS-CoV2 wastewater-based epidemiology. *Sci. Total Environ.*, 777, 146239, 2021, https://doi.org/10.1016/j. scitotenv.2021.146239.
31. Tripathy, S. and Singh, S.G., Label-Free Electrochemical Detection of DNA Hybridization: A Method for COVID-19 Diagnosis. *Trans. Indian. Natl. Acad. Eng.*, 5, 2, 205–209, 2020, 10.1007/ s41403-020-00103-z.
32. Mahari, S., Roberts, A., Shahdeo, D., Gandhi, S., eCovSens-Ultrasensitive Novel In-House Built Printed Circuit Board Based Electrochemical Device for Rapid Detection of nCovid-19 antigen, a spike protein domain 1 of SARS-CoV-2, 2020, bioRxiv, 2020.04.24.059204, 10.1101/2020.04.24.059204.
33. Berggren, C., Bjarnason, B., Johansson, G., Capacitive Biosensors. *Electroanalysis*, 13, 3, 173–180, 2001, https://doi.org/10.1002/1521-4109 (200103)13:3<173::AID-ELAN173>3.0.CO;2-B.
34. Rashed, M.Z., Kopechek, J.A., Priddy, M.C., Hamorsky, K.T., Palmer, K.E., Mittal, N., Valdez, J., Flynn, J., Williams, S.J., Rapid detection of SARS-CoV-2 antibodies using electrochemical impedance-based detector. *Biosens. Bioelectron.*, 171, 112709, 2021, 10.1016/j.bios.2020.112709.
35. Di, B., Hao, W., Gao, Y., Wang, M., Wang, Y.D., Qiu, L.W., Wen, K., Zhou, D.H., Wu, X.W., Lu, E.J., Liao, Z.Y., Mei, Y.B., Zheng, B.J., Che, X.Y., Monoclonal antibody-based antigen capture enzyme-linked immunosorbent assay reveals high sensitivity of the nucleocapsid protein in acute-phase sera of severe acute respiratory syndrome patients. *Clin. Vaccine Immunol.*, 12, 1, 135–40, 2005, 10.1128/CDLI.12.1.135-140.2005.
36. Fehr, A.R. and Perlman, S., Coronaviruses: an overview of their replication and pathogenesis. *Methods Mol. Biol.*, 1282, 1–23, 2015, 10.1007/978-1-4939-2438-7_1.
37. Diao, B., Wen, K., Zhang, J., Chen, J., Han, C., Chen, Y., Wang, S., Deng, G., Zhou, H., Wu, Y., Accuracy of a nucleocapsid protein antigen rapid test in the diagnosis of SARS-CoV2 infection. *Clin. Microbiol. Infect.*, 27, 2, 289.e1–289.e4, 2021, https://doi.org/10.1016/j.cmi.2020.09.057.
38. Eissa, S. and Zourob, M., Development of a Low-Cost Cotton-Tipped Electrochemical Immunosensor for the Detection of SARS-CoV2. *Anal. Chem.*, 93, 3, 1826–1833, 2021, 10.1021/acs.analchem.0c04719.
39. Fabiani, L., Saroglia, M., Galata, G., De Santis, R., Fillo, S., Luca, V., Faggioni, G., D'Amore, N., Regalbuto, E., Salvatori, P., Terova, G., Moscone, D., Lista, F., Arduini, F., Magnetic beads combined with carbon black-based screen-printed electrodes for COVID-19: A reliable and miniaturized electrochemical immunosensor for SARS-CoV2 detection in saliva. *Biosens. Bioelectron.*, 171, 112686, 2021, 10.1016/j.bios.2020.112686.
40. Castillo-Henriquez, L., Brenes-Acuna, M., Castro-Rojas, A., Cordero-Salmeron, R., Lopretti-Correa, M., Vega-Baudrit, J.R., Biosensors for the Detection of Bacterial and Viral Clinical Pathogens. *Sensors (Basel)*, 20, 23, 2020, 10.3390/s20236926.

41. Henritzi, D., Hoffmann, B., Wacheck, S., Pesch, S., Herrler, G., Beer, M., Harder, T.C., A newly developed tetraplex real-time RT-PCR for simultaneous screening of influenza virus types A, B, C and D. *Influenza Other Respir. Viruses*, 13, 1, 71–82, 2019, 10.1111/irv.12613.
42. Wu, K., Liu, J., Saha, R., Su, D., Krishna, V. D., Cheeran, M. C. J., & Wang, J. P., Magnetic particle spectroscopy for detection of influenza A virus subtype H1N1. *ACS Appl. Mater. Interfaces*, 12, 12, 13686–13697, 2020.
43. Ribeiro, B.V., Cordeiro, T.A.R., Oliveira e Freitas, G.R., Ferreira, L.F., Franco, D.L., Biosensors for the detection of respiratory viruses: A review. *Talanta Open*, 2, 100007, 2020, https://doi.org/10.1016/j.talo.2020.100007.
44. Lee, W.S., Ahn, J., Jung, S., Lee, J., Kang, T., Jeong, J., Biomimetic Nanopillar-Based Biosensor for Label-Free Detection of Influenza A Virus. *Biochip J.*, 15, 3, 260–267, 2021, 10.1007/s13206-021-00027-y.
45. Achadu, O.J., Takemura, K., Khoris, I.M., Park, E.Y., Plasmonic/magnetic molybdenum trioxide and graphitic carbon nitride quantum dots-based fluoroimmunosensing system for influenza virus. *Sens. Actuators, B*, 321, 128494, 2020, 10.1016/j.snb.2020.128494.
46. Barman, S. and Sadhukhan, M., Facile bulk production of highly blue fluorecent graphitic carbon nitride quantum dots and their application as highly selective and sensitive sensors for the detection of mercuric and iodide ions in aqueous media. *J. Mater. Chem.*, 22, 41, 21832–21837, 2012, 10.1039/C2JM35501A.
47. Pang, X., Bian, H., Wang, W., Liu, C., Khan, M.S., Wang, Q., Qi, J., Wei, Q., Du, B., A bio-chemical application of N-GQDs and g-C3N4 QDs sensitized TiO2 nanopillars for the quantitative detection of pcDNA3-HBV. *Biosens. Bioelectron.*, 91, 456–464, 2017, https://doi.org/10.1016/j. bios.2016.12.059.
48. Sun, B., Dong, J., Cui, L., Feng, T., Zhu, J., Liu, X., Ai, S., A dual signal-on photoelectrochemical immunosensor for sensitively detecting target avian viruses based on AuNPs/g-C(3)N(4) coupling with CdTe quantum dots and *in situ* enzymatic generation of electron donor. *Biosens. Bioelectron.*, 124-125, 1–7, 2019, 10.1016/j.bios.2018.09.100.
49. Xiong, M., Rong, Q., Meng, H.M., X.B., Two-dimensional graphitic carbon nitride nanosheets for biosensing applications. *Biosens. Bioelectron.*, 89, Pt 1, 212–223, 2017, 10.1016/j.bios.2016.03.043.
50. Courtney, S.J., Stromberg, Z.R., Myers, Y.G.A., Jacobsen, D., Stromberg, L.R., Lenz, K.D., Theiler, J., Foley, B.T., Gans, J., Yusim, K., Kubicek- Sutherland, J.Z., Optical Biosensor Platforms Display Varying Sensitivity for the Direct Detection of Influenza RNA. *Biosensors (Basel)*, 11, 10, 2021, 10.3390/bios11100367.
51. Stromberg, Z.R., Theiler, J., Foley, B.T., Myers, Y.G.A., Hollander, A., Courtney, S.J., Gans, J., Deshpande, A., Martinez-Finley, E.J., Mitchell, J., Mukundan, H., Yusim, K., Kubicek-Sutherland, J. Z. Fast Evaluation of Viral Emerging Risks (FEVER): A computational tool for biosurveillance,

diagnostics, and mutation typing of emerging viral pathogens. *PloS Global Public Health*, 2, 2, e0000207, 2022, 10.1371/journal.pgph.0000207.
52. Park, G., Kim, H.O., Lim, J.W., Park, C., Yeom, M., Song, D., Haam, S., Rapid detection of influenza A (H1N1) virus by conductive polymer-based nanoparticle via optical response to virus-specific binding. *Nano Res.*, 15, 3, 2254–2262, 2022, 10.1007/s12274-021-3772-6.
53. Silva, M.M.S., Dias, A.C.M.S., Silva, B.V.M., Gomes-Filho, S.L.R., Kubota, L.T., Goulart, M.O.F., Dutra, R.F., Electrochemical detection of dengue virus NS1 protein with a poly(allylamine)/carbon nanotube layered immunoelectrode. *J. Chem. Technol. Biotechnol.*, 90, 1, 194–200, 2015, https://doi.org/10.1002/jctb.4305.
54. Uram, J.D., Ke, K., Hunt, A.J., Mayer, M., Submicrometer pore-based characterization and quantification of antibody-virus interactions. *Small*, 2, 8-9, 967–72, 2006, 10.1002/smll.200600006.
55. Zhang, L., Zhang, Z., Lv, Y., Chen, X., Wu, Z., He, Y., Zou, Y., Reduced Graphene Oxide Aerogels with Uniformly Self-Assembled Polyaniline Nanosheets for Electromagnetic Absorption. *ACS Appl. Nano Mater.*, 3, 6, 5978–5986, 2020, 10.1021/acsanm.0c01115.
56. Chen, H., Park, S.G., Choi, N., Moon, J., II, Dang, H., Das, A., Lee, S., Kim, D.G., Chen, L., Choo, J., SERS imaging-based aptasensor for ultrasensitive and reproducible detection of influenza virus A. *Biosens. Bioelectron.*, 167, 112496, 2020, 10.1016/j.bios.2020.112496.
57. Lou, B., Liu, Y., Shi, M., Chen, J., Li, K., Tan, Y., Chen, L., Wu, Y., Wang, T., Liu, X., Jiang, T., Peng, D., Liu, Z., Aptamer-based biosensors for virus protein detection. *TrAC Trends Anal. Chem.*, 157, 116738, 2022, 10.1016/j.trac.2022.116738.
58. Wang, F., Gopinath, S.C., Lakshmipriya, T., Aptamer-Antibody Complementation On Multiwalled Carbon Nanotube-Gold Transduced Dielectrode Surfaces To Detect Pandemic Swine Influenza Virus. *Int. J. Nanomed.*, 14, 8469–8481, 2019, 10.2147/IJN.S219976.
59. Kukushkin, V., II, Ivanov, N.M., Novoseltseva, A.A., Gambaryan, A.S., Yaminsky, I.V., Kopylov, A.M., Zavyalova, E.G., Highly sensitive detection of influenza virus with SERS aptasensor. *PloS One*, 14, 4, e0216247, 2019, 10.1371/journal.pone.0216247.
60. Divya, Dkhar, D.S., Kumari, R., Mahapatra, S., Kumar, R., Chandra, P., Ultrasensitive Aptasensors for the Detection of Viruses Based on Opto-Electrochemical Readout Systems. *Biosensors*, 12, 2, 81, 2022, https://www.mdpi.com/2079-6374/12/2/81.
61. Wu, K., Liu, J., Saha, R., Su, D., Krishna, V.D., Cheeran, M.C., Wang, J.P., Magnetic Particle Spectroscopy for Detection of Influenza A Virus Subtype H1N1. *ACS Appl. Mater. Interfaces*, 12, 12, 13686–13697, 2020, 10.1021/acsami.0c00815.
62. Farre, C., Viezzi, S., Wright, A., Robin, P., Lejal, N., Manzano, M., Vidic, J., Chaix, C., Specific and sensitive detection of Influenza A virus using a

biotin-coated nanoparticle enhanced immunomagnetic assay. *Anal. Bioanal. Chem.*, 414, 1, 265–276, 2022, 10.1007/s00216-020-03081-x.
63. Ghithan, J.H., Moreno, M., Sombrio, G., Chauhan, R., O'Toole, M.G., Mendes, S.B., Influenza virus immunosensor with an electro-active optical waveguide under potential modulation. *Opt. Lett.*, 42, 7, 1205–1208, 2017, 10.1364/OL.42.001205.
64. Singhal, C., Khanuja, M., Chaudhary, N., Pundir, C.S., Narang, J., Detection of chikungunya virus DNA using two-dimensional MoS(2) nanosheets based disposable biosensor. *Sci. Rep.*, 8, 1, 7734, 2018, 10.1038/ s41598-018-25824-8.
65. Bartholomeeusen, K., Daniel, M., LaBeaud, D.A., Gasque, P., Peeling, R.W., Stephenson, K.E., Ng, L.F.P., Ariën, K.K., Chikungunya fever. *Nat. Rev. Dis. Primers*, 9, 1, 17, 2023, 10.1038/s41572-023-00429-2.
66. Chong, H.Y., Leow, C.Y., Abdul Majeed, A.B., Leow, C.H., Flavivirus infection-A review of immunopathogenesis, immunological response, and immunodiagnosis. *Virus Res.*, 274, 197770, 2019, 10.1016/j.virusres. 2019.197770.
67. Nazmi, A., Dutta, K., Hazra, B., Basu, A., Role of pattern recognition receptors in flavivirus infections. *Virus Res.*, 185, 32–40, 2014, 10.1016/j. virusres.2014.03.013.
68. Faria, A.M. and Mazon, T., Early diagnosis of Zika infection using a ZnO nanostructures-based rapid electrochemical biosensor. *Talanta*, 203, 153–160, 2019, 10.1016/j.talanta.2019.04.080.
69. George, A., Amrutha, M.S., Srivastava, P., Sai, V.V.R., Sunil, S., Srinivasan, R., Label-Free Detection of Chikungunya Non-Structural Protein 3 Using Electrochemical Impedance Spectroscopy. *J. Electrochem. Soc.*, 166, 14, B1356, 2019, 10.1149/2.1081914jes.
70. Nasrin, F., Chowdhury, A.D., Ganganboina, A.B., Achadu, O.J., Hossain, F., Yamazaki, M., Park, E.Y., Fluorescent and electrochemical dual-mode detection of Chikungunya virus E1 protein using fluorophore-embedded and redox probe-encapsulated liposomes. *Microchim. Acta*, 187, 12, 674, 2020, 10.1007/s00604-020-04656-2.
71. Song, J., Mauk, M.G., Hackett, B.A., Cherry, S., Bau, H.H., Liu, C., Instrument-Free Point-of-Care Molecular Detection of Zika Virus. *Anal. Chem.*, 88, 14, 7289–94, 2016, 10.1021/acs.analchem.6b01632.
72. Camacho, S.A., Sobral-Filho, R.G., Aoki, P.H.B., Constantino, C.J.L., Brolo, A.G., Zika Immunoassay Based on Surface-Enhanced Raman Scattering Nanoprobes. *ACS Sensors*, 3, 3, 587–594, 2018, 10.1021/acssensors.7b00639.
73. Sanchez-Purra, M., Carre-Camps, M., de Puig, H., Bosch, I., Gehrke, L., Hamad-Schifferli, K., Surface-Enhanced Raman Spectroscopy-Based Sandwich Immunoassays for Multiplexed Detection of Zika and Dengue Viral Biomarkers. *ACS Infect. Dis.*, 3, 10, 767–776, 2017, 10.1021/ acsinfecdis.7b00110.
74. Fatin, M.F., Ruslinda, A.R., Md Arshad, M.K., Tee, K.K., Ayub, R.M., Hashim, U., Kamarulzaman, A., Gopinath, S.C.B., HIV-1 Tat biosensor: Current

development and trends for early detection strategies. *Biosens. Bioelectron.*, 78, 358–366, 2016, 10.1016/j.bios.2015.11.067.

75. Farzin, L., Shamsipur, M., Samandari, L., Sheibani, S., HIV biosensors for early diagnosis of infection: The intertwine of nanotechnology with sensing strategies. *Talanta*, 206, 120201, 2020, 10.1016/j.talanta.2019.120201.
76. Hu, Y., Li, H., Li, J., A novel electrochemical biosensor for HIV-related DNA detection based on toehold strand displacement reaction and cruciform DNA crystal. *J. Electroanal. Chem.*, 822, 66–72, 2018, https://doi.org/10.1016/j.jelechem.2018.05.011.
77. Yeter, E.C., Sahin, S., Caglayan, M.O., Ustundag, Z., An electrochemical label-free DNA impedimetric sensor with AuNP-modified glass fiber/carbonaceous electrode for the detection of HIV-1 DNA. *Chem. Pap.*, 75, 1, 77–87, 2021, 10.1007/s11696-020-01280-5.
78. Fu, X., Cheng, Z., Yu, J., Choo, P., Chen, L., Choo, J., A SERS-based lateral flow assay biosensor for highly sensitive detection of HIV-1 DNA. *Biosens. Bioelectron.*, 78, 530–537, 2016, 10.1016/j.bios.2015.11.099.
79. Jia, Z., Ma, Y., Yang, L., Guo, C., Zhou, N., Wang, M., He, L., Zhang, Z., NiCo(2)O(4) spinel embedded with carbon nanotubes derived from bimetallic NiCo metal-organic framework for the ultrasensitive detection of human immune deficiency virus-1 gene. *Biosens. Bioelectron.*, 133, 55–63, 2019, 10.1016/j.bios.2019.03.030.
80. Yang, S.P., Chen, S.R., Liu, S.W., Tang, X.Y., Qin, L., Qiu, G.H., Chen, J.X., Chen, W.H., Platforms Formed from a Three-Dimensional Cu-Based Zwitterionic Metal-Organic Framework and Probe ss-DNA: Selective Fluorescent Biosensors for Human Immunodeficiency Virus 1 ds-DNA and Sudan Virus RNA Sequences. *Anal. Chem.*, 87, 24, 12206–14, 2015, 10.1021/acs.analchem.5b03084.
81. Jin, F., Li, H., Xu, D., Enzyme-free fluorescence microarray for determination of hepatitis B virus DNA based on silver nanoparticle aggregates-assisted signal amplification. *Anal. Chim. Acta*, 1077, 297–304, 2019, 10.1016/j.aca.2019.05.066.
82. Lu, X., Dong, X., Zhang, K., Han, X., Fang, X., Zhang, Y. A gold nanorods-based fluorescent biosensor for the detection of hepatitis B virus DNA based on fluorescence resonance energy transfer. *Analyst*, 138, 2, 642–50, 2013, 10.1039/c2an36099c.
83. Li, M., Cushing, S.K., Liang, H., Suri, S., Ma, D., Wu, N., Plasmonic nanorice antenna on triangle nanoarray for surface-enhanced Raman scattering detection of hepatitis B virus DNA. *Anal. Chem.*, 85, 4, 2072–8, 2013, 10.1021/ac303387a.
84. Antipchik, M., Korzhikova-Vlakh, E., Polyakov, D., Tarasenko, I., Reut, J., Opik, A., Syritski, V., An electrochemical biosensor for direct detection of hepatitis C virus. *Anal. Biochem.*, 624, 114196, 2021, 10.1016/j.ab.2021.114196.

85. Goletti, D., Petruccioli, E., Joosten, S.A., Ottenhoff, T. H. Tuberculosis Biomarkers: From Diagnosis to Protection. *Infect. Dis. Rep.*, 8, 2, 6568, 2016, 10.4081/idr.2016.6568.
86. Jiang, Y., Liu, H., Wan, K., MPT64 polymorphisms of Mycobacterium tuberculosis strains suggest ongoing immune evasion. *Tuberculosis (Edinb)*, 94, 6, 712–4, 2014, 10.1016/j.tube.2014.08.013.
87. Seifert, M., Vargas, E., Ruiz-Valdepenas Montiel, V., Wang, J., Rodwell, T.C., Catanzaro, A., Detection and quantification of Mycobacterium tuberculosis antigen CFP10 in serum and urine for the rapid diagnosis of active tuberculosis disease. *Sci. Rep.*, 11, 1, 19193, 2021, 10.1038/ s41598-021-98471-1.
88. Bhandari, S., Chapter 2 - Polyaniline: Structure and Properties Relationship, in: *Polyaniline Blends, Composites, and Nanocomposites*, P.M. Visakh, C.D. Pina, E. Falletta, (Eds.), pp. 23–60, Elsevier, USA, 2018.
89. Chen, Y., Guo, S., Zhao, M., Zhang, P., Xin, Z., Tao, J., Bai, L., Amperometric DNA biosensor for Mycobacterium tuberculosis detection using flower-like carbon nanotubes-polyaniline nanohybrid and enzyme-assisted signal amplification strategy. *Biosens. Bioelectron.*, 119, 215–220, 2018, 10.1016/j.bios.2018.08.023.
90. Bai, L., Chen, Y., Bai, Y., Chen, Y., Zhou, J., Huang, A., Fullerene-doped polyaniline as new redox nanoprobe and catalyst in electrochemical aptasensor for ultrasensitive detection of Mycobacterium tuberculosis MPT64 antigen in human serum. *Biomaterials*, 133, 11–19, 2017, 10.1016/j.biomaterials.2017.04.010.
91. Chen, Y., Li, Y., Yang, Y., Wu, F., Cao, J., Bai, L., A polyaniline-reduced graphene oxide nanocomposite as a redox nanoprobe in a voltammetric DNA biosensor for Mycobacterium tuberculosis. *Microchim. Acta*, 184, 6, 1801–1808, 2017, 10.1007/s00604-017-2184-5.
92. Omar, R.A., Verma, N., Arora, P.K., Development of ESAT-6 Based Immunosensor for the Detection of Mycobacterium tuberculosis. *Front. Immunol.*, 12, 653853, 2021, 10.3389/fimmu.2021.653853.
93. Mohd Azmi, U.Z., Yusof, N.A., Kusnin, N., Abdullah, J., Suraiya, S., Ong, P.S., Ahmad Raston, N.H., Abd Rahman, S.F., Fathil, M.F.M., Sandwich Electrochemical Immunosensor for Early Detection of Tuberculosis Based on Graphene/Polyaniline-Modified Screen-Printed Gold Electrode. *Sensors (Basel)*, 18, 11, 2018, 10.3390/s18113926.
94. He, F., Xiong, Y., Liu, J., Tong, F., Yan, D., Construction of Au-IDE/CFP10-ESAT6 aptamer/DNA-AuNPs MSPQC for rapid detection of Mycobacterium tuberculosis. *Biosens. Bioelectron.*, 77, 799–804, 2016, 10.1016/ j.bios.2015.10.054.
95. Mohd Azmi, U.Z., Yusof, N.A., Abdullah, J., Alang Ahmad, S.A., Mohd Faudzi, F.N., Ahmad Raston, N.H., Suraiya, S., Ong, P.S., Krishnan, D., Sahar, N.K., Portable electrochemical immunosensor for detection of Mycobacterium tuberculosis secreted protein CFP10-ESAT6 in clinical sputum samples. *Microchim. Acta*, 188, 1, 20, 2021, 10.1007/s00604-020-04669-x.

96. Chan, K.F., Lim, H.N., Shams, N., Jayabal, S., Pandikumar, A., Huang, N.M., Fabrication of graphene/gold-modified screen-printed electrode for detection of carcinoembryonic antigen. *Mater. Sci. Eng. C*, 58, 666–74, 2016, 10.1016/j.msec.2015.09.010.
97. Perumal, V., Saheed, M.S.M., Mohamed, N.M., Saheed, M.S.M., Murthe, S.S., Gopinath, S.C.B., Chiu, J.M., Gold nanorod embedded novel 3D graphene nanocomposite for selective bio-capture in rapid detection of Mycobacterium tuberculosis. *Biosens. Bioelectron.*, 116, 116–122, 2018, 10.1016/j.bios.2018.05.042.
98. Dhiman, A., Kalra, P., Bansal, V., Bruno, J.G., Sharma, T.K., Aptamer- based point-of-care diagnostic platforms. *Sens. Actuators, B*, 246, 535–553, 2017, https://doi.org/10.1016/j.snb.2017.02.060.
99. Li, L., Yuan, Y., Chen, Y., Zhang, P., Bai, Y., Bai, L., Aptamer based voltammetric biosensor for Mycobacterium tuberculosis antigen ESAT-6 using a nanohybrid material composed of reduced graphene oxide and a metal-organic framework. *Microchim. Acta*, 185, 8, 379, 2018, 10.1007/ s00604-018-2884-5.
100. Chen, Y., Liu, X., Guo, S., Cao, J., Zhou, J., Zuo, J., Bai, L., A sandwich-type electrochemical aptasensor for Mycobacterium tuberculosis MPT64 antigen detection using C(60)NPs decorated N-CNTs/GO nanocomposite coupled with conductive PEI-functionalized metal-organic framework. *Biomaterials*, 216, 119253, 2019, 10.1016/j.biomaterials.2019.119253.
101. Vermisoglou, E., Panacek, D., Jayaramulu, K., Pykal, M., Frebort, I., Kolar, M., Hajduch, M., Zboril, R., Otyepka, M., Human virus detection with graphene-based materials. *Biosens. Bioelectron.*, 166, 112436, 2020, 10.1016/j.bios.2020.112436.
102. Koncki, R., Recent developments in potentiometric biosensors for biomedical analysis. *Anal. Chim. Acta*, 599, 1, 7–15, 2007, 10.1016/j. aca.2007.08.003.
103. Yunus, M.H., Yusof, N.A., Abdullah, J., Sulaiman, Y., Ahmad Raston, N.H., Md Noor, S.S., Simultaneous Amperometric Aptasensor Based on Diazonium Grafted Screen-Printed Carbon Electrode for Detection of CFP10 and MPT64 Biomarkers for Early Tuberculosis Diagnosis. *Biosensors (Basel)*, 12, 11, 2022, 10.3390/bios12110996.
104. Yunus, M.H., Yusof, N.A., Ismail, S., Md Noor, S.S., Mohammad, F., Sulaiman, Y., Ahmad Raston, N.H., Abdullah, J., Soleiman, A.A., Surface- Enhanced Carboxyphenyl Diazonium Functionalized Screen-Printed Carbon Electrode for the Screening of Tuberculosis in Sputum Samples. *Nanomaterials (Basel)*, 12, 15, 2022, 10.3390/nano12152551.
105. Bizid, S., Mlika, R., Haj Said, A., Chemli, M., Korri-Youssoufi, H., Investigations of poly(p-phenylene) modified with ferrocene and their application in electrochemical DNA sensing. *Sens. Actuators, B*, 226, 370–380, 2016, https://doi.org/10.1016/j.snb.2015.11.137.
106. Granot, E., Basnar, B., Cheglakov, Z., Katz, E., Willner, I., Enhanced Bioelectrocatalysis Using Single-Walled Carbon Nanotubes (SWCNTs)/ Polyaniline Hybrid Systems in Thin-Film and Microrod Structures

Associated with Electrodes. *Electroanalysis*, 18, 1, 26–34, 2006, https://doi.org/10.1002/elan.200503403.

107. Benvidi, A., Rajabzadeh, N., Mazloum-Ardakani, M., Heidari, M.M., Comparison of impedimetric detection of DNA hybridization on chemically and electrochemically functionalized multi-wall carbon nanotubes modified electrode. *Sens. Actuators, B*, 207, 673–682, 2015, https://doi.org/10.1016/j.snb.2014.10.043.
108. Bizid, S., Blili, S., Mlika, R., Haj Said, A., Korri-Youssoufi, H., Direct E-DNA sensor of Mycobacterium tuberculosis mutant strain based on new nanocomposite transducer (Fc-ac-OMPA/MWCNTs). *Talanta*, 184, 475–483, 2018, 10.1016/j.talanta.2018.03.025.
109. Bizid, S., Blili, S., Mlika, R., Haj Said, A., Korri-Youssoufi, H., Direct Electrochemical DNA Sensor based on a new redox oligomer modified with ferrocene and carboxylic acid: Application to the detection of Mycobacterium tuberculosis mutant strain. *Anal. Chim. Acta*, 994, 10–18, 2017, 10.1016/j.aca.2017.09.022.
110. Miodek, A., Mejri, N., Gomgnimbou, M., Sola, C., Korri-Youssoufi, H., E-DNA sensor of Mycobacterium tuberculosis based on electrochemical assembly of nanomaterials (MWCNTs/PPy/PAMAM). *Anal. Chem.*, 87, 18, 9257–64, 2015, 10.1021/acs.analchem.5b01761
111. Tufa, L.T., Oh, S., Tran, V.T., Kim, J., Jeong, K.-J., Park, T.J., Kim, H.-J., Lee, J., Electrochemical immunosensor using nanotriplex of graphene quantum dots, Fe3O4, and Ag nanoparticles for tuberculosis. *Electrochim. Acta*, 290, 369–377, 2018, https://doi.org/10.1016/j.electacta.2018.09.108.
112. Appaturi, J.N., Pulingam, T., Thong, K.L., Muniandy, S., Ahmad, N., Leo, B.F., Rapid and sensitive detection of Salmonella with reduced graphene oxide-carbon nanotube based electrochemical aptasensor. *Anal. Biochem.*, 589, 113489, 2020, 10.1016/j.ab.2019.113489.
113. Muniandy, S., Teh, S.J., Appaturi, J.N., Thong, K.L., Lai, C.W., Ibrahim, F., Leo, B.F., A reduced graphene oxide-titanium dioxide nanocomposite based electrochemical aptasensor for rapid and sensitive detection of Salmonella enterica. *Bioelectrochemistry*, 127, 136–144, 2019, https://doi.org/10.1016/j.bioelechem.2019.02.005.
114. Hu, J., Tang, F., Wang, L., Tang, M., Jiang, Y.-Z., Liu, C., Nanozyme sensor based-on platinum-decorated polymer nanosphere for rapid and sensitive detection of Salmonella typhimurium with the naked eye. *Sens. Actuators, B*, 346, 130560, 2021, https://doi.org/10.1016/j. snb.2021.130560.
115. Kumar, D.N., Baider, Z., Elad, D., Blum, S.E., Shtenberg, G., Botulinum Neurotoxin-C Detection Using Nanostructured Porous Silicon Interferometer. *Chemosensors*, 9, 8, 228, 2021, https://www.mdpi. com/2227-9040/9/8/228.
116. Afkhami, A., Hashemi, P., Bagheri, H., Salimian, J., Ahmadi, A., Madrakian, T., Impedimetric immunosensor for the label-free and direct detection of botulinum neurotoxin serotype A using Au nanoparticles/

graphene-chitosan composite. *Biosens. Bioelectron.*, 93, 124–131, 2017, 10.1016/j.bios.2016.09.059.
117. Krampa, F.D., Aniweh, Y., Kanyong, P., Awandare, G.A., Recent Advances in the Development of Biosensors for Malaria Diagnosis. *Sensors (Basel)*, 20, 3, 2020, 10.3390/s20030799.
118. Wu, F., Singh, J., Thomas, P.A., Ge, Q., Kravets, V.G., Day, P.J., Grigorenko, A.N., Ultrasensitive and rapid detection of malaria using graphene-enhanced surface plasmon resonance. *2D Mater.*, 7, 4, 045019, 2020, 10.1088/2053-1583/aba88e.
119. Gupta, D.K., Dembele, L., Voorberg-van der Wel, A., Roma, G., Yip, A., Chuenchob, V., Kangwanrangsan, N., Ishino, T., Vaughan, A.M., Kappe, S.H., Flannery, E.L., Sattabongkot, J., Mikolajczak, S., Bifani, P., Kocken, C.H., Diagana, T.T., The Plasmodium liver-specific protein 2 (LISP2) is an early marker of liver stage development. *Elife*, 8, 2019, 10.7554/eLife.43362.
120. Nam, H.W., Song, K.J., Ahn, H.J., Yang, Z., Chong, C.K., Cho, P.Y., Ahn, S.K., Kim, T.S., Probability of antibody formation against circumsporozoite protein of Plasmodium vivax among Korean malaria patients. *Korean J. Parasitol.*, 52, 2, 143–9, 2014, 10.3347/kjp.2014.52.2.143.
121. Mouatcho, J.C. and Goldring, J.P.D., Malaria rapid diagnostic tests: challenges and prospects. *J. Med. Microbiol.*, 62, Pt 10, 1491–1505, 2013, 10.1099/jmm.0.052506-0.
122. Barber, B.E., William, T., Grigg, M.J., Piera, K., Yeo, T.W., Anstey, N.M., Evaluation of the sensitivity of a pLDH-based and an aldolase-based rapid diagnostic test for diagnosis of uncomplicated and severe malaria caused by PCR-confirmed Plasmodium knowlesi, Plasmodium falciparum, and Plasmodium vivax. *J. Clin. Microbiol.*, 51, 4, 1118–23, 2013, 10.1128/JCM.03285-12.
123. Dzakah, E.E., Kang, K., Ni, C., Wang, H., Wu, P., Tang, S., Wang, J., Wang, J., Wang, X., Plasmodium vivax aldolase-specific monoclonal antibodies and its application in clinical diagnosis of malaria infections in China. *Malar. J.*, 12, 199, 2013, 10.1186/1475-2875-12-199.
124. Raman, J., Ashok, C.S., Subbayya, S., II, Anand, R.P., Selvi, S.T., Balaram, H., Plasmodium falciparum hypoxanthine guanine phosphoribosyltransferase. Stability studies on the product-activated enzyme. *FEBS J.*, 272, 8, 1900– 11, 2005, 10.1111/j.1742-4658.2005.04620.x.
125. Singh, N.K., Thungon, P.D., Estrela, P., Goswami, P., Development of an aptamer-based field effect transistor biosensor for quantitative detection of Plasmodium falciparum glutamate dehydrogenase in serum samples. *Biosens. Bioelectron.*, 123, 30–35, 2019, 10.1016/j.bios.2018.09.085.
126. Regiart, M., Gimenez, A.M., Marques, R.F., Soares, I.S., Bertotti, M., Microfluidic device based on electrodeposited Nanoporous Gold/Carbon Nanotubes for Plasmodium vivax detection. *Sens. Actuators, B*, 340, 129961, 2021, https://doi.org/10.1016/j.snb.2021.129961.

127. Obisesan, O.R., Adekunle, A.S., Oyekunle, J.A.O., Sabu, T., Nkambule, T.T., II, Mamba, B.B., Development of Electrochemical Nanosensor for the Detection of Malaria Parasite in Clinical Samples. *Front. Chem.*, 7, 89, 2019, 10.3389/fchem.2019.00089.
128. Burcu Bahadir, E. and Kemal Sezginturk, M., Applications of electrochemical immunosensors for early clinical diagnostics. *Talanta*, 132, 162–74, 2015, 10.1016/j.talanta.2014.08.063.
129. Dip Gandarilla, A.M., Regiart, M., Bertotti, M., Correa Gloria, J., Morais Mariuba, L.A., Brito, W.R., One-step enzyme-free dual electrochemical immunosensor for histidine-rich protein 2 determination. *RSC Adv.*, 11, 1, 408–415, 2020, 10.1039/d0ra08729g.
130. Gandarilla, A.M.D., Gloria, J.C., Barcelay, Y.R., Mariuba, L.A.M., Brito, W.R., Electrochemical immunosensor for detection of Plasmodium vivax lactate dehydrogenase. *Mem. Inst. Oswaldo Cruz*, 117, e220085, 2022, 10.1590/0074-02760220085.
131. Lafleur, J.P., Jonsson, A., Senkbeil, S., Kutter, J.P., Recent advances in lab-on-a-chip for biosensing applications. *Biosens. Bioelectron.*, 76, 213–33, 2016, 10.1016/j.bios.2015.08.003.
132. Ruiz-Vega, G., Arias-Alpizar, K., de la Serna, E., Borgheti-Cardoso, L.N., Sulleiro, E., Molina, I., Fernandez-Busquets, X., Sanchez-Montalva, A., Del Campo, F.J., Baldrich, E., Electrochemical POC device for fast malaria quantitative diagnosis in whole blood by using magnetic beads, Poly-HRP and microfluidic paper electrodes. *Biosens. Bioelectron.*, 150, 111925, 2020, 10.1016/j.bios.2019.111925.
133. Nezami, A., Nosrati, R., Golichenari, B., Rezaee, R., Chatzidakis, G., II, Tsatsakis, A.M., Karimi, G., Nanomaterial-based aptasensors and bio-affinity sensors for quantitative detection of 17β-estradiol. *TrAC, Trends Anal. Chem.*, 94, 95–105, 2017, https://doi.org/10.1016/j. trac.2017.07.003.
134. Bayat, P., Nosrati, R., Alibolandi, M., Rafatpanah, H., Abnous, K., Khedri, M., Ramezani, M., SELEX methods on the road to protein targeting with nucleic acid aptamers. *Biochimie*, 154, 132–155, 2018, 10.1016/j. biochi.2018.09.001.
135. Lo, Y., Cheung, Y.W., Wang, L., Lee, M., Figueroa-Miranda, G., Liang, S., Mayer, D., Tanner, J.A., An electrochemical aptamer-based biosensor targeting Plasmodium falciparum histidine-rich protein II for malaria diagnosis. *Biosens. Bioelectron.*, 192, 113472, 2021, 10.1016/j.bios.2021.113472.
136. Singh, N.K., Arya, S.K., Estrela, P., Goswami, P., Capacitive malaria aptasensor using Plasmodium falciparum glutamate dehydrogenase as target antigen in undiluted human serum. *Biosens. Bioelectron.*, 117, 246–252, 2018, https://doi.org/10.1016/j.bios.2018.06.022.
137. Chakma, B., Jain, P., Singh, N.K., Goswami, P., Development of Electrochemical Impedance Spectroscopy Based Malaria Aptasensor Using HRP-II as Target Biomarker. *Electroanalysis*, 30, 8, 1847–1854, 2018, https://doi.org/10.1002/elan.201800142.

138. Loyez, M., Wells, M., Hambye, S., Hubinon, F., Blankert, B., Wattiez, R., Caucheteur, C., PfHRP2 detection using plasmonic optrodes: performance analysis. *Malar. J.*, 20, 1, 332, 2021, 10.1186/s12936-021-03863-3.
139. Sharma, A., Mishra, R.K., Goud, K.Y., Mohamed, M.A., Kummari, S., Tiwari, S., Li, Z., Narayan, R., Stanciu, L.A., Marty, J.L., Optical Biosensors for Diagnostics of Infectious Viral Disease: A Recent Update. *Diagnostics (Basel)*, 11, 11, 2021, 10.3390/diagnostics11112083.
140. Sharma, S., Kumar, A., Singh, K.S., Tyagi, H.K., 2D photonic crystal based biosensor for the detection of chikungunya virus. *Optik*, 237, 166575, 2021, https://doi.org/10.1016/j.ijleo.2021.166575.
141. Chaudhary, V.S., Kumar, D., Kumar, S., Gold-Immobilized Photonic Crystal Fiber-Based SPR Biosensor for Detection of Malaria Disease in Human Body. *IEEE Sens. J.*, 21, 16, 17800–17807, 2021, 10.1109/ JSEN.2021.3085829.
142. Hadinegoro, S.R., The revised WHO dengue case classification: does the system need to be modified? *Paediatr. Int. Child Health*, 32, Suppl 1(s1), 33–8, 2012, 10.1179/2046904712Z.00000000052.
143. Khristunova, E., Dorozhko, E., Korotkova, E., Kratochvil, B., Vyskocil, V., Barek, J., Label-Free Electrochemical Biosensors for the Determination of Flaviviruses: Dengue, Zika, and Japanese Encephalitis. *Sensors (Basel)*, 20, 16, 2020, 10.3390/s20164600.
144. Wong, W.R., Krupin, O., Sekaran, S.D., Mahamd Adikan, F.R., Berini, P., Serological Diagnosis of Dengue Infection in Blood Plasma Using Long-Range Surface Plasmon Waveguides. *Anal. Chem.*, 86, 3, 1735– 1743, 2014, 10.1021/ac403539k.
145. Alcon, S., Talarmin, A., Debruyne, M., Falconar, A., Deubel, V., Flamand, M., Enzyme-linked immunosorbent assay specific to Dengue virus type 1 non-structural protein NS1 reveals circulation of the antigen in the blood during the acute phase of disease in patients experiencing primary or secondary infections. *J. Clin. Microbiol.*, 40, 2, 376–81, 2002, 10.1128/ JCM.40.02.376-381.2002.
146. Suthanthiraraj, A., P.P., Sen, A.K., Localized surface plasmon resonance (LSPR) biosensor based on thermally annealed silver nanostructures with on-chip blood-plasma separation for the detection of denguenon-structural protein NS1 antigen. *Biosens. Bioelectron.*, 132, 38–46, 2019, 10.1016/j. bios.2019.02.036.
147. Wasik, D., Mulchandani, A., Yates, M., V. Point-of-Use Nanobiosensor for Detection of Dengue Virus NS1 Antigen in Adult Aedes aegypti: A Potential Tool for Improved Dengue Surveillance. *Anal. Chem.*, 90, 1, 679– 684, 2018, 10.1021/acs.analchem.7b03407.
148. Solanki, S., Soni, A., Pandey, M.K., Biradar, A., Sumana, G., Langmuir–Blodgett Nanoassemblies of the MoS2–Au Composite at the Air–Water Interface for Dengue Detection. *ACS Appl. Mater. Interfaces*, 10, 3, 3020–3028, 2018, 10.1021/acsami.7b14391.

149. Kim, J., Byun, S., Smith, A.J., Yu, J., Huang, J., Enhanced Electrocatalytic Properties of Transition-Metal Dichalcogenides Sheets by Spontaneous Gold Nanoparticle Decoration. *J. Phys. Chem. Lett.*, 4, 8, 1227–32, 2013, 10.1021/jz400507t.
150. Lee, J.A., Hwang, S., Kwak, J., Park, S., II, Lee, S.S., Lee, K.-C., An electrochemical impedance biosensor with aptamer-modified pyrolyzed carbon electrode for label-free protein detection. *Sens. Actuators, B*, 129, 1, 372–379, 2008, https://doi.org/10.1016/j.snb.2007.08.034.
151. Omar, N.A.S., Fen, Y.W., Abdullah, J., Chik, C. E. N. C. E., Mahdi, M.A., Development of an optical sensor based on surface plasmon resonance phenomenon for diagnosis of dengue virus E-protein. *Sens. Bio-Sens. Res.*, 20, 16–21, 2018, https://doi.org/10.1016/j.sbsr.2018.06.001.
152. Cui, J., Gao, L., Chen, S., Huang, Z., Wang, X., Electrochemical voltammetric behaviors of synthetic dengue virus RNAs at ITO sensing electrode. *J. Electroanal. Chem.*, 851, 113463, 2019, https://doi.org/10.1016/j.jelechem.2019.113463.
153. Tripathy, S., Joseph, J., Pothuneedi, S., Das, D., Vanjari, S.R.K., Rao, A., Singh, S.G., A miniaturized electrochemical platform with an integrated PDMS reservoir for label-free DNA hybridization detection using nano-structured Au electrodes. *Analyst*, 144, 23, 6953–6961, 2019, 10.1039/ c9an01076a.
154. Odeh, A.A., Al-Douri, Y., Voon, C.H., Mat Ayub, R., Gopinath, S.C.B., Odeh, R.A., Ameri, M., Bouhemadou, A., A needle-like Cu2CdSnS4 alloy nanostructure-based integrated electrochemical biosensor for detecting the DNA of Dengue serotype 2. *Microchim. Acta*, 184, 7, 2211–2218, 2017, 10.1007/s00604-017-2249-5.
155. Ariffin, E.Y., Tan, L.L., Abd Karim, N.H., Heng, L.Y., Optical DNA Biosensor Based on Square-Planar Ethyl Piperidine Substituted Nickel(II) Salphen Complex for Dengue Virus Detection. *Sensors (Basel)*, 18, 4, 2018, 10.3390/s18041173.
156. Mariappan, M., Suenaga, M., Mukhopadhyay, A., Maiya, B.G., Synthesis, structure, DNA binding and photonuclease activity of a nickel(II) complex with a N,N′-Bis(salicylidene)-9-(3,4-diaminophenyl)acridine ligand. *Inorg. Chim. Acta*, 390, 95–104, 2012, https://doi.org/10.1016/j. ica.2012.04.016.

8

Metal Nanoparticle-Based Impedimetric Biosensors for Rapid Detection of Bacterial Pathogen in Aquaculture

Subbiah Padmapriya[1], Muthusamy Anand[1*], Kannan Rangesh[1] and Muthuchamy Maruthupandy[2]

[1]*Department of Marine and Coastal Studies, Madurai Kamaraj University, Madurai, Tamil Nadu, India*
[2]*Department of Health Sciences, Dong-A University, Busan, South Korea*

Abstract
Typical methods for identifying bacterial pathogens generally necessitate a core laboratory along with skilled technicians and can take several hours to days. Recent advances in scientific and engineering, however, have resulted in a significant bacterial pathogen identified. Analytical biosensors, which are frequently utilized in medical contexts, are now being extended to enable swift and on-site surveillance of pathogenic bacteria in aquaculture settings. Biotic stresses of viral, parasitic, bacterial, and fungal origin are major constraints to increasing aquaculture production. Evaluation as well as swift detection of microbial contaminants are critical in identifying and mitigating biotic stresses, which have become one among the most difficult aspects in aquaculture. Usual analytical procedures, which include titrimetric analysis, spectrophotometric, mass spectrometric, spectroscopic, and chromatographic techniques, are time-consuming along with, in some instances, inaccessible when needed. Impedimetric biosensor development for rapid, selective, and sensitive detection is a broad and dynamic field of aquaculture research. This chapter fills a knowledge gap in the detection of bacterial contaminants of biotic origin in aquaculture using electrochemical impedance spectroscopy (EIS) techniques utilizing metal and metal oxide nanoparticles. The discussion centers around the analytical attributes of the biosensor, encompassing

**Corresponding author*: anandm21@yahoo.com

Sathish-Kumar Kamaraj, Arun Thirumurugan, Muthuchamy Maruthupandy, Mercedes Guadalupe López Pérez and Shanmuga Sundar Dhanabalan (eds.) Nanoscience and Nanotechnology for Smart Prevention, Diagnostics and Therapeutics: Fundamentals to Applications, (183–216) © 2024 Scrivener Publishing LLC

sensitivity, specificity, and usability. Additionally, a concise juxtaposition with current bacterial detection methods is provided.

***Keywords*:** Biosensors, bacterial pathogens, impedimetric detection, aquaculture

8.1 Introduction

The fastest-growing primary industry is aquaculture, which helps supply the expanding demand for animal proteins and lipids. It has provided easily digestible protein, essential fats, and adequate micronutrients to billions of people around the world. In rural areas, aquaculture holds the promise of becoming a source of both employment and food. It also has a high chance of bringing in money from exports to help the country's gross domestic product (GDP). Over the past seven decades, the total production of industry of fisheries and aquatic farming (excluding algae) has grown dramatically, from 19 metric tonnes (live weight the equivalent) in the year 1950 to an all-time high of around 179 million tonne in 2018, a growth rate of 3.3% per year. Production then slightly decreased in 2019 (a decrease of 1% from 2018), before barely recovering (by 0.2%) to reach 178 million tonnes in 2020 [1]. Total first sale price of agriculture and fishery productions of marine creatures in the year 2020 was US$406 billion, with aquaculture accounting for US$ 265 billion [2]. However, excessive breeding nutritious fish caused in outbreaks of several bacterial illnesses, forcing the aquaculture industry to suffer annual economic losses in the billions of dollars worldwide. Pathogen-caused disease outbreaks are a major threat to the field of aquaculture because they cost the world economy $6 billion USD each year [3].

8.1.1 Sources of Contaminants in Aquaculture and Its Impacts

Multiple abiotic and biotic stressors have recently threatened its prodigious growth potential (Figure 8.1). As an alternative, it appears that growing illness susceptibility, the establishment of novel diseases, and the subpar effectiveness of conventional diagnostic techniques are the key challenges to working on sustainability. Significant disparities in the quality of fish raised in aquaculture are caused by the possibility of different aquaculture procedures as well as the potential for environmental damage. Products from the fishing sector have also been found as potential sources of natural toxins, parasites, heavy metals, and other contaminants, as well as pathogenic bacteria as *Vibrio* sp and *Salmonella* sp. Food sickness has been

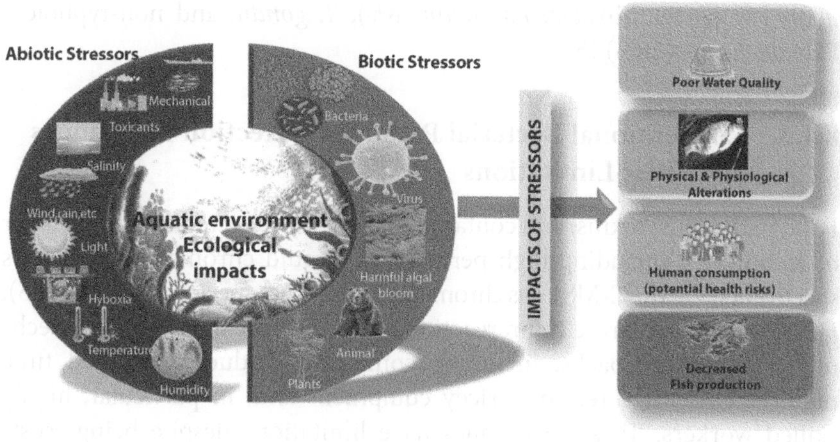

Figure 8.1 Source of contaminants and its impacts.

linked to the consumption fish like mackerel since it contains high levels of histamine (50 mg/100 g fish) [4]. Since there is a higher potential that the ponds would become polluted by the environment, many of these risks are particularly serious when it comes to fish products from aquaculture. Regulation agencies like the US and the European Union Food and Drug Administration (FDA) have set strict standards for the quality of processed fish products sold by exporting nations due to the rising demand for the commodity and the necessity to safeguard consumers from food-borne illnesses. The development of quick and accurate quality evaluation techniques is required to keep these things at an acceptable quality level.

8.1.2 The Most Prevalent Categories of Potential Pathogens

Aquatic pathogenic bacteria have garnered a lot of attention recently for their role in diseases and productivity losses. Numerous aquatic pathogenic bacteria have been discovered to have an impact on the well-being and security of the global populations of people, animals, and plants. A recent Centers for Disease Control and Prevention (*CDC*) report found that over a 2-year period, 31 major pathogens were to blame for 1,351 fatalities, 55,961 hospitalizations, and 9.4 million instances of aquatic illness. The majority of illnesses (58%) were brought on by norovirus, which came in first place behind non-typhoidal *Salmonella* sp. (11%), *Campylobacter* sp. (9%), and *Clostridium perfringens* (10%). The most common causes of hospitalization were nontyphoidal *Salmonella* sp. (35%), *Campylobacter* sp. (15%), *norovirus* (26%), *Toxoplasma gondii* (8%), *T. gondii*, and non-typhoidal *Salmonella* sp. (28%) [5].

8.1.3 Conventional Bacterial Pathogen Detection Techniques and Their Limitations

Food pathogens, toxins, and contaminants can be detected using a number of techniques, including high-performance liquid chromatography-mass spectrometry (HPLC-MS), gas chromatography-mass spectrometry (GCMS), HPLC, and polymerase chain reaction (PCR) (Figure 8.2) [6]. These techniques have drawbacks, including complexity, arduousness, and time requirements; they rely on pricey equipment and, in particular, highly skilled workers. These techniques have limitations despite being trustworthy, precise, and highly sensitive for identifying food contaminants and toxicants. As a result, it is challenging to meet the expectations of quick, on-the-spot screening of numerous samples in impoverished and underdeveloped areas without detecting technologies and qualified workers. The complex food matrix makes it necessary to develop some quick, easy-to-use, and sensitive methods for detecting minute amounts of contaminants in food in order to ensure food safety. In the establishment of

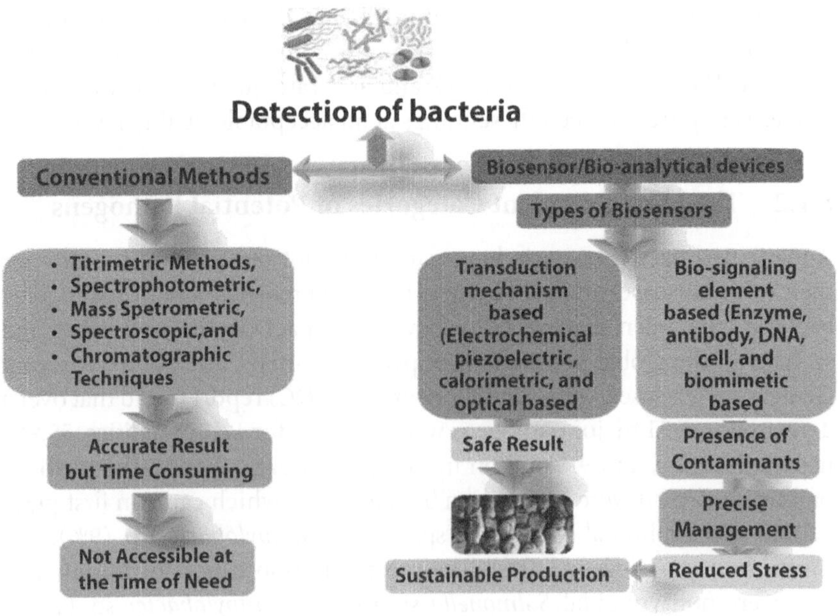

Figure 8.2 The conventional and biosensor techniques for detecting bacteria.

rapid techniques for the identification of aquatic pathogens, it is still necessary to take into account the differentiating between dead and living cells, automation, expense, simplicity, training, and reliability. As one type of electrochemical biosensor, it has been demonstrated that the impedance technique is a viable technique for detecting aquatic pathogenic bacteria because of their portability, quickness, sensitivity, and, most importantly, the fact that it can be used for on-site detection.

Impedance detection techniques are classified into two kinds based on the existence or absence of particular bio-recognition components. The first kind measures the resistance variation brought on by the target binding to bioreceptors (nucleic acids and antibodies immobilized on the electrode surface), whereas the second kind depends on metabolites produced by bacterial cells. Impedance biosensors for detecting bacteria have been thoroughly reviewed over the years, as this chapter summarises.

8.1.4 Nanotechnology Influenced Impedance Biosensor for Detection of Aquatic Pathogens

Infectious diseases pose a potential threat to global economy and public health. Fast and precise pathogen identification is key to preventing disease transmission. The traditional approaches to detecting these microorganisms are time-consuming, expensive, and ineffective for on-site monitoring. A quick, accurate, and point-of-care diagnostic can be offered by biosensors. Nanomaterials have become significant actors in the sector of biosensors because of their exceptional electrical, chemical, and optical properties. In this chapter, mainly focused on metal nanoparticles used in electrochemical impedance biosensors that can identify infectious diseases will be discussed, along with how they improved the sensitivity and speed of the various sensing platforms. Due to their portability, speed, sensitivity, affordability, ease of miniaturization, and label-free operation, electrochemical biosensors based on impedance technique have developed as a promising technique of identifying bacteria that cause disease. However, the fact that they can be applied to on-the-spot detection is more significant.

Impedance biosensor, which relies based on impedance changes in culture media as a result of bacterial growth as changes in conductance, as a result of charged ions and compounds produced by biological metabolism, or as a result of bacterial cell adhesion towards the electrode surface as interfacial capacitance, is a field of study that has generated a lot of literature [7]. Traditional impedance biosensor is not a selective technique, it should be highlighted. Utilizing selective culture media may help achieve some selectivity. However, as demonstrated in this chapter, functionalizing the electrodes with high-affinity recognition element that specifically bind target cells, such as antibodies, aptamers, proteins, and so on, allows for a significant improvement in the method's selectivity.

Additionally, as discussed in the chapter of this book titled *Metal nanoparticles based impedimetric biosensors for speedy detection of bacterial pathogen in aquaculture*, different nanoparticle-enabled impedimetric biosensors using different biorecognition elements for quick detection of diverge bacterial pathogens may aid to lower detection thresholds and increase the selectivity of the method. Impedance biosensors track changes in their surface's electrical properties (either resistance or capacitance), which are influenced by interactions between the analyte present in the biorecognition element and the sample solution attached to the surface of the sensor. Planar metal electrodes can be used to measure the faradic impedance of a test solution in the occurrence of redox pair.

This sensor's architecture enables the use of several bio-recognition components for label-free bacterial detection. The subsequent parts of this chapter will go into further depth on all significant facets of electrochemical impedance spectrometry (EIS)-based pathogen identification.

8.2 Nanoparticles

The most significant nanoparticles (NPs) are metal, metal oxide, magnetic, quantum dot, semiconductor, and ceramic NPs, which range in size from 1 to 100 nm [8]. These NPs can be found as spheres, needles, rods, and plates, among other shapes. The surface-to-volume ratio and the distance of the energy level are generally affected by the size of the NPs. Many of the NPs properties change as a result of these two variables. Due to their distinctive properties, NPs play important roles in many electrochemical processes, but their primary performances can be broadly categorized as follows: (1) electrochemical reaction catalysis, (2) an increased transfer of electrons, (3) biological molecule immobilization, and (4) Labeling biological molecules is step 4 [9]. They are distinguished by their different shapes, sizes, and characteristics. Due to their large surface area, they are also considered cutting-edge materials. The majority of biosensor are constructed with highly conductive nanomaterials. Four dimensions are used to categorize nanomaterials: zero dimension (0D), one dimension (1D), two dimension (2D), and three-dimension (3D) [10] 0Ds nanomaterials, like quantum dots, have all dimensions that are lesser than 100 nm (QDs).

Nanowires, nanotubes, and nanorods are typical examples of 1Ds nanomaterials, and all three have sizes under 100 nm. In the case of the 2Ds, where a minimum of one dimension is less than 100 nm, we find shapes like the ribbon, the plates and sheets, and the triangles [8]. All dimensions of a 3Ds nanomaterial are greater than 100 nm, including those of objects like stars, corns, dumbbells, flowers, boxes, and cubes. Only a subset of the aforementioned nanomaterials has seen actual application in biosensor development. Both organic (which typically includes carbon nanoparticles) and inorganic (which typically includes semiconductors and metal nanoparticles) classifications are commonly used to describe nanomaterials. The majority of existing NPs together with nanostructured materials (NSMs) can be classified into one of four broad material classes, which are as follows:

i. Carbon-based nanomaterials

These nanomaterials (NMs) exhibit various shapes, such as ellipses, hollow tubes, and spheres, often incorporating carbon as a component. Among carbon-based NMs, there are notable representatives like fullerenes (C60), carbon nanofibers, carbon nanotubes (CNTs), carbon black, graphene (Gr), and carbon onions. These carbon-based nanomaterials are synthesized through major industrial processes, like arc discharge, laser ablation, and chemical vapor deposition (CVD), with the exception of carbon black.

ii. Inorganic-based nanomaterials

Metal as well as metal oxide NPs and NSMs are among the inorganic-based nanomaterials (NMs) that are included in this category. These NMs can be synthesized into semiconductors, metals (Au or Ag NPs), and metal oxides (TiO_2 and ZnO NPs) (silicon and ceramics).

iii. Organic-based nanomaterials

Excluding carbon-based and inorganic-based nanomaterials, this category includes NMs that are primarily made of organic materials. Organic NMs can be converted into the proper structures, such as dendrimers, micelles, liposomes, and polymer NPs, through the design and self-assembly of molecules with the help of noncovalent (weak) interactions.

iv. Composite-based nanomaterials

Composite NMs are NSMs and multiphase NPs with one nanoscale phase that may put NPs together with other NPs or NPs integrated with greater or bulkier materials (for example, hybrid nanofibers) or more complex structures, such as metal-organic frameworks. The bulk materials for the

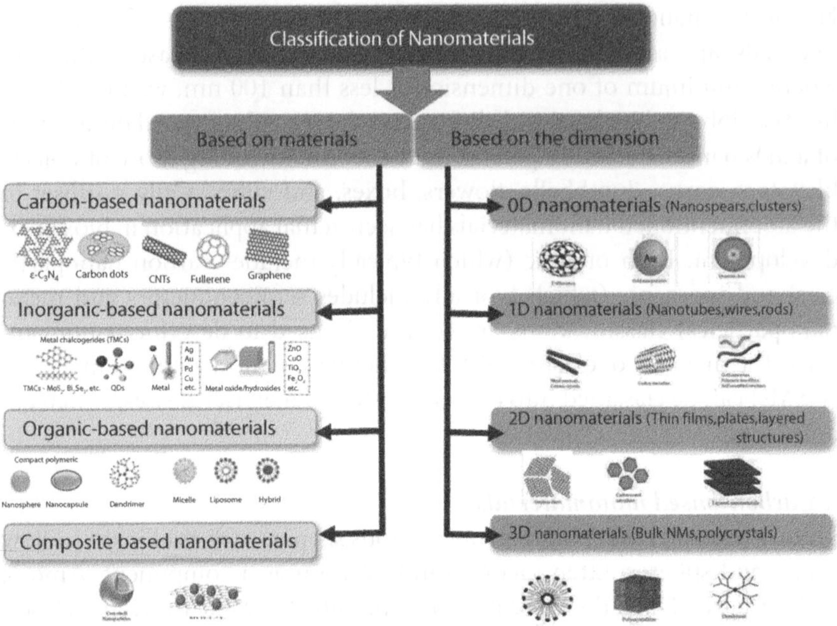

Figure 8.3 Classification of nanomaterials.

composites can be any type of ceramic, metal, or polymer, along with any combination of carbon-, metal-, or organic-based nanomaterials. Nanomaterials are categorized according to their composition and size as shown in Figure 8.3.

8.2.1 Metal and Metal Oxide Nanoparticles

(i) Silver nanoparticles (Ag NPs)

With diameters ranging from 1 to 100 nm, silver nanoparticles are suitable for biomedical applications. The ratio of surface to volume shockingly rises in impressive increments as Ag NP diameter is reduced, and this effect causes observable changes in organic, physical, and compound exercises. Ag NPs are currently being tested in novel and cutting-edge symptomatic equipment, like biosensors and immunosensors (Figure 8.4).

(ii) Gold nanoparticles (Au NPs)

As can be inferred from their extraordinary optical or electrical properties, gold NPs are widely used in the sector of infection discovery.

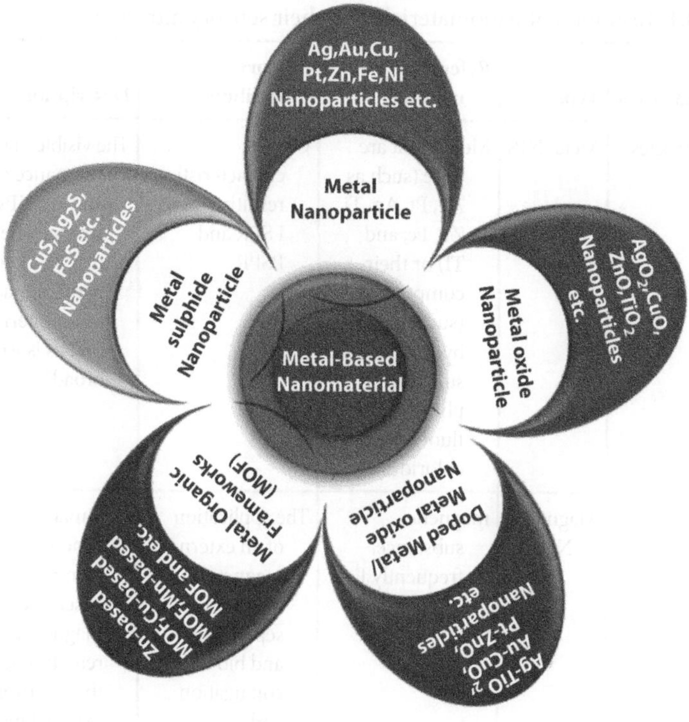

Figure 8.4 Different types of metal-based nanomaterials.

(iii) Magnetic nanoparticles (MNPs)

Magnetic NPs are a diverse group of particles with their own associated properties, such as catalysis, attractive liquids, and magnetic resonance imaging.

(iv) Zinc oxide nanoparticles (ZnO NPs)

ZnO is the zinc compound that is present in nanostructures in the greatest concentration. Transducers, surface acoustic wave instruments, gas sensors, and photonic devices are just a few of the uses for ZnO that come to mind. In some unusual detectors called mechanochemical sensors, ZnO, a material with piezoelectric properties, is utilized.

(v) Aluminum nanoparticles (Al NPs)

Among the various varieties of Al NPs, nanoporous morphology is the most well-known and desirable for biosensing researchers. This nanostructure is appropriate for use in innovative methods because it is made up of a number of significant materials with a large surface area and optical, as

Table 8.1 Examples of nanomaterials and their sensory attributes.

Nanomaterials	Type	Reference material	Sensory attributes	Description
Nanoparticles	Metal NPs	Metals that are pure (such as Au, Pt, Ag, Ti, Zn, Fe, and Tl) or their compounds (such as oxides, hydroxides, sulphides, phosphates, fluorides, and chlorides)	Optical characteristics resulting from LSPR and PSPR	The visible plasmon resonance of Au and Ag NPs is strong, whereas the UV absorption band of many other intermediate metals is weak and broad.
	Magnetic NPs	magnetic substance, frequently Fe, Ni, and Co	The application of an external magnetic field for simple separation and bio-conjugation and immobilization to prepare samples for detection	minimal toxicity, superb compatibility, and absence of residual magnetism retention following the removal of an external magnetic field, iron oxide nanoparticles (IO NPs), especially superparamagnetic Fe_3O_4 (magnetite) NPs, have found widespread utilization in the food industry.
	Quantum dots	Semiconductors, metals, magnetic materials	Fluorescent behavior	Typically, QDs emit light between 450 and 850 nm. They become more sensitive to tracking after Stokes'

well as physical properties, like chemical and thermic stability, which are essentially viable in bio conditions like the human body.

(vi) Copper nanoparticles (Cu NPs)

Copper NPs have drawn a lot of interest due to their extraordinary capability and potential to replace more expensive NPs. Modern nanotechnology can now be used to study the infectious activity of Cu NPs. One major drawback of these NPs is their inherent propensity to oxidize in ambient conditions. On the other hand, copper NPs offer well-suited properties, leading to the development of numerous biosensors aimed at viral diagnosis.

8.2.2 Influence of Nanomaterials on Biosensor Performance

As a result of their high electrocatalytic activity and their propensity to promote the electron transfer/charge directly in mediator-less bio-sensing systems, metal as well as metal oxide nanostructures are the materials that are most often employed for biosensor modification. In order to facilitate quicker electron transfer kinetics between the active sites of immobilized bio receptors and the electrode, ZnO, CuO, NiO, TiO_2, and Fe_3O_4 have all been commonly utilized in electrochemical bio-sensing. This has resulted to a synergistic enhancement in sensing performance [11].

Because of their resistance to oxidation processes and their high level of biocompatibility, precious metal nanostructures have been used for the purpose of electrode modification. These nanostructures have been used with precious metals such as Au, Pt, Ag, and Pd. Either these nanostructures could be organized on the exterior of the electrode, or they could be combined with other components (such polymeric or sol-gel materials) in the matrix of the electrode. In the end, the benefits of those one-of-a-kind qualities of nanomaterials enable the manufacturing of nano-biosensor chips and portable sensing devices for speedy and accurate multi-target in complicated biological and environmental matrices using nano-biosensor devices (Table 8.1).

8.3 Biosensor

8.3.1 Design and Principle

A biosensor is a tool or probe that combines an electronic component with a biological element, like an enzyme or antibody, to produce a

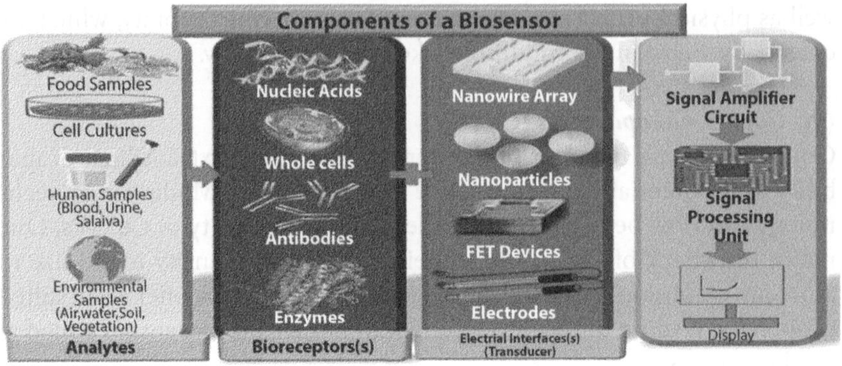

Figure 8.5 The constituting elements of a biosensor.

quantifiable signal. The electronic component can detect and record physiological changes as well as identify environmental chemicals or biological substances. Biosensors are available in a variety of sizes and shapes, and they have the ability to measure and detect even very low concentrations of certain pathogens, toxic chemicals, and pH levels. A typical biosensor consists of an analyte, a bioreceptor, a transducer, electronics, and a display shown in Figure 8.5.

8.3.2 Attributes of Biosensors

There are some static and dynamic requirements that must be met in order to create a highly effective and capable biosensor system. The performance

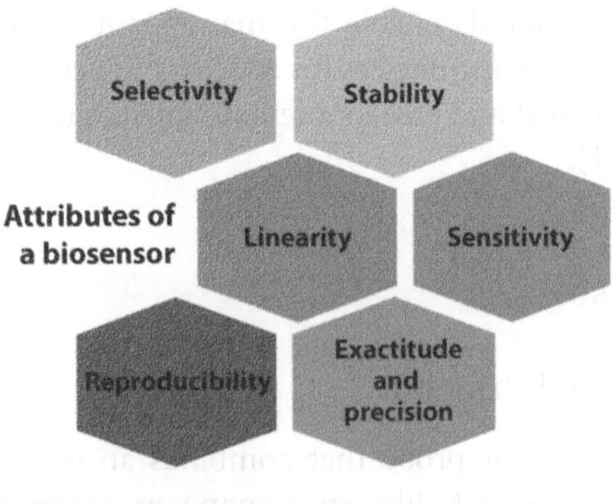

Figure 8.6 The basic attributes of a biosensor.

of the biosensors can be optimized for commercial uses based on (Figure 8.6) specifications.

8.3.3 Classification of Biosensors

The classification of biosensors is a large and interdisciplinary field. The classification of biosensors is based on a number of factors, and Figure 8.7 shows the overall classification scheme.

Bioreceptors are regarded as the main element in the construction of biosensors, as was previously discussed. Biosensors are divided into five different classifications based on the type of bioreceptor employed: immunosensors, enzymatic biosensors, aptamer or nucleic acid-based biosensors, nanobiosensor and microbial or whole-cell biosensors [12]. Immunosensors are the most prevalent type of biosensor and have high specificity and sensitivity, making them particularly useful in diagnosis. The second division depends on the transducer, and the sensors are divided into the following groups: electrochemical (which is additionally divided into amperometric, potentiometric, impedance, and conductometric groups), electronic biosensor, optical, thermal biosensor, and mass-based or gravimetric [13].

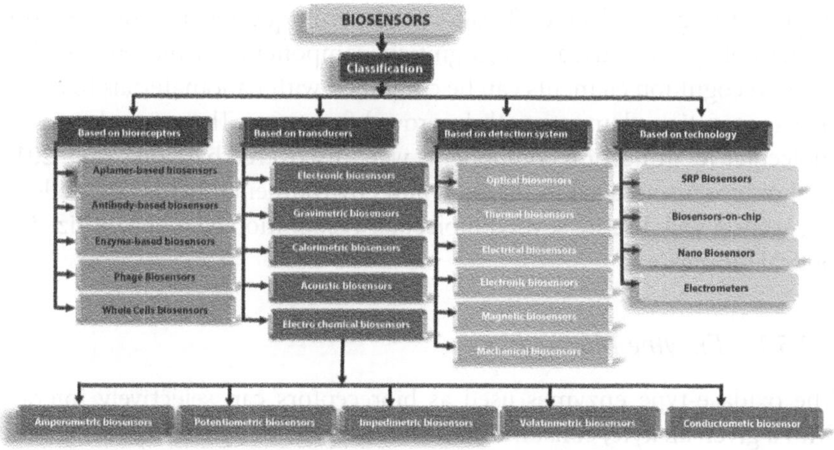

Figure 8.7 Classification of biosensor based on bioreceptor, transducer, detection system and technology.

8.3.4 Bioreceptors or Biosensing Elements

In a bioreceptor, this type of molecule uses a biochemical detection mechanism. Two categories of bioreceptors exist: catalytic biosensors encompass enzymes, cells, or tissues, while affinity biosensors involve nucleic acids and antibodies.

8.3.5 Bacteria Detection Using Molecular Recognition Elements

Biosensors are devices that can detect the presence of biological substances by converting their concentrations to electrical signals. Two essential elements that make up a typical biosensing system are a device that selectively affirms a biomolecule and one that reads out the corresponding signal [14]. The bioreceptor of a biosensor, often referred to as the recognition component, enables a precise and sensitive analysis. A biosensor's molecular recognition components can include bacteriophages, aptamers, antibodies, and DNA sequences. The particular biosensing device's functional strategy has a significant bearing on the kind of bioreceptor selected detection of a specific substance. Antibodies use the sensitivity and specificity of interactions between biomolecular antibodies and antigen-producing immune cells as recognition elements [15]. Functional monomers are employed by molecularly imprinted polymers (MIPs) to initiate polymerization around a template, resulting in the formation of synthetic recognition sites that precisely mimic the template. Aptamers have the ability to specifically bind to a target molecule. Because of their unique reliance on the host, bacteriophages are used as recognition components in biosensors [12]. These recognition elements can be combined with nanomaterials to create new recognition elements with improved functions. Three novel types of molecular identification components were discovered by Dorst *et al.* 2010 [16] in biosensors used in the food and environmental fields. The following section will discuss several molecular recognition elements utilized in optical biosensors.

8.3.5.1 Enzyme Bioreceptor

The oxidase-type enzymes used as bioreceptors can selectively interact with a given analyte, consume dissolved oxygen, and yield the easily measurable compound H_2O_2. Beta-galactosidase and horseradish peroxidase are the two most frequently used enzymes. The labeling of the antibodies with these enzymes enables the detection of pathogenic bacteria in food.

The advantages of this bioreceptor include its commercial enzyme availability, high purity, specificity, suitability for various transduction techniques, and its capability to detect a wide range of analytes.

8.3.5.2 Cells as Bioreceptor

Bacteria, yeast, or higher eukaryotic cells, such as those from vertebrates or mammals, can be used as these sensors. Living cells have sensitivity, functional evaluation for biochemical agents, as well as an extremely low detection threshold that make them advantageous as recognition elements. In 2010 [17], a cell-based biosensor was created by Banerjee and Bhunia (2010) [18] for the quick detection of a variety of pathogenic bacteria, including *Listeria monocytogenes, enterotoxigenic Bacillus* sp, *Vibrio* sp, *Micrococcus* sp, and *Serrati* sp, as well as toxins from *Staphylococcus aureus, Clostridium perfringens, Stoichactis helianthus, L. monocytogenes,* and *Bacillus* sp.

8.3.5.3 Antibody Bioreceptor

Biosensors use antibodies, which are a type of universal bioreceptor. In response to antigenic stimulation, B-lymphocytes produce these proteins. In order to detect chemicals, biomolecules, microorganisms, etc., an antigen-specific antibody must fit its particular antigen.

8.3.5.4 Nucleic Acid Biosensor

These types of biosensors, which use DNA, RNA, and peptide nucleic acids as a bioreceptor, are gaining popularity due to their high sensitivity and selectivity as a result of strong base pair affinities. Numerous studies have focused on nucleic acid-based biosensors for the detection of aquatic pathogens like *Escherichia coli, Salmonella* sp., *C. jejuni*, etc. (Figure 8.8).

8.3.5.5 Bacteriophage Bioreceptor

Phages are bacterial viruses that range in size from 20 to 200 nm [19]. They attach to their target bacteria using particular receptors found on the surface of the host cells and introduce their genetic material. These viruses use receptors on their tail to identify the target bacteria. They are desirable candidates for use as probes in sensor devices because they can target particular bacteria. By using a variety of sensing platforms, phages are used

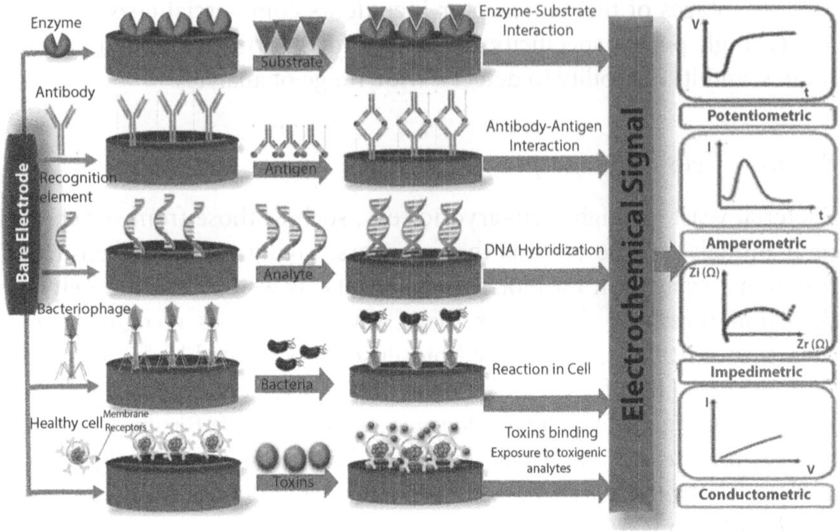

Figure 8.8 Diagram of impedimetric biosensors fabricated by different biorecognition elements: antibody-based sensor, enzyme, bacteriophage, and cell-based sensor.

as a biorecognition element for the exposure of different pathogens like *Staphylococcus aureus, Escherichia coli,* and *Bacillus anthracis* spores.

8.3.5.6 Nanobiosensors Based on MIPs

MIPs are polymers that specifically bind to and recognize targets or structural analogs. MIPs aid in the development of artificial recognition elements in sensing components that are stable and "solid-state like." Amino acid derivatives' imprinting on microporous polymers was first discovered by Mosbach and colleagues in 1984, after which the term "imprinted polymer" became popular among scientists [20]. The progress of the types, uses, and history of MIPs was done by Bel Bruno [21]. Uzun and colleagues [22] have reported the integration of Molecularly Imprinted Polymer (MIP)-based sensors with innovative approaches such as computational design, advanced polymerization techniques, and optimized material compositions. MIPs have been applied in optical and electrochemical sensors, according to Ahmad *et al.* [23], MIPs in biosensors can collect target materials on the sensing surface to increase sensitivity and can also separate analysis objects from food samples to improve selectivity. Improved response times can be achieved by using molecularly imprinted nanospheres in a label-free optical biosensor, and MIPs and nanomaterials combined can also demonstrate high sensitivity and selectivity. For instance, a gram-negative bacterial biofilm containing the signal molecule N-acylhomoserine lactones (AHLs) was quickly and accurately detected by a CdSe/ZnS QDs@MIP probe [23]. In that study, a surface imprinting method was used to create QDs@MIP. With an increase in AHL concentration from 0.54 to 0.88 nM over 30 seconds, the constructed platform improved. Li and colleagues created a dual recognition system using aptamers and MIPs. Molecularly Imprinted Polymers (MIPs) were synthesized using electro polymerization, involving a carbon dot (C-dots)-labeled DNA aptamer, o-aminophenol, and lincomycin. This process occurred on a graphene oxide-modified gold electrode (Au-GO). Furthermore, a surface plasmon resonance (SPR)-based platform using an Au slide-MIP configuration was developed for the detection of *S. paratyphi* [24].

8.4 Transducer Component

The transducer is necessary for the biosensor. Electrochemical, calorimetric (thermometric), mass (piezoelectric or surface acoustic wave devices), and optical transducer types are the four main types used in biosensor devices.

8.4.1 Electrochemical Transducers

Typically, electrochemical equipment measure impedance or conductivity changes, or they measure the current at a fixed voltage, the voltage at zero current (potentiometry), (Amperometry), or both.

8.4.2 Optical Transducers

When microorganisms bind to transducer-mounted, immobilized receptors, surface, optical transducers can be labeled optical biosensors (fluorescence, luminescence, transmission, and assess even the slightest variations in the refractive index or thickness scattering) and label-free optical biosensors (ellipsometry, surface plasmon resonance, resonant mirror sensor, interferometer), divided into two categories.

8.4.3 Mass-Based Transducers

For pathogen detection, mass-sensitive sensors use piezoelectric crystals. According to Shabani *et al.*, this type of transducer measures the change in frequency as bacteria bind to the transducer's receptor [26].

8.4.4 Electrochemical Biosensor

Electrochemical biosensors are outpacing other biosensors in popularity due to their low price, excellent selectivity and sensitivity, ability to operate within turbid media, and promise for miniaturization for the assessment of food-borne viruses Palchetti and Mascini [27]. A chemical reaction between immobilized biological molecules and the intended analyte is used in electrochemical biosensors to change observable electrical properties of solution, such as an electric current or potential, by producing or consuming ions. The most extensively researched and applied class of biosensors are electrochemical ones, which rely on the electrochemical

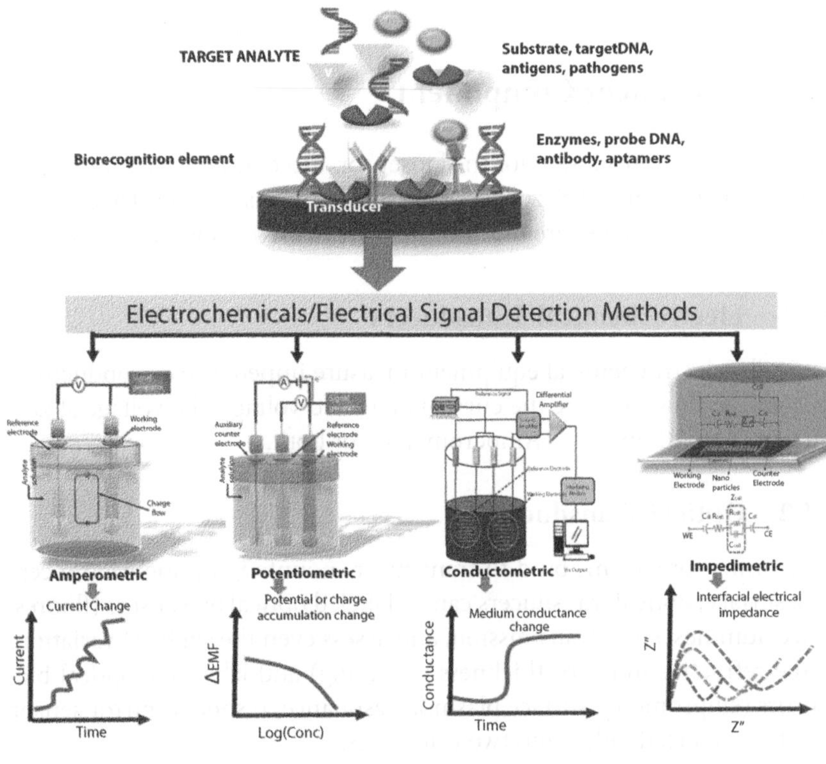

Figure 8.9 A schematic representation of electrochemical immunosensor methods, encompassing amperometric/voltammetric, potentiometric, and conductometric biosensors. Additionally, it showcases the equivalent circuit for the impedimetric biosensor, denoting various components such as C_{dl} (double-layer capacitance of the electrodes), Z_{cell} (impedance due to the nanoparticles), R_{sol} (solution resistance), C_{de} (electrode capacitance), R_{cell} (resistance), and C_{cell} (capacitance in parallel). The figure illustrates the corresponding readings obtained from this configuration.

characteristics of the analyte and transducer. High levels of sensitivity, selectivity, and detection are displayed by electrochemical biosensors. This biosensor works by generating detectable electrochemical signals in the kind of current, voltage, impedance, and capacitance on the transducer surface between the analyte and bioreceptor. Electrochemical biosensors are divided into the following categories depending on the transduction principle: (a) amperometric; (b) potentiometric; (c) conductometric; (d) impedimetric; and (e) voltammetric (shown in Figure 8.9).

(a) Potentiometric biosensors

Potentiometric biosensors measure the accumulated charge on the working electrode due to interactions between the analyte and the bioreceptor, occurring under no current conditions. Ion-selective and electrode ion-sensitive field-effect transistors are used to convert a biological process into a potential signal.

(b) Amperometric biosensors

These sensors have two or three electrodes and function in amplitude measurements. Such sensors measure the current produced by electrochemical reduction or oxidation of electroactive species at the working electrode when they apply a constant potential to the working electrode with respect to the reference electrode. The amount of current produced on the surface of the working electrode is directly related to the concentration of analyte present in the solution. Compared to potentiometric biosensors, this method offers sensitive, quick, exact, and linear response, making it more suitable for mass production. However, these sensors' shortcomings include their poor selectivity and interference from other electroactive materials.

(c) Conductometric biosensors

Conductometric biosensors measure the difference in conductance (change in the conductivity properties of the analyte) between two electrodes as a result of an electrochemical reaction. Typically, the observation of metabolic processes in living biological systems is made possible by the use of conductometric and impedimetric biosensors.

(d) Voltammetric biosensors

Voltammetric biosensors use controlled variations in applied potential to measure the current while detecting analytes. These sensors can detect numerous analytes at once and offer extremely sensitive measurements.

(e) Impedimetric biosensors

The electrical impedance that is created a small sinusoidal wave occurs at the electrode/electrolyte interface upon the application of an excitation signal is measured by impedimetric biosensors (optional). In this method, a sensor electrode is used with a minor amplitude AC voltage, and an impedance analyzer is used to measure the current response in- and out-of-phase frequency dependence.

8.5 Mechanisms for Impedance-Based Detection of Microorganisms

Considering these electrically linked attributes of bacterial cells, three strategies having been identified for the establishment of impedance-based techniques for the identification and measurement of bacterial cells.

8.5.1 Detection Based on Bacterial Metabolism

The foundation of this strategy is the quantification of fluctuations in electrical impedance within a growth medium or reaction mixture brought on by proliferation of bacterial organisms.The main source of the impedance changes is the discharge of ionic byproducts derived from living cells into the medium, which includes both minor ions (like K+ and Na+ ion channels) and the primary energy metabolism (catabolism). Two electrodes dipped in the reactant solution or the growth medium are typically used to measure the impedance change over time. Impedance growth curves are created by graphically plotting the measured electrical signals over the vertical axis in relation to the incubation durations on the horizontal axis. According to Yang *et al.*, the impedance growth curve's shape accurately depicts the different stages of bacterial growth [28]. Using silicon-based microfluidic biochips, Gomez *et al.* studied the impedance-based microbial detection (*L. monocytogen*), where cells are concentrated and collected from a liquid sample, and their growth and metabolism are detected as they are incubated with a volume of 400 pL inside the chip [29]. The microbial sample is concentrated and collected using dielectrophoresis on bacterial cells, and metabolism is detected using impedance measurements. Additionally, they showed how the time required to find the bacteria was significantly reduced by this technique.

8.5.2 Detection Reliant on the Insulating Attributes of the Cell Membrane

The electrical properties of the adjacent components that make up bacterial cells are very different. The bilayer of lipid and numerous proteins that make up the bacterial cell membrane is primarily in charge of moving nutrients, ions, and waste materials across the membrane. This membrane has a conductivity of about 10^{-7} S/m and is generally very well insulated. These immobilized bioreceptors, which include antibodies, nucleic acids, bacteriophages, and lectins, bind to specific bacteria in this type of impedimetric biosensor. Bacterial cells attached to an electrode surface can effectively reduce the electrode area that the current reaches and thereby increase the interface impedance because their cell membranes are highly insulating [28]. A throwaway screen-printed carbon electrodes (SPCE) immunosensor for the identification of ochratoxin A in food samples was created by Malvano *et al.* in 2016 [30].

For the impedimetric assessment of *Escherichia coli* K12 cells, Moghtadar *et al.* (2016) [31] used pencil graphite electrodes deposited by gold nanorods. On these gold nanorod electrodes, bacteriophages had attached themselves. As *Escherichia coli* binds to the bacteriophage, the value of interfacial charge-transfer resistance (Rct) rises, reaches a maximum in about 25 to 30 minutes, and then falls as bacterial lysis happens due to phage invasion on the surface of the electrode. *S. aureus*, a non-target bacterium, showed no time-dependent changes in contrast. enabling detection threshold of 10^3 CFU/mL in roughly 100 L of bacterial solution, this method for detecting bacteria is very straightforward and reasonably priced.

8.5.3 Ionic Cytoplasm Substance Release-Based Detection

DNA, granules for storing nutrients, ribosomes, and numerous dissolved charged molecules make up the interior of bacteria, making it highly conductive (1 S/m). The impedance can be altered by disruption or the ions being released intracellularly into a medium with low conductivity. The detection of bacterial cells is done by using this change in impedance. For quality control and pathogen monitoring, several commercially available systems are used, including the Malthus ATanalyzer (Malthus Instruments), Bactometer (bioMerieux), and Bac Trac™.

8.6 Metal Nanoparticles Enabled Immunosensing to Identify Bacterial Pathogens

Electrochemical immunosensors are now more sensitive due to the extraordinary capacity of Au nanoparticles to amplify electrical signals by facilitating electron transfer from redox reactions inside the device. As a result, these sensors stand out as one of the most effective methods for creating highly sensitive and reliable detection devices. By expanding the electrode's surface area with Au NPs, more capturing antibody can be immobilized. Ag NPs were additionally employed for development of numerous electrochemical immunosensors in a manner similar to how the electric conductivity properties of the Au NPs and increased available surface area for immobilization (Figure 8.10).

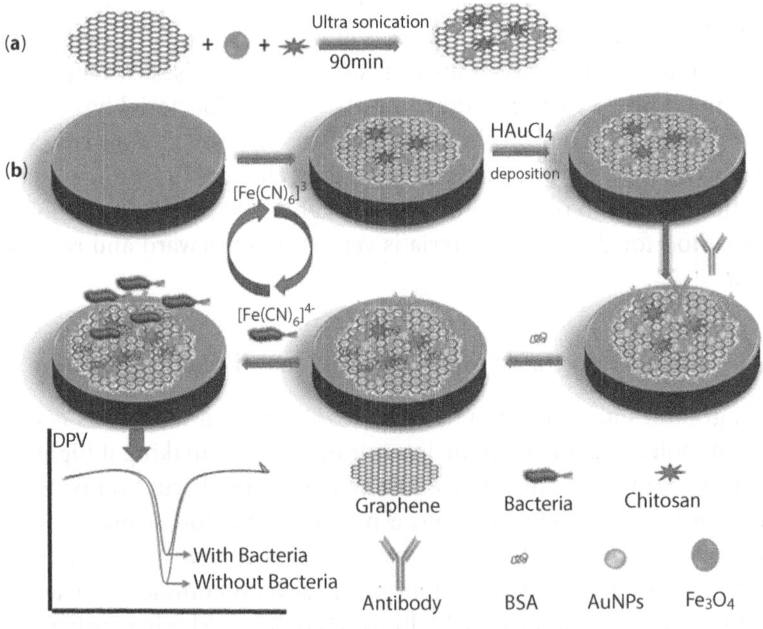

Figure 8.10 Diagram demonstrating the development of the fabrication of the (a) Fe3O4@graphene nanocomposite and the (b) electrochemical immunosensor, respectively. Differential pulse voltammetry = DPV. I = current (a); E = potential (V). Adopted source from Feng *et al.* [25].

8.6.1 *Escherichia coli*

T7 bacteriophages with the lacZ operon were adeptly used by Nugen *et al.* to infect *Escherichia coli* and cause excessive beta-galactosidase (-gal) expression [32]. The 4-aminophenyl-galactopyranoside (PAPG) would be the substrate that the β-gal would use to catalyse, releasing the electroactive species p-aminophenol that could be detected electrochemically. In order to identify *E. coli*, label-free electrochemical biosensors were also created. Malhotra *et al.* constructed an immune-electrode that realized that *E. coli* O157: H7 cells can be distinguished from the non-pathogenic *Escherichia coli* and other bacterial cells using graphene covered copper(II)-assisted cysteine hierarchical structure (rGO-CysCu) [48]. The immunoelectrochemical detection of *Escherichia coli* O157:H7 at levels of 90 CFU/mL was carried out using a neutral red functionalized reduced graphene oxide (rGO-NR Au@Pt) nanocomposite was used to modify the gold@platinum core@shell NP, which also exhibited high electrocatalytic activity [33].

Wang *et al.* [34] development of an innovative 3D Ag nanoflower served as the foundation for yet another label-free electrochemical biosensor. Upon capturing *Escherichia coli* O157:H7 using the biosensor, $[Fe(CN)_6]^{3/4}$ was employed as the redox probe to discern changes in resistance.

8.6.2 *Vibrio cholera*

Human cholera, one of the earliest and most prevalent epidemic diseases, is a result of the pathogen *Vibrio cholerae*. Numerous pandemics have been brought on by *Vibrio cholerae*, which is primarily known for its severe diarrhoea, water loss, vomiting, and a high mortality. As a result, it is included in the list of infectious diseases subject to international quarantine. In order to capture the targets, antibodies were covalently immobilized on a microelectrode modified with a CeO2 nanowire to create a label-free biosensor that only requires one step to detect *V. cholerae*. By using $[Fe(CN)_6]^{3-/4-}$ as the redox probe in an impedance analysis, the resulting biosensor was discovered [35].

8.6.3 *Bacillus cereus*

Close contact between humans and the species of the Bacillus genus known as *Bacillus cereus* (*B. cereus*) can result in food poisoning. This kind of diarrhoea can be brought on by numerous foods, especially leftovers that have been improperly stored in the refrigerator. Abdominal discomfort, nausea, and diarrhoea, which are symptoms of *Bacillus cereus*, are very similar to those of *Clostridium perfringens*. Additionally, it is more challenging to differentiate from other transient indications of deterioration (like those brought on by *Staphylococcus aureus*). Therefore, it is crucial to create a trustworthy method for identifying *Bacillus cereus* in food.

Monoclonal *B. cereus* antibodies were immobilized on gold nanoparticles with double-layer to capture the target, and also chitosan was employed to connect the sensing component with GCE in Liang *et al.* [36] novel *B. cereus* electrochemical sensor. The sensor showed rapid bacterial contamination detection, high sensitivity, and long-term stability. Amine *et al.* [37] created a *Bacillus anthracis* spores electrochemical biosensor without labels by modifying the electrode with pyrrole and using $[Fe(CN)_6]^{3/4}$ as a redox probe.

8.6.4 *Staphylococcus aureus*

Gram-positive *Staphylococcus aureus (S. aureus)* bacteria are widely distributed that can cause sepsis, pneumonia, pseudomembranous colitis, pericarditis, and other serious purulent infections in people. Here, in the United States and Canada, respectively, *Staphylococcus aureus* enterotoxin-related food poisoning causes among 33% and 45% of all food poisoning caused by bacteria. Additionally, there have been numerous instances of poisoning in China.

By adding insulating 6-mercaptohexadecanoid acid self-assembled molecular single layer (SAM) onto a gold electrode and attaching Particular antibodies against *S. aureus* for the next immunoreaction, it was possible to make an immunosensor impedimetric for measuring stressed and revived *S. aureus* [38].

Gram-positive bacteria belonging to the *Micrococcaceae* family include *S. aureus*. This pathogen has the potential to cause bone infections, heart valve infections, and pneumonia, all of which have high rates of mortality as well as morbidity.

Wei *et al.* [38] described an electrochemical sensor for the detection of *S. aureus* using reduced graphene oxide-gold nanoparticles (rGO-AuNP) nanocomposite and single-stranded DNA as an aptamer. Mansour *et al.* [39] method of detecting *S. aureus* by impedance spectroscopy also involved observing the change in resistance before and after the *S. aureus*, recognized by anti-*S. aureus*, immobilized on gold electrode. Ferri-/ferrocyanide was used as a redox probe. Further applications of the developed biosensor included the detection of stressed and revived pathogens. Connolly *et al.* [40] recently developed a low-cost screen-printed electrode for a *S. aureus* biosensor using impedance spectroscopy. The targets were

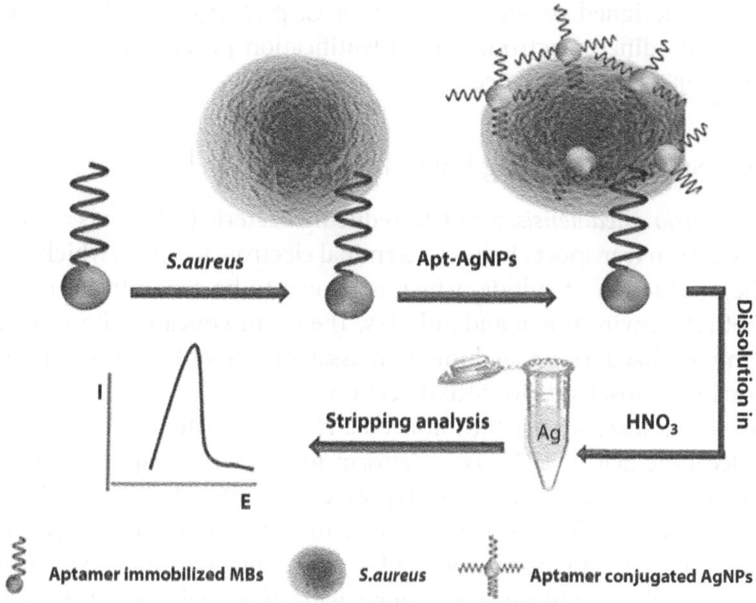

Figure 8.11 Illustration depicting the utilization of aptamer-conjugated silver nanoparticles (Ag NPs) for the electrochemical detection of *S. aureus*. Adapted source from Abbaspour *et al.* [42].

incubated in electrode-equipped chambers, and the results were examined using a novel method of analysis. Although impedance spectroscopy offers a label-free method, it still has a higher detection limit than other electrochemical biosensors. Methicillin-resistant *S. aureus* was obtained from patient nose swabs and captured and detected using a microfluidic device and magnetic nanoparticles with antibody functionalization [41]. Alkaline phosphatase-functionalized strain-specific antibodies for electrochemical detection is used to identify *S. aureus*, as shown in Figure 8.11 [42].

8.6.5 *Clostridium perfringens*

The predominant prevalent type of Clostridium found in clinically present genital gangrene pathogens is *Clostridium perfringens*. *C. perfringens* has the ability to degrade muscle sugar and connective tissue and then release a significant amount of gas. This severely emphysematizes the tissue, reduces blood flow, and eventually results in a significant area of necrotic tissue. The bacteria were given the name *C. perfringens* because it can form a capsule inside the body. Wang *et al.* [43] introduced an electrochemical biosensor designed to identify DNA of *C. perfringens,* utilizing CeO2/chitosan-modified electrodes. The identification process involved monitoring changes in impedance.

8.6.6 Sulfate-Reducing Bacteria

Desulforibrio caledoiensis, a sulfate-reducing bacteria (SRB), has a sulphate in its electron transport chain as a terminal electron acceptor, which results in the production of sulfide, which is known to be a significant problem for both the environment and industry. The lectin-concanavalin A (ConA) was immobilized for an agglutination assay by one of the earliest impedimetric immunosensors for SRB detection [44].

An immunosensor for SRB utilized a Ni foam coated with 11-mercaptoundecanoic acid (MPA) as a platform for bacterial trapping. Reduced graphene sheets (RGS) were also reported to be included in the SRB biosensor. These two-dimensional nanostructures are low manufacturing cost, a redox catalyst, and are biocompatible. The detection range of a chitosan nanocomposite film biosensor platform with RGS doping was from 1.8 10^2 to 1.8 10^7 CFU/mL. This RGS-based immunosensor produced results more quickly than previous attempts at SRB detection, and although the sensitivity was not enhanced, the assay could be conducted with a sample size as small as 10 L [45].

8.6.7 The Concurrent Detection of Several Pathogens

8.6.7.1 Streptococcus pyogenes, Salmonella typhimurium, and Pseudomonas aeruginosa

A Gram-positive bacteria called *Streptococcus pyogenes* causes cellulitis, scarlet fever (rash), pharyngitis, impetigo, and erysipelas. The described immunosensor underwent modification by depositing a polytyramine (Ptyr) layer on Dropsens gold SPEs. Afterward, biotin-neutravidin was used to attach biotin-tagged antibodies. The sample volume is 10 L, and the single-shot incubation method yielded a LOD of 10^2 cells in 30 minutes [46].

A Gram-negative bacteria called *Pseudomonas aeruginosa* is frequently discovered in tainted water. Human infections caused by this pathogen can show up as, among other things, respiratory infections, systemic infections, and urinary tract infections. An impedance-based immunosensor for *P. aeruginosa* has been immobilizing polyclonal antibodies directed against this pathogen on top of a screen-printed carbon electrode (SPCE) [54].

Table 8.2 Examples of metal nanoparticle-based impedimetric biosensor.

Analyst	Detection type	Materials	Performance	Reference
E. coli	Impedimetric	rGO-CysCu/Gold electrode	Linear range: 100–10^8 CFU/mL LOD: 3.8 CFU/mL Assay time: > 1 h	[48]
E. coli	Impedimetric	BSA-conjugated 3D Ag nanoflowers	Linear range: 3.0×10^2–3.0×10^8 CFU/Ml LOD: 100 CFU/mL	[34]
Vibrio cholerae	Impedimetric	CeO_2 nanowire-modified microelectrode	Linear range: 1.0×10^2–1.0×10^4 CFU/mL	[35]
B. cereus	Impedimetric	GNPs-sDNA-(nheA)/PGE	Sensitivity: 10^0 CFU/mL LOD: 9.4×10^{-12} mol/L	[49]
S. aureus	Impedimetric	Aptamer/rGO-AuNP/GCE	Linear range:10–10^6 CFU/mL LOD: 10 CFU/mL (S/N=3) Assay time: < 1 h	[50]
S. aureus	Impedimetric	MPA/gold electrode	Linear range: 10^1–10^7 CFU/mL LOD: 10 CFU/mL	[39]
S. aureus	Impedimetric	screen printed electrode	Linear range: 3.6×10^7–9.3×10^7 CFU/mL Assay time: approximately 30 min	[51]

(*Continued*)

Table 8.2 Examples of metal nanoparticle-based impedimetric biosensor. (*Continued*)

Analyst	Detection type	Materials	Performance	Reference
DNA of *C. perfringens*	Electro-chemiluminescence	gold electrode (rolling circle amplification)	LOD: 10^{-15} M Assay time: approximately 1 h	[33]
C. perfringens	Impedimetric	CeO2/chitosan/GCE	Linear range: 1.0 $\times 10^{-14}$–1.0 $\times 10^{-7}$ mol/L LOD: 7.06 $\times 10^{-15}$ mol/L	[43]
E. coli	Electrochemical immunosensors	AuAgNPs	10^2 CFU/mL	[46]
Salmonella typhimurium	Electrochemical immunosensors	AuNPs	7.7 cells/mL	[47]
E. coli O157:H7	Electrochemical immunosensors	rGO-NR Au@Pt	91 CFU/mL	[52]
E. coli O157:H7	Electrochemical immunosensors	AuNPs	457 CFU/mL (in minced beef) 309 CFU/mL (in tap water)	[53]

De Oliveira *et al.* [47] reported a disposable microfluidic device with a limit of detection as low as 7 cells/mL for detecting *Salmonella typhimurium* using a magneto-immunoassay with both Au NPs and magnetic particles linked to specific antibodies (Table 8.2).

8.7 Conclusion

Nanomaterials exhibit unique sensory attributes, with metal nanoparticles (primarily Au and Ag) and carbon nanotubes (CNTs) being extensively employed in this domain. The emergence of novel nanostructures like graphene and its derivatives is poised to drive future advancements, owing to their distinctive surface characteristics. Although nano-based impedimetric biosensors offer a variety of benefits, they also have some drawbacks. There have been hundreds of publications along with two decades of research; however, no impedance-based biosensor has been successful in the marketplace so far. The development of commercial nanomaterial-influenced impedimetric biosensors to identify the aquatic pathogenic bacteria should therefore continue. This will call for improved stability, decreased volume, increased sensitivity, and reduced costs. Enhancing the capability to detect intricate matrices, minimizing expenses, and devising portable functionalities are prospective objectives for the future advancement of nanosensor design and fabrication. Furthermore, economic factors ought to be taken into account when these sensors are put on the market and used on a larger scale by the food industry and regulatory groups that keep an eye on the quality and safety of food.

Acknowledgement

The Rashtriya Uchchatar Shiksha Abhiyan (MKU-RUSA) Project of Madurai Kamaraj University provided full support for the development of this book chapter under the terms of the sanction order PF.No.003/RUSA/MKU/2020-2021, dated 24.12.2020.

References

1. FAO, *The State of World Fisheries and Aquaculture 2022. Towards Blue Transformation*, FAO, Rome, 2022, https://doi.org/10.4060/cc0461en.
2. Leung, T.L.F. and Bates, A.E., More rapid and severe disease outbreaks for aquaculture in the tropics: Implications for food security. *J. Appl. Ecol.*, 50, 1, 215–222, 2012. doi:10.1111/1365-2644.12017.
3. Stentiford, G.D., Sritunyalucksana, K., Flegel, T.W., Williams, B.A.P., Withyachumnarnkul, B., Itsathitphaisarn, O. *et al.*, New Paradigms to Help Solve the Global Aquaculture Disease Crisis. *PloS Pathog.*, 13, 2, e1006160, 2017.
4. Venugopal, V., Biosensors in fish production and quality control. *Biosens. Bioelectron.*, *17*, 3, 147–157, 2002.
5. Scallan, E., Hoekstra, R.M., Angulo, F.J., Tauxe, R.V., Widdowson, M.-A., Roy, S.L., Griffin, P.M., Foodborne Illness Acquired in the United States—Major Pathogens. *EID*, *17*, 1, 7–15, 2011, doi: 10.3201/eid1701.p11101.
6. Guan, J. and Levin, R.E., Quantitative detection of *Escherichia coli* O157:H7 in ground beef by the polymerase chain reaction. *Food Microbiol.*, *19*, 2–3, 159–165, 2002. doi:10.1006/fmic.2001.0477.
7. Katz, E. and Willner, I., Probing biomolecular interactions at conductive and semiconductive surfaces by impedance spectroscopy: Routes to impedimetric immunosensors, DNA-Sensors, and enzyme biosensors. *Electroanalysis*, *15*, 11, 913–947, 2003. doi:10.1002/elan.200390114.

8. Yadav, M., Singh, G., Lata, S. Revisiting Some Recently Developed Conducting Polymer@Metal Oxide Nanostructures for Electrochemical Sensing of Vital Biomolecules: A Review. *J. Anal. Test.*, 6, 274–295, 2022.
9. Ding, L., Bond, A.M., Zhai, J., Zhang, J., Utilization of nanoparticle labels for signal amplification in ultrasensitive electrochemical affinity biosensors: a review. *Anal. Chim. Acta*, 797, 1–12, 2013. doi:10.1016/j.aca.2013.07.035.
10. Tiwari, J.N., Tiwari, R.N., Kim, K.S., Zero-dimensional, one-dimensional, two-dimensional and three-dimensional nanostructured materials for advanced electrochemical energy devices. *Prog. Mater. Sci.*, *57*, 4, 724–803, 2012.
11. Tombelli, S., Minunni, M., Mascini, M., Analytical applications of aptamers. *Biosens. Bioelectron.*, 20, 12, 2424–2434, 2005.
12. Shukla, S.K., Advances in Biomembranes and Lipid Self-Assembly. *Polymeric Micellar Structures for Biosensor Technology*, 24, 143–161, 2016. doi:10.1016/bs.abl.2016.04.005.
13. Malhotra, B. D. and Ali, M. A., Nanomaterials in Biosensors: Fundamentals and Applications. *Nanomater. Biosens.*, 1–74, 2017, https://doi.org/10.1016/B978-0-323-44923-6.00001-7.
14. Naresh, V. and Lee, N., A Review on Biosensors and Recent Development of Nanostructured Materials-Enabled Biosensors. *J. Sens.*, *21*, 4, 1109, 2021.
15. Hoyos-Nogués, M., Gil, F.J., Mas-Moruno, C., Antimicrobial Peptides: Powerful Biorecognition Elements to Detect Bacteria in Biosensing Technologies. *Molecules (Basel, Switzerland)*, 23, 7, 1683, 2018, https://doi.org/10.3390/molecules23071683.
16. Van Dorst, B., Mehta, J., Bekaert, K., Rouah-Martin, E., De Coen, W., Dubruel, P., Robbens, J., Recent advances in recognition elements of food and environmental biosensors: A review. *Biosens. Bioelectron.*, *26*, 4, 1178–1194, 2010.
17. Velusamy, V., Arshak, K., Korostynska, O., Oliwa, K., Adley, C., An overview of foodborne pathogen detection: In the perspective of biosensors. *Biotechnol. Adv.*, *28*, 2, 232–254, 2010.
18. Banerjee, P. and Bhunia, A.K., Cell-based biosensor for rapid screening of pathogens and toxins. *Biosens. Bioelectron.*, *26*, 1, 99–106, 2010.
19. Jamal, M., Bukhari, S.M., Andleeb, S., Ali, M., Raza, S., Nawaz, M.A., Shah, S.S., Bacteriophages: an overview of the control strategies against multiple bacterial infections in different fields. *Tetrahedron Lett.*, 59, 2, 123–133, 2019.
20. Andersson, L., Sellergren, B., Mosbach, K., Imprinting of amino acid derivatives in macroporous polymers. *Tetrahedron Lett.*, *25*, 45, 5211–5214, 1984.
21. BelBruno, J. J., Molecularly Imprinted Polymers. *Chem. Rev.*, 119, 1, 94–119, 2019, https://doi.org/10.1021/acs.chemrev.8b00171.
22. Uzun, L. and Turner, A. P., Molecularly-imprinted polymer sensors: realising their potential. *Biosens. Bioelectron.*, 76, 131–144, 2016, https://doi.org/10.1016/j.bios.2015.07.013.

23. Ahmad, O.S., Bedwell, T.S., Esen, C., Garcia-Cruz, A., Piletsky, S.A., Molecularly Imprinted Polymers in Electrochemical and Optical Sensors. *Trends Biotechnol.,* 37, 3, 294–309, https://doi.org/10.1016/j.tibtech.2018.08.009.
24. Wu, L., Li, X., Miao, H., Xu, J., Pan, G., State of the art in development of molecularly imprinted biosensors. *View, 3*, 3, 20200170, 2022.
25. Feng, K., Li, T., Ye, C., Gao, X., Yue, X., Ding, S., Zhang, J., A novel electrochemical immunosensor based on Fe3O4@graphene nanocomposite modified glassy carbon electrode for rapid detection of Salmonella in milk. *Int. J. Dairy Sci., 105*, 3, 2108–2118, 2022.
26. Shabani, E., Dowlatshahi, S., Abdekhodaie, M.J., Laboratory detection methods for the human coronaviruses. *Eur. J. Clin. Microbiol., 40*, 2, 225–246, 2021.
27. Palchetti, I. and Mascini, M., Nucleic acid biosensors for environmental pollution monitoring. *Analyst, 133*, 7, 846–854, 2008.
28. Yang, L. and Bashir, R., Electrical/electrochemical impedance for rapid detection of foodborne pathogenic bacteria. *Biotechnol. Adv., 26*, 2, 135–150, 2008.
29. Gómez-Sjöberg, R., Morisette, D. T., Bashir, R., Impedance microbiology-on-a-chip: Microfluidic bioprocessor for rapid detection of bacterial metabolism. *J. Microelectromech. Syst.*, 14, 4, 829–838, 2005, https://doi.org/10.1109/JMEMS.2005.845444.
30. Malvano, F., Albanese, D., Crescitelli, A., Pilloton, R., Esposito, E., Impedimetric label-free immunosensor on disposable modified screen-printed electrodes for ochratoxin A. *Biosensors, 6*, 3, 33, 2016.
31. Moghtader, F., Congur, G., Zareie, H.M., Erdem, A., Piskin, E., Impedimetric detection of pathogenic bacteria with bacteriophages using gold nanorod deposited graphite electrodes. *RSC Adv., 6*, 100, 97832–97839, 2016.
32. Chen, J., Alcaine, S.D., Jiang, Z., Rotello, V.M., Nugen, S.R., Detection of Escherichia coli in Drinking Water Using T7 Bacteriophage-Conjugated Magnetic Probe. *J. Anal. Chem., 87*, 17, 8977–8984, 2015.
33. Zhu, F., Zhao, G., Dou, W., Electrochemical sandwich immunoassay for *Escherichia coli* O157:H7 based on the use of magnetic nanoparticles and graphene functionalized with electrocatalytically active Au@Pt core/shell nanoparticles. *Microchim. Acta*, 185, 10, 455, 2018, https://doi.org/10.1007/s00604-018-2984-2.
34. Huang, H., Liu, M., Wang, X., Zhang, W., Yang, D.P., Cui, L., Wang, X., Label-free 3D Ag nanoflower-based electrochemical immunosensor for the detection of Escherichia coli O157: H7 pathogens. *NRL, 11*, 1, 1–8, 2016.
35. Tam, P.D. and Thang, C.X., Label-free electrochemical immunosensor based on cerium oxide nanowires for *Vibrio cholerae* O1 detection. *Mat. Sci. Eng. C-Mater. Biol. Appl.*, 58, 953–959, 2016, https://doi.org/10.1016/j.msec.2015.09.027.

36. Kang, X., Pang, G., Chen, Q., Liang, X., Fabrication of Bacillus cereus electrochemical immunosensor based on double-layer gold nanoparticles and chitosan. *Sens. Actuators, B, 177*, 1010–1016, 2013.
37. Ait Lahcen, A., Arduini, F., Lista, F., Amine, A., Label-free electrochemical sensor based on spore-imprinted polymer for Bacillus cereus spore detection. *Sens. Actuators B: Chem.*, 276, S0925400518314618, 114–120, 2018, doi: 10.1016/j.snb.2018.08.031.
38. Jia, F., Duan, N., Wu, S., Ma, X., Xia, Y., Wang, Z., Wei, X., Impedimetric aptasensor for *Staphylococcus aureus* based on nanocomposite prepared from reduced graphene oxide and gold nanoparticles. *Microchim. Acta*, 181, 9–10, 967– 974, 2014, doi: 10.1007/s00604-014-1195-8.
39. Bekir, K., Barhoumi, H., Braiek, M., Chrouda, A., Zine, N., Abid, N., Mansour, H.B., Electrochemical impedance immunosensor for rapid detection of stressed pathogenic Staphylococcus aureus bacteria. *ESPR, 22*, 20, 15796–15803, 2015.
40. Ward, A.C., Hannah, A.J., Kendrick, S.L., Tucker, N.P., MacGregor, G., Connolly, P., Identification and characterisation of Staphylococcus aureus on low cost screen printed carbon electrodes using impedance spectroscopy. *Biosens. Bioelectron.*, 110, 65–70, 2018, https://doi.org/10.1016/j.bios.2018.03.048.
41. Nemr, C.R., Smith, S.J., Liu, W., Mepham, A.H., Mohamadi, R.M., Labib, M., Kelley, S.O., Nanoparticle-mediated capture and electrochemical detection of methicillin-resistant Staphylococcus aureus. *J. Anal. Chem., 91*, 4, 2847–2853, 2019.
42. Abbaspour, A., Norouz-Sarvestani, F., Noori, A., Soltani, N., Aptamer-conjugated silver nanoparticles for electrochemical dual-aptamer-based sandwich detection of staphylococcus aureus. *Biosens. Bioelectron, 68*, 149–155, 2015, doi: 10.1016/j.bios.2014.12.040.
43. Qian, X., Qu, Q., Li, L., Ran, X., Zuo, L., Huang, R., Wang, Q., Ultrasensitive electrochemical detection of Clostridium perfringens DNA based morphology-dependent DNA adsorption properties of CeO2 nanorods in dairy products. *J. Sens., 18*, 6, 1878, 2018.
44. Wan, Y., Zhang, D., Hou, B., Monitoring microbial populations of sulfate-reducing bacteria using an impedimetric immunosensor based on agglutination assay. *Talanta*, 80, 1, 218–223, 2009, https://doi.org/10.1016/j.talanta.2009.06.057.
45. Wan, Y., Lin, Z., Zhang, D., Wang, Y., Hou, B., Impedimetric immunosensor doped with reduced graphene sheets fabricated by contollable electrodeposition for the non-labelled detection of bacteria. *Biosens. Bioelectron.*, 26, 5, 1959–1964, 2011.
46. Ahmed, A., Rushworth, J.V., Wright, J.D., Millner, P.A., Novel impedimetric immunosensor for detection of pathogenic bacteria *Streptococcus pyogenes in* human saliva. *Anal. Chem.*, 85, 24, 12118–12125, 2013.

47. Oliveira, T.R., Martucci, D.H., Faria, R.C., Simple disposable microfluidic device for Salmonella typhimurium detection by magneto-immunoassay. *Sens. Actuators B-Chem.*, 255, 684–691, 2018.
48. Pandey, C.M., Tiwari, I., Singh, V.N., Sood, K.N., Sumana, G., Malhotra, B.D., Highly sensitive electrochemical immunosensor based on graphene-wrapped copper oxide-cysteine hierarchical structure for detection of pathogenic bacteria. *Sens. Actuators B: Chem.*, 238, 1060–1069, 2017, doi: 10.1016/j. snb.2016.07.121.
49. Izadi, Z., Sheikh-Zeinoddin, M., Ensafi, A.A., Soleimanian-Zad, S., Fabrication of an electrochemical DNA-based biosensor for Bacillus cereus detection in milk and infant formula. *Biosens. Bioelectron.*, 80, 582–589, 2016, https://doi.org/10.1016/j.bios.2016.02.032.
50. Zhang, Z., Zhou, J., Du, X., Electrochemical biosensors for detection of food-borne pathogens. *Micromachines*, *10*, 4, 222, 2019.
51. Khue, V.Q., Huy, T.Q., Phan, V.N., Tuan-Le, A., Le, D.T.T., Tonezzer, M., Hanh, N.T.H., Electrochemical stability of screen-printed elec- trodes modified with Au nanoparticles for detection of methicillin-resistant Staphylococcus aureus. *Mater. Chem. Phys.*, *255*, 123562, 2020.
52. Russo, L., Leva Bueno, J., Bergua, J.F., Costantini, M., Giannetto, M., Puntes, V., de La Escosura-Muñiz, A., Merkoçi, A., Low-Cost Strategy for the Development of a Rapid Electrochemical Assay for Bacteria Detection Based on AuAg Nanoshells. *ACS Omega*, 3, 18849–18856, 2018, https://doi.org/10.1021/acsomega.8b02458.
53. Hassan, A. R., de la Escosura-Muñiz, A., Merkoçi, A., Highly sensitive and rapid determination of Escherichia coli O157:H7 in minced beef and water using electrocatalytic gold nanoparticle tags. *Biosens. Bioelectron.*, 67, 511–515, 2015, https://doi.org/10.1016/j.bios.2014.09.019.
54. Bekir K, *et al.* An investigation of the well-water quality: immunosensor for pathogenic Pseudomonas aeruginosa detection based on antibody-modified poly(pyrrole-3 carboxylic acid) screen-printed carbon electrode. *Environ. Sci. Pollut. Res.*, 22, 23, 18669–18675, 2015.

9

Properties and Applications of Dendrimers: A New Class of Polymers

Chenthis Kanisha Chelliah[1], Manavalan Murugan[2*], Vinod S. Undal[3†], Manish R. Ahir[4], Chackaravarthy Gnanasekaran[5], Ramachandran Govindan[5], Rajivgandhi Govindan[6‡] and Franck Quero[6]

[1]*Department of Nanotechnology, Noorul Islam Centre for Higher Education, Kumaracoil, Kanyakumari, Tamil Nadu, India*
[2]*Department of Biomedical Engineering, Noorul Islam Centre for Higher Education, Kumaracoil, Kanyakumari, Tamil Nadu, India*
[3]*Department of Botany, Ghulam Nabi Azad College, Barshitakali Dist-Akola, Maharashtra, India*
[4]*Department of Microbiology, Ghulam Nabi Azad College, Barshitakali Dist-Akola, Maharashtra, India*
[5]*Marine Pharmacology and Toxicology Laboratory, Department of Marine Science, Bharathidasan University, Tiruchirappalli, Tamil Nadu, India*
[6]*Laboratorio de Nanocelulosa y Biomateriales, Departamento de Ingeniería Química, Biotecnología y Materiales, Facultad de Ciencias Físicas y Matemáticas, Universidad de Chile, Santiago, Chile*

Abstract

Recent years, nanocarriers are heightened worldwide in the research field of drug delivery approaches due to the ability to improve the nanomaterial or other drugs. The ability to transport the drug molecules twice at their site of action and it is a promising factor in drug discovery research. Importantly, nanocarriers are excellent carrier molecules having excellent biocompatibility, reduced toxicity, biodegradability, prolonged drug circulation, enhanced solubility, responsive stimuli nature and target site drug delivery. Among the nanocarriers, dendrimers are the recently emerged successful nanocarriers which have been used in various

**Corresponding authors*: muruganbt@niuniv.com
†*Corresponding author*: molbio.vinod@gmail.com
‡*Corresponding author*: rajivgandhimicro@yahoo.com

Sathish-Kumar Kamaraj, Arun Thirumurugan, Muthuchamy Maruthupandy, Mercedes Guadalupe López Pérez and Shanmuga Sundar Dhanabalan (eds.) Nanoscience and Nanotechnology for Smart Prevention, Diagnostics and Therapeutics: Fundamentals to Applications, (217–242) © 2024 Scrivener Publishing LLC

research filed including biomedicine, biopharmaceutical, agrochemical industry, food industry and textile industry. Recent years, dendrimers based drug delivery molecules are heightened worldwide and also act as alternative drug delivery molecules, because of enhanced stability, solubility, encapsulation efficiency, maintaining of therapeutic concentration and controlled drug delivery system. Based on the above said information, the current study is highly concentrated on dendrimers research, especially; polymer based dendrimers and their progress, properties and applications in various fields.

Keywords: Dendrimers, polymers, drug delivery, drug release capacity, biological applications, tumor therapy, gene therapy, cancer therapy

9.1 Introduction

The word 'dendrimers' arrives from two Greek remarks "dendron" which indicates tree and "meros" which denotes fractions, thus referring to the characteristic tree-like form of these compounds [1]. It is a globular macromolecules containing 1-100 nm range and their morphology has three diverse domain including 1. Inner core, 2. Hyper branched mantle, 3. Corona with marginal reactive functional clusters [2]. Last 25 years, the development of polymers based research is very high in research field. In polymers research, more than 50,000 patents and published literature evidences are available till-date. Based on the mono disperse nature, elevated symmetrical and exterior polyvalence of the dendrimer has diverse nature. The frequent enlargement effects throughout dendrimer production lead to superior production and amount of divisions, ultimately forming a three-dimensional globular makeup. The separate synthetic procedure formulates dendrimer possess a definite interior–shell architecture and contracted polydispersity. The formation of dendrimers having multiple motifs in nature and these are most frequent one in application filed based on the functions and enhancement. The dendrimers are enormously divided, globular, multivalent, mono-dispersed fragment molecules with synthetic flexibility and numerous probable appliances [3].

Fragment rich dendrimers are synthetic polymers assembled by branched units that arisen from a crucial point and received the anionic, neutral and cationic terminal functionalities in more number on exterior portion which leads to a formation of hydrophilic or hydrophobic nature [4]. Nanometric fragment of the dendrimers are readily balanced, globular, mono dispersed nature and also homogenous nature. Based on the low solubility and short life, most of the drugs are ineffective and restricted. Subsequently, there is an emerging need to develop a novel route with better efficiency and deliver the drug completely in to the targeted places. More drugs are failure due to their inability to deliver the drug in target sites and less drug releasing efficiency and very low number of successful rate. So, the current drug delivery model is overcome these problems and concentrating on drug delivery efficiency and releasing efficiency. The nanoparticles are most supported to deliver the drug supply and this platform used to transfer the fundamental properties of drug fragments through nature of solubility, short life time, biocompatibility and its release characteristics [5]. Recent years, more nanoparticle mediated therapeutic products are opened in the market and few are in under medical and preclinical evaluation. The Dendrimers are very different compared with polymers due to its nature of dimension, so it is frequently utilized by nanomedicine investigation. It is found to be valuable as release and transport the drugs, genes and other materials have exhibited a few dendrimers. Sometimes, it is used their own frequent approach through biological properties including antifungal, antibacterial and cytotoxic properties [6]. Based on the low solubility, immovable problem and high toxicity nature of most of the drugs could not be exploited. So, the dendrimers are the best choice to overcome this problem and improved their clinical evaluations and applications [7].

9.2 Archives of Dendrimers

Recent years, the advantages of plant and animals natures are frequently helped to generate the dendritic approaches to improve the drug nature and properties as applied in respiratory tract of the animals. As same as, the soil plants are also more helped to dendritic features to synergy and reason for the foliage to sunlight. In addition, the depth stage of land needs more requirements to expose a big functional surface when accumulating the water from the soil. Likewise, dendrimers acted as a linked material and accumulating the molecules and divergent the nanoparticles [8]. The dendritic blue print composition and their differentiation are previously reported by Fritz Vögtle and coworkers in 1978, followed by Denkewalter at 1981, Donald Tomalia at Dow Chemical's at 1983 and George Newkome at 1985. The synthetic convergent approach is reported by Frechet at 1990 and many researchers are followed in the same kind of research. In their research, the dissimilar properties and applications of dendrimers investigations are heightened. Finally, the researchers are concluded that the characteristic and various other applications of the research still not understood well. In 1979 to 1985, Donald A. Tomalia and co-workers are synthesized the dendrimers at the Dow Laboratories and it is great breakthrough worldwide [9]. Researchers made thin, vacant core and tendril polymers consisting with divided superficial format, which divide one from another, it is constrained and predictable approaches named as Tomalia and called as dendrimers.

9.3 Dendrimers as Drug Delivery Vehicles

The various beneficial mediators suffer from a range of restrictions resembling little solubility in aqueous nature and short life time. Hence, dendrimers are more important to blue print delivery method, which able to transport a drugs competently [10]. The traditional medicine release methods are valuable to a little degree in overcoming these concerns, however they have unsuccessful to verify their efficacy in transporting several medicines. The drug delivery approach through nanoparticle provide new platform to alter the necessary of remedy molecules including solubility, half life time, bioavailability, biodegradability, biocompatibility and discharged characteristic nature [5].

The dendrimers have materialized as an essential cluster of nano controlled transporters for the improvement of nano drug to care for a range of diseases. Because, the dendrimers have structural differences and more flexibility nature which can be employed to transport the drugs and genes in more dissimilar means. The hydrophopic interior and hydrophilic periphery of the dendrimers could behave similar to unimolecular micelles and it has been employed based on the solubilize the hydrophobic drugs by entrapping them in trans molecular opening [11]. Sometimes, the cationic dendrimers are used as a nonviral gene transporter through conjugation process [12, 13]. Previously, Liu *et al.* [14] worked with polyethylene glycol, polysaccharides which is polymers performed the conjugation of dendrimers.

9.4 Interactions Between Drug Molecules and Dendrimers

With organic composites a dendrimer can intermingle in numerous manners, such as drugs. The physical and the chemical are the two main type's interactions of dendrimer. They could involve uncomplicated encapsulation, hydrophobic hydrogen bonding. The physical interaction is a non-bonding interaction a mid-particular clusters within dendrimer and the drug, as well as refers to the uncomplicated encapsulation of the molecule [10]. The vacant interior cavities have hydrophobic assets, which build probable its interaction by means of weakly soluble drugs through hydrophobic interactions [15]. Recently, the hydrophilic natures of the polyester-co-polymer mediated dendrimers are prepared with more bio-compatible nature including butanetetracarboxylic and aspartic acid. In addition, the new dendrons of polyethylene glycol, monomethacrylate and dihydroxybenzoic acid are developed. The loading of rhodamine and β-carotene to polyester-co-polyether dendrimers can be attained in acceptable style and these dendrimers can be effective drug release systems [16].

9.5 Properties of Dendrimers

The properties of dendrimers are made by various properties, in particular the chemical affectability, solubility and glass conversion temperature of the materials are very important dendrimers. This is for the reason that the radially enlarge in number of terminal clusters through the raise in production number of the dendrimers in contrast to linear polymers having simply twice terminal groups [17]. Based on the globular outline of the dendrimers, the physical and chemical performance of the dendrimers are available with linear macromolecules. Based on the production numbers belongs to the enlarged characteristic nature, the soluble dendrimers information are available in Table 9.1. Nevertheless, the reactivity and solubility of the dendrimers particularly in case of globular shape molecules by the nature of the marginal clusters [17].

Typically, dendrimers have small intrinsic viscosity [17]. Some of the dendrimers are having considerable importance that related to more contrast with conventional polymers including particular number of marginal groups, formal cell take-up, the competence between the physical affection to ligands, numerous active spots and structural adopting. These properties contain in comparable molecular homogeneity, multifunctional exterior and the existence of inner cavities in addition to further precise properties to be appropriate for a variety of high technology uses [18].

9.6 Factors Affecting the Properties of Dendrimers

Exposed dendritic structural design of dendrimer supplies prospective of the drugs are reviewed based on the liability and week soluble nature

Table 9.1 Properties of dendrimer.

S. no.	Properties	Dendrimer
1	Organization	Compact and Globular
2	Outline	Spherical
3	Structural design	Normal
4	Structural control	Extremely high
5	Synthesis	Stepwise growth
6	Crystallinity	Noncrystalline and amorphous materials Lower glass temperatures
7	Reactivity	High
8	Aqueous solubility	High
9	Nonpolar solubility	High
10	Stickiness	Nonlinear association with molecular mass
11	Ionic conductivity	High
12	Compressibility	Low
13	Polydispersity	Monodisperse

of the drugs. More researchers are reported that the encapsulation and complex of the various drug molecules are depends on the dendrimers. Importantly, the drug delivery and releasing the drugs based on the solubilization are very important in the dendrimers. It is depending on various parameters including pH, production of dendrimers, interior and exterior dendrimer, and concentration of dendrimers in liquid solution can be influence the solubilization and all the parameters are available below (Figure 9.1).

9.6.1 Consequence of pH

Investigation of the configurationally organization of PAMAM dendrimers as a role of pH, with concerning molecular activities indicated that the dendrimer have an comprehensive structure, dependent on an extremely controlled organization at little down pH (pH<4). With indicated pH, the inner is receiving more and more "vacant" as the production quantity enlarges with consequence of detestation among the certainly stimulating amines together with dendrimer exterior, however the third amines in center. At this pH, the conformation has a superior level of back-folding as a result of the weak "inter-dendron" repulsive forces [19].

9.6.2 Effect of Solvent

The ability to prepare the dendrimers from solvent to solvate is the important process and extremely important constraint and it is the most reliable confirmation for dendrimers. With the declining solvent superiority, i.e. diminishing salvation, the dendrimers of the entire productions usually

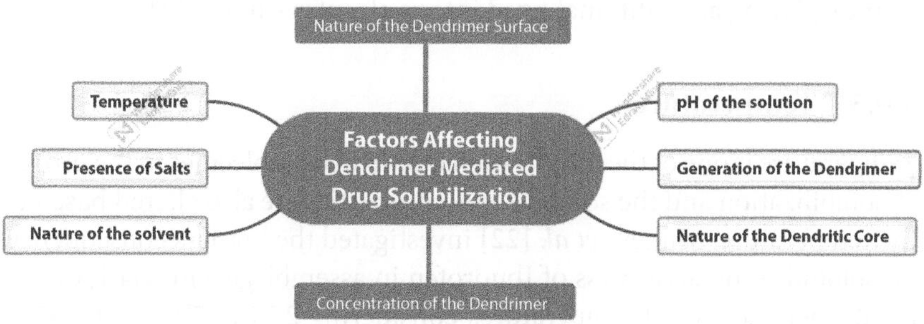

Figure 9.1 Influence of dendrimers medicated drug solubilization by various factors.

experience a bigger amount of back-folding [20]. However, being extra elastic, the little production dendrimers demonstrate the maximum affinity headed for back-overlaps, with the outcome of deprived solvation contrasted to the superior making dendrimers. The PPI based dendrimers by *nuclear magnetic resonance* investigation suggested that the benzene of nonpolar liquid nature and partially solvate dendrons are more supporting to the exchange of intramolecular of the dendrimers.

9.6.3 Effect of Salt

Salt concentration based ionic potency of dendrimers has very resistant consequence on charged PPI dendrimers, and arrangement of the dendrimers, it depends on the highest level of the back folding to certain analogous. These all are depend on the pH or decreased solvation. At little salt circumstances, the revolting forces among the charged dendrimer fragments outcome in an extensive arrangement in order to decrease charge repulsion in the makeup [19].

9.6.4 Effect of Concentration

Dendrimers arrangement are not only depends on their outer shells of the small fragments including salts, protons, solvents. Also, dendrimers depends on some of the larger substances including molecular thickness which arranged each other with larger consequences. The X-ray scattering analysis is used to monitor the PPI based dendrimers in polar liquid nature including methanol and it detected that the molecular identification of the dendrimers upon growing concentration turn into more and more reduced. These molecular reduction could shrink revolting strengths among dendrimer fragments with boosting capability of a dendrimers near displaying an additional fixed intermolecular stuffing [21].

9.6.5 Temperature

Temperature is one of the important factor in drug molecules in the process of solubilization and the solubilization properties are also altered based on the temperature. Milhem *et al.* [22] investigated the hotness consequences on solubilization activeness of Ibuprofen in assemblage with G4 PAMAM dendrimers at several temperatures considering 27°C, 35°C, 40°C, 45°C, and 50°C. When the time of temperature increasing is exaggerated, the solvability of the drug changed their nature, so the specific temperature is needed to get the exact drug efficiency. Wang *et al.* [23], investigated the capability of magnetic poly-(methyl acrylate-divinyl benzene) microspheres through amine ended dendrimers on the periphery for adsorbing hexavalent chromium and documented that the adsorption competence of the microspheres enlarged through increasing hotness. The metal ion chromium is the important study, also very tough and utilized by various processes for adsorption process than desolation. So, the frequency of the temperature-based solubilization model of the dendrimers study is still not understood well and also need more investigation, and more attempts are secured to interpret the concerned performances.

9.7 Reasons Influencing Drug Solubilization and Release

It supplies chances for interaction with labile or inadequately soluble drugs by the open dendritic design of dendrimer. The dendrimer based encapsulation and complexation process of the drug molecules have been optimized and discovered by several researchers. Despite the facts, drug solubilization, and delivered the drugs with dendrimers which striking prospective pertinent to a drug in broad range are used. Then, the numerous aspects of the dendrimers based rug preparation including temperature, pH, composite solution, production, exterior, interior, and concentration. The solubilization based dendrimers preparation model was available in Figure 9.1. In the functions of dendrimers, the hydrophobes exchange and exterior amino groups are concentrated more in the progress of drug solubility, delivered the drugs, pH, temperature and some other sources. Based on the pH, the marginal and internal dendrimers are constructed. It accounted a relative enhance in the loading competence of flurbiprofen in water by means of Polyamidoamine (PAMAM) dendrimer, maximum at pH 10, transitional at pH 7, and the minimum at pH 2 [24]. A number of investigations have accounted a significant function of pH in the solubility of different drugs, including ibuprofen, ketoprofen, and nicotinic acid [25, 26], when used in combination with PAMAM dendrimers. Moreover, it has been documented that the potential mechanisms behind the solubility improvement of ibuprofen and ketoprofen are their ionization and electrostatic exchanges with dendrimer exterior amines, which are pH reliant [22].

9.8 Current Marketing Status of Dendrimers

Based on the recent reports, the solubilization of drugs using dendrimers and market able applications of the end product is not exposed well due to the decreased rate of drug release efficiency and improvement. As per the previous information, the drug booster level of the dendrimers research is not succeed in market and pharmaceutical industry, however, the more number of other dendrimers based products are available in the market for used as a various applications. The most available market product of the dendrimer made with mucoadhesive gel named as VivaGel® used for bacterial infection of vaginosis. Also, it is reported that the VivaGel® is an excellent anti-viral and anti-bacterial agent. In bacterial infection, vaginosis is one of the most dangerous infection, and the dendrimer based existing VivaGel® as excellent inhibitor for this dangerous infections and also, the mechanism of VivaGel® is not defined properly with any researchers. Also, the outcome of the VivaGel® is also not validated by existing reports. As per report, the markedly available VivaGel® is the only product developed based on the dendrimers. In addition, some of the other dendrimer-based products are also available, such as SuperFect® and PrioFect®, and these are all used for the molecular and cellular studies. Another one of Stratus® CS is used for the cardiac biomarker in the form of analytical system. Additionally, Alert Ticket®-a is used in the diagnostic system for anthrax, Starburst® and Priostar® are commercially available dendrimers and also available in market for applied in various applications as well as various discipline.

Nearly, 40 different polymer conjugated medicines are progressed in medical and premedical phase stage. There are 19% of phase II and III reports are currently available and 26$ are in pipeline which captured by phase I and II. There are various other dendrimers based drugs are in progress to apply in various infections. This domain is progressively advantage to different expertise and considerable approaches with medicinal properties. It is important system used in the field of drug delivery, and one of the key areas to detect the diseases such as MRI imaging, targeted therapy, tissue regeneration, cell repair and vaccine improvements.

9.9 Structure and Chemistry of Dendrimers

The interior portions of the dendrimers are arranged by central atoms or cluster of the atoms which used to initiate the fragments in dendrimers

(Figure 9.2). With the middle makeup, the divisions of further atoms known 'dendrons' develop with a range of chemical reactions. There carry on to be a debate concerning the accurate configuration of dendrimers, in exacting whether they are entirely extended with highest compactness at the exterior or whether the terminal-groups fold back into a tightly filled center [27].

The dendrimers be able to synthesized through a range of control not achievable with most linear polymers, leading to closely monodisperse, globular macromolecules through a huge number of marginal groups as distinguish (Figure 9.3), the organization of a few dendrimer duplicate components, for instance, the 1,3-diphenylacetylene part grown by Newkome [28]. The dendrimers be an innovative group of polymeric belongings. The chemistry of this new biomolecule is one of the most striking and rapidly emergent field of innovative chemistry [29]. The chemistry of dendrimer, as further specific exploring areas, has its individual expressions and abbreviations. Moreover, an additional short arrange mental classification is useful to illustrate the various chemical actions carried out at the dendrimer exterior. Dendrigrafts are a division of dendritic polymers similar to dendrimers, can be assembled through a sound specific molecular

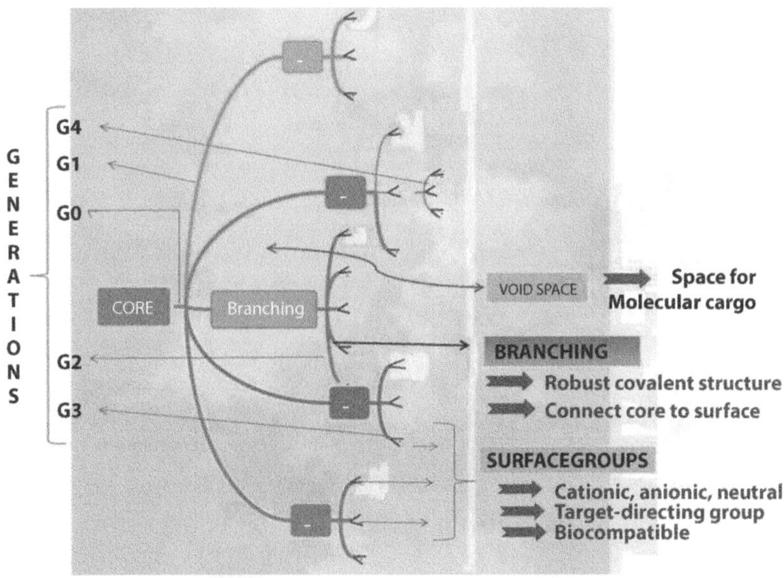

Figure 9.2 Schematic representation of the dendrimer arrangement. Adapted from Henri Brunner [30].

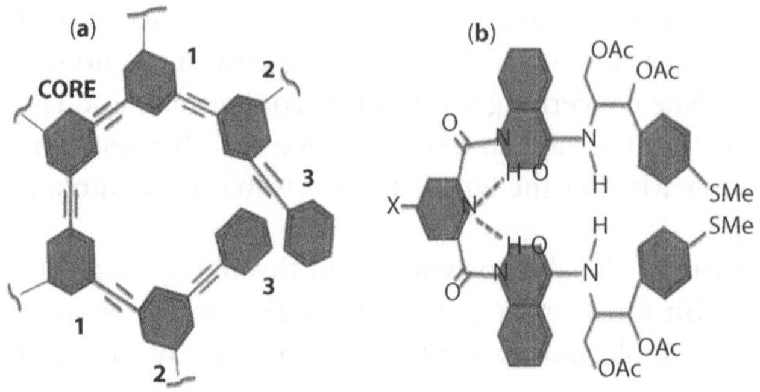

Figure 9.3 Types of dendrimers. (a) More type dendrimers consisting of phenyl acetylene subunits. (b) Parquette-type dendrons (adapted: [27]). Therapeutic, Diagnostic and Environmental Applications of Dendrimer

constitution, i.e., being monodisperse [3]. For the host-guest chemistry and is principally sound operational to connect in multivalent communications, the distinctive configuration of dendrimers supply unique prospective. At the similar instance, one of the primary planned appliances of dendrimers biomolecules was as container composites, in which little substrates are unite within the interior voids [31]. The practical confirmation

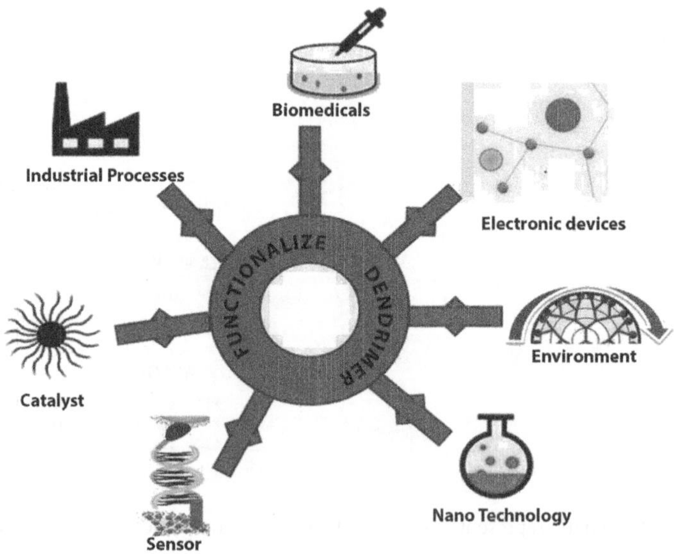

Figure 9.4 Diagrammatic representation of utilizations of functionalized dendrimers.

for unimolecular micelle properties was recognized several years past in cooperation, hyperbranched polymers [32] and dendrimers [20].

The dendrimer embraces its distinctive structural uniqueness like nanoscopic dimension, spheroidal margin, maximum branching, cavernous inner, etc. and stimulating properties, like little viscosity, elevated solubility, maximum reactivity, in arrangement with the maximum functionalities of the dendritic polymers possible that they have wide number of potential relevance in diverse fields [33]. The multiple ample range high technology used for the dendrimers are available in Figure 9.4, since, their unparalleled molecular stability, multifunctional exterior and distinctive interior cavities.

9.10 Dendrimers in Various Fields

It must meet other criteria while competent drug delivery systems, such as: (i) biological adaptability, safe polymer/pendant utility; (ii) constitutional organization more dimension with outline of medicine or imaging-mediators storage space; (iii) defined, tiny-box with/otherwise framework assets and elevated medicine-cargo competence; (iv) not have immunogenicity; (v) correct cellular union and internalization; (vi) sufficient bio-elimination or decomposable; (vii) precise scaffolding and/or exterior flexible utility for cell particular targeting component; (viii) restricted or stimuli-responsive drug delivery attributes; (ix) fragment stage extraction and safety near drug in opposition to antidiscrimination throughout transportation to destination position; (x) smallest wide constitutional with blood–plasma attaching assets; (xi) simplicity near reliable, replicable, medical rating synthesis.

9.10.1 Dendrimers in Biomedical Field

Recent years, the dendrimers are used in various biological applications with excellent outputs. As polymer natures of the dendrimers is most frequent user and consist of analogous to protein and enzymes used for various functions. In dendrimers arrangement, more molecules are encapsulated into their inner voids. Dendrimers are used as an effective biological mediators, and it is most advantages to biomedical applications. Recent years, PAMAM is the excellent dendrimers used for detect the targeted cancer cells. The targeting group can be conjugated with dendrimers which is acted as a host and allowed to imaging mediator to selective bond to particular spot including receptors on tumor cells to develop the recognition.

The frequently used drug of cisplatin is the outside group of a carboxylate terminated PAMAM dendrimers used as a boost in the solubility of cisplatin. Importantly, it is boost the cisplatin ten times higher compared with any other liberated drugs [34].

9.10.2 Magnetic Resonance Imaging Contrast Agents of Dendrimers

One of the important roles of dendrimer is metal chelates which helped to mediators for magnetic resonance imaging contrast. It has huge properties and used in image model of distinguishing media. It is stronger after trail with more trials and shown better efficiency than any other agents [35].

9.11 Dendrimers in Antitumor Therapy

The dendrimer fractions were used as investigative reagent for tumor recognition by magnetic resonance imaging and as contrast mediator; by changeable dimension and hydrophilicity and by collective with tumor targeting antibodies, these complexes can be utilized as specific imaging reasons [36]. The drug must be safe, under no irradiative surroundings, consequently act as prodrug when not irradiated. The dendrimers enclose photo sensitizes named 5-aminolevulinic acid has been connect to the outside of dendrimers and studied as a mediator for photodynamic treatment of tumorigenic keratinocytes [37].

9.12 Dendrimers as Gene Transfer Reagents

As transfection, eukaryotic cells are a technique for efficient alterations in the heritable material of cells [38]. The utilization of dendrimers for transfection was primarily documented by the group of Haesnler [39]. The company named Quiagen evolved an industrial transfection arrangement based on PAMAM dendrimers followed by the work of Boas *et al.* [38]. The dendrimers are dynamically under inspection for the release of DNA and organic molecule medicines, predominantly for cancer treatment. The amino ended PAMAM or PPI dendrimers as nonviral gene transmit mediators, supplying the transfection of DNA by endocytosis into the cell nucleus [40]. Additionally, a sequence of amphiphilic dendrimer based on the stiff diphenyl ethyne core was manufactured and their actions as transfection mediator were recognized [41]. In this investigation, dendritic amidoamine side series of diverse generations were covalently united to the chitosan which was chosen to combine the biological actions of chitosan in gene release, antibacterial activity and lesion curative activity with the delivery earnings establish for dendrimers.

9.13 Drug Delivery of Dendrimers

The centre and exterior portion of the dendrimers are the suitable place for loaded the drug fragments efficiently. Encapsulations of the well-known antitumor medicine cisplatin contained by PAMAM dendrimer afford conjugates which can slow down the liberate and provide superior accumulation in solid tumors and reduce toxicity than free cisplatin [42]. The encapsulation of silver salts with PAMAM dendrimers generate conjugates that can reveal slow silver discharge rates along with antimicrobial actions against various gram positive bacteria [43]. The dendrimers through complexation or encapsulation, drug fragments can be induced. Therapeutic mediators can be connected to a dendrimer for the release of drugs, for example, dendrimers in Boron Neutron Capture Therapy (BNCT-2005).

9.14 Targeted Drug Delivery of Dendrimers

Dendrimers exterior has excellent drug delivery capacity and it has multifunctional effect with high potentialities for drug releases. In addition, it can be deliver the drug completely with more efficiency against various infections. When the dendrimers are used in double fold dilution, the drug will act prolonged nature into the affected cells and enhanced its pharmaceutical efficiency effectively. Before, the natures of the drugs are creating the problem due to the low efficiency, scarcity of drug delivery and more toxicity. In case of the use of dendrimers, it acted as an improved bioavailability, long-term treatment, soluble nature, and minimized toxicity [44].

9.15 Transdermal Drug Delivery of Dendrimers

Last few decades, the transdermal drug release approaches are frequently used against various infections. Transdermal drug delivery system was applied, in order to improve the effectiveness of the drug. Drug delivery through skin to reach an organized consequence of drug is known as transdermal drug deliverance. Transdermal liberation provide controlled, sound management of the drug which broaden the efficacy of drug having small half-life during the reservoir of drug accessible in the delivery system and its controlled discharge account [45] which are extremely proficient in management of acute and chronic rheumatoid and osteoarthritis, could be improving the drug penetration through the skin as distribution enhancers.

9.16 Dendrimers in Vaccine Development

Vaccination is the best choice to eradicating the communicable disease worldwide due to the emergence of multi drug resistant bacteria. Once the bacteria get multi drug resistant behavior, it is very challenging to inhibit through the vaccination process as well as tragedies for prevent the dissemination of the pathogens which led to very harmful to human. As a result, there is always an emergent requirement for innovative productions of vaccines with best possible qualities to target these communicable diseases [46]. The dendrimers have capability to act as competent resistant stimulating compounds (adjuvants) that can improve the effectiveness of vaccines, due to their most advantageous physical characteristics. Moreover to generate conjugates with immune stimulators and/or antigens, dendrimers can supply well-defined multivalent scaffolds [47]. Upon administration without any side effect, these conjugates believed to be outstanding transporters of minute antigens. In this respect, peptides dendrons make up solely from lysine have achieve much attention from the investigators [48, 49]. The production of earliest glycopeptide dendrimer-type AIDS vaccine form consisting of a V3 loop peptide–succinyl–maltose–proline–poly(lysine) dendrimer scaffold have further documented by Baigude *et al.* [50]. This vaccine model has been declared to have superior aqueous solubility, small cell toxicity, and antigenicity of dendrimer itself due to the existence of sugar moieties [51]. The numerous investigations have been carried out with dendrimeric species particularly PAMAM dendrimers as adjuvants against diverse pathogens. Superior effectiveness and decline in undesirable consequence on host exhibited that dendrimers in conjugation with vaccines/antigens can be used effectively [52].

9.17 Application of Dendrimers

9.17.1 Molecular Probes of Dendrimers

Based on the distinct morphology and peculiar characteristic nature of the dendrimers having attractive fragments which adopted by molecular probes. For the production of integrated molecular probe, the immobilization of the sensor units are needed on exterior places of the dendrimers and it is a competent study due to their larger surface are and high surface density.

9.17.2 X-Ray Contrast of Dendrimers

X-ray diffraction is one of the instruments used to the entire field such as physical, chemical, biological and pharmaceutical. Especially, it is used to detect various infectious diseases in medical filed. In biomedical field, more accurate result of X-ray is observed such as tumor, kidney cells, urinary infections and infracts. Here, the active potential polymeric X-ray contrast agent is dendrimer. It is used in the form of different metallic complexes including bismuth and tin oxide.

9.17.3 Dendrimers as MRI Contrast Agents

Previously, Balogh *et al.* [53] and Co-workers discovered a new class of high molecular weight of the dendrimers which is used in MRI contrast mediators in medical filed. Followed, the Gd-III-DTPA-based PAMAM dendrimers are discovered by Wiener and Co-workers. It is used to identify the action of drug in the specific target sites of the infectious places through MRI contrast mediators as synthetic polymer differences mediators' mode. Generated a float united Gd(III)–DTPA PAMAM nanomaterial, that enlarged continuing loosening time near cancer tissue stating premier attraction folate sensor.

9.17.4 Dendrimers Used as a Boron Neutron Capture Therapy

Based on Boron capture reaction, the Boron therapy is concerned with the management of cancers [54]. The usefulness of Poly (amidoamine) nanomaterial's during exploring within tumor liberation nearby mediators for NCT be incredible in biology and health discipline. With the investigation of operative G5 Poly(amidoamine) nanomaterials plus combining cetuximab precise to EGF sense organ to starburst type of nanopolymer so as to passed approximately 1100 boron particles. However with living body outcomes exposed so as to gathering before complex be 10 assets superior within brain cancer mass into correlation toward well brain tissues. The investigation be opening headed for express efficiency of a boronated monoclonal proteins intended for boron NCT of an intracerebral glioma [55]. The nanopolymer dependent boron NCT in associating through EGF sensor intention fractions be able to augment the boron consumption within cancer mass [56].

9.17.5 Application of Dendrimers in Environment

The dendrimers have huge number of exterior functional clusters through which they can connect to diverse surfaces, such dendrimer supported surfaces could act as valuable adsorbents for diagnostic and electro-diagnostic dealings in environmental investigations. Wang *et al.* [57] reported the utilization of functionalized SBA-15 mesoporous silica gel through Poly(Amidoamine) dendrimer (PAMAM-SBA-15) for the elimination of Cu(II), Pb(II) and Cd(II) ions from aqueous mixtures. Likewise, Carmo and Piamac counted their investigation of Copper adsorption on Chloropropyl silica gel surface functionalization with G-3 Poly (Propyleneimine) hexadeca amine dendrimer (Dab-Am-16). Moreover approach, Defever *et al.* [58], documented the remediation of Naphthalene by 3rd – 6th generation Poly (Amidoamine) dendrimer and graphene oxide (GrO). The information reveals the prospective applications of these nanomaterials in eliminating poly aromatic hydrocarbon (PAH) contaminants from water.

9.18 Noxious Outline Concerning Dendrimers

The noxiousness, as well as all the further properties of a compound, is straightly connected to its configuration. The particular elements composing a dendrimer (center, division, exterior assemblages) contribute to amplify or constraints of its toxicity. The extensive dendrimer-express investigation carried out nowadays contributed to the characterization of these shows potential compounds for therapeutic appliances. Considering with numerous major progresses that dendrimers afforded, concerning pharmacokinetic and pharmacodynamics properties, toxicity elements were also evaluated. The dendrimers with cationic exterior groups tend to interact through lipid bilayer, raise the permeability and declines the reliability of biological membrane which is analogous to macromolecules. The system involved is the outflow of cyotosolic proteins such as luciferase and lactate dehydrogenase, on interaction of dendrimers through the cell membrane that lastly causes its interruption and cell lysis [59]. Moreover, the *in vitro* cytotoxicity of cationic melamine dendrimers having exterior groups such as amine, guanidine, carboxylate, sulfonate or phosphonate were documented and it was reported that cationic dendrimers were much more cytotoxic than anionic or PEGylated dendrimers [60].

However Jones *et al.* conducted nano-toxicological investigation on amine-terminated PAMAM dendrimers that exposed that intravenous administration was deadly to mice and reason for disseminated intravascular coagulation-like situation. With flow cytometry and microscopic examination it was verified that cationic fluorescein isothiocyanate labeled G7 PAMAM dendrimers caused platelet interruption, whereas neutral (hydroxyl terminated) and anionic (carboxyl terminated) PAMAM dendrimers did not modify platelet morphology or their function [61].

9.19 Dendrimers and Transport System

With the substantial exterior part with existence of useful clusters contained, that elevated solvable inside water, biocompatibility, with flexibility, the nanomaterial have been utilized additional with medicine carriers. Because of their diminutive facet, these dendrimers have unlocked the novel aspect headed for idea of restricted discharge. The transport of both hydrophobic and hydrophilic molecules is an additional enormous benefits before the super molecule being employed for drug transporters [62].

The micro polymer be able to interrelate through fragments actually or elementally. As soon as drugs be actually connected near polymers, these never go through modifications. On the other hand, though this exchanges be simple headed for outline, there be slight regulation above these discharge. Additionally, as mentioned dual formations are actually combined, the medicine: polymer proportion be generally little. The elemental dealings be capable of occurring during triple diverse manners. Primary of these be throughout a straight union among exterior of a nanomaterial with the medicine fragment. These be capable of furthermore connections with a bond, which be utilize while medicine do not have purposeful assemblages that be able to connect to nanopolymer or once these association be essential near converting solvability constraints otherwise discharge outline. The other rationale to utilize the diverse configuration is declining an amount of medicine fragments close to exterior of a dendrimer. Therefore attain the extra considerable quantity of medicine hurdle near the nanopolymer. At last, molecule be able to prepared the central fraction of a medicine, since it will be unrestricted within particular cell network [63].

Between the dealings that can happen be easy encapsulation, which takes place gratitude to globular profile of nanopolymer with the inner hollow space so as to there. Once within this void, the medicine be able to interrelate through particular clusters within polymer makeup, like oxygen or nitrogen particles, permitting H connections to outline. On the other hand, never attachment desires near take place as H union, with aquaphobic exchanges be capable of generating easy encapsulation, however substantial encapsulation be furthermore achievable. These interior compartment have aquaphobic assets, building the probable headed for carrying fragments that be never simply solvable [64]. The elementally nanopolymer be capable of carrying medicines by electrostatically attaching them and several efficient assemblages establish going on exterior, like amines or carboxyl clusters. The same as an benefit after the exchange happens, it can exist distinguished to the solvability of aquaphobic fragments be enlarged; in counting to that, a medicines never drop compound or corrective reliability. Instances of the exchanges have be observe by a few NSAIDs so as to have carboxylic elements in their composition, like ibuprofen or piroxicam, which outline a steady composite through a dendrimer electrostatic communications to facilitate formation.

A further mode to link drug fragment to dendrimers to carry them is by fusion these two makeup through covalent attachments. It be essential just before the existence of useful assemblages like paminobenzoic acid otherwise associations with amides or esters, in order to produce this combination. It must be revealed to facilitate twisting of carrying be of unique attention while it be required toward a restricted discharge by the medicines, because the nanopolymer-drug composite disperses with the lesser velocity while evaluating near the open medicines, moreover during marked treatment, as these composites be capable of immersed within particular cell mass. The trouble so as to occur be that the composite being unsolvable; still, mentioned be able to resolved via fusion little PEG series. Several of a medicine with the purpose comprise effectively coupled near these nanomaterial includes cisplatin, methotrexate, naproxen, as well as penicillin. Once what be required toward transmitting medicines through aquaphilic distinctive, be attached toward the nanomaterial polyethylene glycol, generating monomolecular micelles so as to above the surface of polyethylene glycol, solvable hydrophilic fragments. On other hand, within the dendrimer, it can transmit hydrophobic molecules [65].

9.20 Conclusion

Recent years, dendrimers are heightened and leading role in pharmaceutical industry and medicinal industry. It can be utilized as excellent progress drug creation properties, which are called as structural properties including nanoscale level, various dimension, nature of solubility, inner cavity, branching behavior, polyvalency and controlled drug delivery efficiency. In addition, the long life time is more important. All the drugs are connected with dendrimers through covalent bonds, electrostatic exchanges and encapsulations. The interaction between the drug and dendrimers are more important, and benefit to the properties of drugs to improve the nature. Importantly, it is very excellent drug transporters and can be engineered to the release of drugs in target sites with prolonged time without toxicity nature. The important benefit is, dendrimers are active drug transporter molecule with biocompatible agent and transferred the drugs in target sites safely with more accurately. Also, it delivered the drugs in target sites with different ways. It is one of the most promising drug delivery molecule compared with any other nanocarrier molecules.

References

1. Carmo, D.R., Silveira, S.F.T., Laurentiz, S.R., Bicalho, M.L., Filho, D.L., Paim, L.L., and Synthesis and a Preliminary Characterization of Poly(Proppylene) Imine Hexadecylamine Dendrimer (DAB-Am-16) Modified with Methyl Acrylate. *Am. Chem. Sci. J.*, 3, 314–324, 2013.
2. Tomalia, D.A., Birth of a New Macromolecular Architecture: Dendrimers as Quantized Building Blocks for Nanoscale Synthetic Polymer Chemistry. *Prog. Pol. Sci. (Oxford)*, 30, 294–324, 2005.
3. Tomalia, D.A., Birth of a New Macromolecular Architecture: Dendrimer as Quantized Building Blocks for Nanoscale Synthetic Organic Chemistry. *Aldrichimica Acta*, 37, 39–57, 2004.

4. Lyu, Z., Ding, L., Huang, A.Y.T., Kao, C.L., Peng, L., Poly (amidoamine) Dendrimers: Covalent and Supramolecular Synthesis. *Mater. Today Chem.*, 13, 34–48, 2019.
5. Hu, C.M., Aryal, S., Zhang, L., Nanoparticleassisted Combination Therapies for Effective Cancer Treatment. *Ther. Deliv.*, 1, 323–334, 2010.
6. Cheng, Y., Zhao, L., Li, Y., Xu, T., Design of Biocompatible Dendrimers for Cancer Diagnosis and Therapy: Current Status and Future Perspectives. *Chem. Soc. Rev.*, 40, 2673–2703, 2011.
7. Sherje, A.P., Jadhav, M., Dravyakar, B.R., Kadam, D., Dendrimers: A Versatile Nanocarrier for Drug Delivery and Targeting. *Int. J. Pharm.*, 548, 707–720, 2018.
8. Tomalia, D.A., Special Issue: "Functional Dendrimers". *Molecules*, 21, 1035, 2016.
9. Tomalia DA, Baker H, Dewald J, Hall M, Kallos G, Martin S, *et al.* A new class of polymers: starburst-dendritic macromolecules. *Polym J.* 17, 117–132, 1985.
10. Madaan, K., Kumar, S., Poonia, N., Lather, V., Pandita, D., Dendrimers in Drug Delivery and Targeting: Drug-dendrimer Interactions and Toxicity Issues. *J. Pharm. Bioallied Sci.*, 6, 139–150, 2014.
11. Menjoge, A.R., Kannan, R.M., Tomalia, D.A., Dendrimer-based Drug and Imaging Conjugates: Design Considerations for Nanomedical Applications. *Drug Discov. Today*, 15, 171–185, 2010.
12. Parekh, H.S., The Advance of Dendrimers-A Versatile Targeting Platform for Gene/drug Delivery. *Curr. Pharm. Des.*, 13, 2837–2850, 2007.
13. Kurtoglu, Y.E., Mishra, M.K., Kannan, S., Kannan, R.M., Drug Release Characteristics of PAMAM Dendrimer–Drug Conjugates with Different Linkers. *Int. J. Pharm.*, 384, 189–194, 2010.
14. Liu, J., Gray, W.D., Davis, M.E., Luo, Y., Peptide- and Saccharide-Conjugated Dendrimers for Targeted Drug Delivery: A Concise Review. *Interface Focus*, 2, 307–324, 2012.
15. Pillai, O. and Panchagnula, R., Polymers in Drug Delivery. *Curr. Opin. Chem. Biol.*, 5, 447–451, 2001.
16. Jean-d'AmourTwibarine, T. and Grindley, B., Polyester Dendrimers: Smart Carriers for Drug Delivery. *Polymers*, 6, 179–213, 2014.
17. Grayson, S.M. and Fréchet, J.M.J., Convergent Dendrons and Dendrimers: from Synthesis to Applications. *Chem. Rev.*, 101, 3819–3868, 2001.
18. Hawker, C.J., Farrington, P.J., Mackay, M.F., Wooley, K.L., Frechet, J.M.J., Molecular Ball Bearings: The Unusual Melt Viscosity Behavior of Dendritic Macromolecules. *J. Am. Chem. Soc.*, 117, 4409–4410, 1995.
19. Gupta, U., Agashe, H., Jain, N.K., Polypropylene Imine Dendrimer Mediated Solubility Enhancement: Effect of Ph and Functional Groups of Hydrophobes. *J. Pharm. Pharm. Sci.*, 10, 358–367, 2007.
20. Newkome, G.R., Young, J.K., Baker, G.R., Potter, R.L., Audoly, L., Cooper, D., Weis, C.D., Morris, K.F., Johnson, C.S., Cascade Polymers: pH Dependence

of Hydrodynamic Radii of Acid Terminated Dendrimers. *Macromolecules*, 26, 2394–2396, 1993.
21. Barth, R.F., Adams, D.M., Soloway, A.H., Alam, F., Darby, M.V., Boronated Starburst Dendrimer-Monoclonal Antibody Immunoconjugates: Evaluation as a Potential Delivery System for Neutron Capture Therapy. *Bioconjugate Chem.*, 5, 58–66, 1994.
22. Milhem, O.M., Myles, C., McKeown, N.B., Attwood, D., D'Emanuele, A., Polyamidoamine Starburst R Dendrimers as Solubility Enhancers. *Int. J. Pharm.*, 197, 239–241, 2000.
23. Wang, Q., Guan, Y., Liu, X., Ren, X., Yang, M., High-capacity Adsorption of Hexavalent Chromium from Aqueous solution Using Magnetic Microspheres by Surface Dendrimer Graft Modifification. *J. Colloid Interface Sci.*, 375, 160–166, 2012.
24. Asthana, A., Chauhan, A.S., Diwan, P.V., Jain, N.K., Poly(amidoamine) (PAMAM) Dendritic Nanostructures for Controlled Site Specific Delivery of Acidic Anti-inflammatory Active Ingredient. *AAPS Pharm. Sci. Tech.*, 6, 536–542, 2005.
25. Yiyun, C. and Tongwen, X., Solubility of Nicotinic Acid in Polyamidoamine Dendrimer solutions. *Eur. J. Med. Chem.*, 40, 1384–1389, 2005a.
26. Yiyun, C. and Tongwen, X., Polyamidoamine Dendrimers Used as Solubility Enhancers of Ketoprofen. *Eur. J. Med. Chem.*, 40, 1390–1393, 2005b.
27. Zimmerman, S.C., Lawless, L.J., Supramolecular Chemistry of Dendrimers. In: *Dendrimers IV. Topics in Current Chemistry,* Vögtle, F., Schalley, C.A. (eds) 217, Springer, Berlin, Heidelberg, 95–120, 2001.
28. Newkome, G.R., He, E., Moorefield, C.N., Vogtle, F., *Dendrimers and Dendrons. Concepts, Synthesis, Applications*, Wiley VCH, Weiheim, 2001.
29. Boris, D. and Rubinstein, M., A Self-Consistent Mean field Model of a Starburst Dendrimers: Dense core vs. Dense Shells. *Macromolecules*, 29, 7251–7260, 1996.
30. Brunner, H., Dendrizymes: Expanded ligands for enantioselective catalysis. *J. Organometal. Chem.*, 500, 1995, 39-46, 1995.
31. Maciejewski, M., Concepts of Trapping Topologically by Shell Molecules. *J. Macromol. Sci. Chem.*, 17, 689–703, 1982.
32. Kim, Y.H. and Webster, O.W., Water Soluble Hyperbranched Polyphenylene: "A Unimolecular Micelle?". *J. Am. Chem. Soc.*, 211, 437–447, 1990.
33. Visintin, A., Iliev, D.B., Monks, B.G., Halmen, K.A., Golenbock, D.T., MD-2. *Immunobiology*, 211, 437–447, 2006.
34. Csaba, N., Garcia-Fuentes, M., Alonso, M.J., The performance of Nanocarriers for Transmucosal Drug Delivery. *Expert Opin. Drug Deliv.*, 3, 463–478, 2006.
35. Ritchie, D.W. and Kemp, G.J., Protein Docking Using Spherical Polar Fourier Correlations. *Proteins*, 39, 178–194, 2000.
36. Kobayashi, H. and Brechbiel, N.W., Dendrimer- based Macromolecular MRI Contrast Agents. *Mol. Imaging*, 2, 1–10, 2005.

37. Sonke, S. and Tomalia, D.A., Dendrimers in Biomedical Applications Reflections on the Field. *Adv. Drug Del. Rev.*, 57, 2106–2129, 2005.
38. Boas, U., Christensen, J.B., Heegaard, P.M.H., Dendrimers in Medicine and Biotechnology. *R. Soc. Chem.*, 81, 152–154, 2006.
39. Haesnler, J. and Szoka, F.C., Polymeric Gene Delivery: Principles and Applications. *Bioconjugate Chem.*, 4, 372–379, 1993.
40. Eichman, J.D., Bielinska, A.U., Kukowska-Latallo, J.F., Baker, R.J., The Use of PAMAM Dendrimers in the Efficient Transfer of Genetic Material into Cells. *Pharm. Sci. Technol.*, 3, 232– 245, 2000.
41. Ferruti, P., Marchisio, A.M., Duncan, R., Poly (amido-amine): Biomedical Applications. *Macromol. Rapid Commun.*, 23, 332–355, 2002.
42. Joester, D., Losson, M., Pugin, R., Heinzelmann, H., Walter, E., Merkle, H.P., Diederich, F., Amphiphilic Dendrimers: Novel Self-Assembling Vectors for E Gene Delivery. *Angew. Chem. Int. Ed. Engl.*, 42, 1486–1490, 2003.
43. Malik, N., Evagorou, E.G., Duncan, R., Dendrimer–platinate: A Novel Approach to Cancer Chemotherapy. *Anticancer Drugs*, 10, 767–776, 1999.
44. Twyman, L.J., Beezer, A.E., Esfand, R., Hardy, M.J., Mitchell, J.C., The Synthesis of Water Soluble Dendrimers, and their Application as Possible Drug Delivery Systems. *Tetrahedron Lett.*, 40, 1743–1746, 1999.
45. Barata, T.S., Teo, I., Brocchini, S., Zloh, M.M., Shaunak, S., Partially Glycosylated Dendrimers Block MD-2 and Prevent TLR4-MD-2-LPS Complex Mediated Cytokine Responses. *PloS Comput. Biol.*, 7, 1–12, 2011.
46. Heegaard, P.M.H., Boas, U., Sorensen, N.S., Dendrimers for Vaccine and Immunostimulatory Uses. A Review. *Bioconjugate Chem.*, 21, 405–418, 2010.
47. J.P. Tam, Multiple Antigen Peptide System. 5,229,490, US5229490A, 1993.
48. Crespo, L., Sanclimens, G., Pons, M., Giralt, E., Royo, M., Albericio, F., Peptide and Amide Bond-containing Dendrimers. *Chem. Rev.*, 105, 1663–1670, 2005.
49. Wang, C.Y., Looney, D.J., Walfield, A.M. *et al.*, Long-term Hightiter Neutralizing Activity Induced by Octameric Synthetic HIV-1 Antigen. *Science*, 254, 285–288, 1991.
50. Baigude, H., Katsuraya, K., Okuyama, K., Uryu, T. *et al.*, Synthesis of Structurally-Controlled AIDS Vaccine Model with Glycopeptide Dendrimer Scaffolds. *Macromol. Chem. Phys.*, 205, 684–691, 2004.
51. Schluer, A.D. and Rabe, J.P., Dendronized Polymers: Synthesis, Characterization, Assembly at Interfaces, and Manipulation. *Angew. Chem. Int. Ed. Engl.*, 39, 864–883, 2000.
52. Wang R., Charoenvit Y., Corradin G., Porrozzi R., Hunter R.L., Glenn G., Alving C.R., Church P. and Hoffman S.L., Induction of Protective Polyclonal Antibodies by Immunization with Plasmodium yoelii circumsporozoite Protein Multiple Antigen Peptide Vaccine. *J. Immunol.*, 155, 1637–1641, 1995.

53. Balogh, L., Swanson, D.R., Tomalia, D.A., Hagnauer, G.L., McManus, A.T., Dendrimer–Silver Complexes and Nanocomposites as Antimicrobial Agents. *Nano Lett.*, 1, 18–21, 2001.
54. Barth, R.F., Soloway, A.H., Fairchild, R.G., Brugger, R.M., Boron Neutron Capture Therapy for Cancer. *Realities and Prospects. Cancer*, 70, 29953007, 1992.
55. Barth, R.F., Wu, G., Yang, W., Binns, P.J., Riley, K.J., Patel, H. *et al.*, Neutron Capture Therapy of Epidermal growth Factor (+) Gliomas Using Boronated Cetuximab (IMCC225) as a Delivery Agent. *Appl. Radiat. Isot.*, 61, 899903, 2004.
56. Yang, W., Barth, R.F., Wu, G., Tjarks, W., Binns, P., Riley, K., Boron Neutron Capture Yherapy of EGFR or EGFRvIII Positive Gliomas Using Either Boronated Monoclonal Antibodies or Epidermal Growth Factor as Molecular Targeting Agents. *Appl. Radiat. Isot.*, 67, S328331, 2009.
57. Wang, D.J. and Imae, T., Fluorescence Emission from Dendrimer& its pH Dependence. *J. Am. Chem. Soc.*, 126, 13204–13205, 2004.
58. Defever, S.R., Geithner, K.N., Bhattacharya, P., Ding, F., Ke, C.P., Sarupria, S., PAMAM Dendrimer and Grapheme: Materials for Removing Aromatic Contaminants from Water. *Environ. Sci. Technol.*, 49, 4490–4497, 2015.
59. Chen, H.T., Neerman, M.F., Parrish, A.R., Simanek, E.E., Cytotoxicity, Hemolysis, and Acute *in Vivo* Toxicity of Dendrimers Based on Melamine, Candidate Vehicles for Drug Delivery. *J. Am. Chem. Soc.*, 126, 1004410048, 2004.
60. Cheng, Y., Xu, Z., Ma, M., Xu, T., Dendrimers as Drug Carriers: Applications in Different Routes of Drug Administration. *J. Pharm. Sci.*, 97, 123–43, 2008.
61. Jones, C.F., Campbell, R.A., Franks, Z., Gibson, C.C., Thiagarajan, G., Vieira-de-Abreu, A. *et al.*, Cationic PAMAM Dendrimers Disrupt Key Platelet Functions. *Mol. Pharm.*, 9, 15991611, 2012.
62. Gupta, U., Agashe, H.B., Asthana, A., Jain, N.K., A Review of *in Vitro-in Vivo* Investigations on Dendrimers: The Novel Nanoscopic Drug Carriers. *Nanomed. Nanotechnol. Biol. Med.*, 2, 66–73, 2006.
63. Svenson, S., Dendrimers as Versatile Platform in Drug Delivery Applications. *Eur. J. Pharm. Biopharm.*, 71, 445–62, 2009.
64. Kalomiraki, M., Thermos, K., Chaniotakis, N.A., Dendrimers as tunable vectors of drug Delivery Systems and Biomedical and Ocular Applications. *Int. J. Nanomed.*, 11, 1–12, 2015.
65. Klajnert, B. and Bryszewska, M., Dendrimers: Properties and Applications. *Acta Biochim. Pol.*, 48, 1, 199–208, 2001.

10

Microneedle of Drug Delivery Systems

Fouad Damiri[1]*, Hitendra M. Patel[2], Sagar Salave[3], Bharathi K.[3], Nagavendra Kommineni[4], B.H. Jaswanth Gowda[5], Karthika Paul[6,7], Sanju Bala Dhull[8] and Mohammed Berrada[1]

[1]University Hassan II of Casablanca, Faculty of Sciences Ben M'Sick, Department of Chemistry, Laboratory of Biomolecules and Organic Synthesis (BioSynthO), Casablanca, Morocco
[2]Department of Chemistry, Sardar Patel University, Vallabh Vidyanagar, Gujarat, India
[3]Department of Pharmaceutics, National Institute of Pharmaceutical Education and Research (NIPER), Ahmedabad, India
[4]Center for Biomedical Research, Population Council, New York, NY, USA
[5]Department of Pharmaceutics, Yenepoya Pharmacy College & Research Centre, Yenepoya (Deemed to be University), Mangalore, Karnataka, India
[6]Department of Pharmaceutical Analysis, JSS College of Pharmacy, JSS Academy of Higher Education and Research, Sri Shivarathreeshwara nagara, Bannimantap, Mysore, Karnataka, India
[7]Department of Pharmaceutical Analysis, Vivekananda College of Pharmacy, Bangalore, Karnataka, India
[8]Department of Food Science and Technology, Chaudhary Devi Lal University, Sirsa (Haryana), India

Abstract

Over the past few years, there has been significant interest in a developing transdermal delivery technology due to its ability to effectively deliver therapeutic and cosmeceutical products for various purposes. These applications include vaccines, drugs, and biomolecules aimed at addressing skin-related issues. One major advantage of transdermal administration is its ability to bypass the initial metabolic breakdown of the drug and facilitate sustained release. However, the stratum corneum acts as a barrier, limiting drug delivery. To overcome this challenge, microneedles (MNs) have emerged as a painless, minimally invasive, and

*Corresponding author: fouad.damiri@outlook.fr

Sathish-Kumar Kamaraj, Arun Thirumurugan, Muthuchamy Maruthupandy, Mercedes Guadalupe López Pérez and Shanmuga Sundar Dhanabalan (eds.) Nanoscience and Nanotechnology for Smart Prevention, Diagnostics and Therapeutics: Fundamentals to Applications, (243–260) © 2024 Scrivener Publishing LLC

user-friendly transdermal drug delivery system, offering high drug bioavailability. In recent years, various types of MNs have been developed using innovative production techniques. These MNs can be classified morphologically as solid, coated, dissolved, or hollow, depending on the specific transdermal drug delivery method employed, namely "poke and patch," "coat and poke," "poke and release," and "poke and flow," respectively.

Keywords: Microneedles, transdermal drug delivery, hydrogel microneedle

10.1 Introduction

The preferred method of delivering treatment, diagnosis, and therapy for diseases is through oral administration. This route of administration has been extensively investigated by formulation scientists in the initial and subsequent phases of drug delivery research [1–3]. Oral administration of drugs is associated with certain drawbacks including systemic metabolism, limited absorption, and a lack of tissue specificity, leading to reduced efficacy in achieving the desired therapeutic outcomes. Consequently, it becomes imperative to explore alternative dosage forms such as parenteral, transdermal, intravesical, and novel drug delivery systems to overcome these limitations [4–6]. Traditional medications designed for topical administration are utilized to treat a wide range of skin conditions, offering both curative properties and aesthetic benefits in the field of cosmetic dermatology. These established pharmaceutical agents are applied externally to the skin surface, addressing diverse skin ailments and promoting the healing process [7]. Nevertheless, the skin acts as a barrier that prevents efficient and precise delivery of substances [8, 9].

Micro-needles, known as MNs, are small projections ranging from 25 to 2500 μm in height. They come in different forms and can be attached to a base for stability. These MNs serve the purpose of extracting bodily fluids and administering therapeutic substances to cells. Nonetheless, MNs are comprised of multiple micro-projections with varying shapes, connected to a supportive base [10, 11]. The utilization of MN arrays in biological membranes has the potential to generate micrometer-sized channels for transportation [12, 13]. These channels, once formed, are significantly larger in size compared to individual molecules. Consequently, they can facilitate the movement of not only macromolecules but also supramolecular complexes and microparticles [14].

10.2 Mechanism of Drug Delivery

The administration of topical drugs operates through the diffusion process. When utilizing the microneedle drug delivery system, the skin undergoes temporary disruption. This system involves an arrangement of numerous microneedles placed on a small patch, similar to a standard transdermal patch found in the market [7, 15]. The purpose is to deliver an adequate dosage of medication, enabling the desired therapeutic outcome. By piercing through the stratum corneum, which serves as a barrier layer, the microneedle device allows for direct placement of the drug into the epidermis or upper dermis. Subsequently, the drug enters the systemic circulation, reaching its target site and eliciting a therapeutic response. The process of drug delivery via microneedles is illustrated in Figure 10.1.

10.3 Types and Fabrication of Microneedle

Microneedles can be categorized into four distinct types based on its design and intended use. Solid microneedles are used extensively for skin pre-treatment prior to drug administration. Insertion of solid microneedles on the surface of the skin enables formation of micron-scale pores that

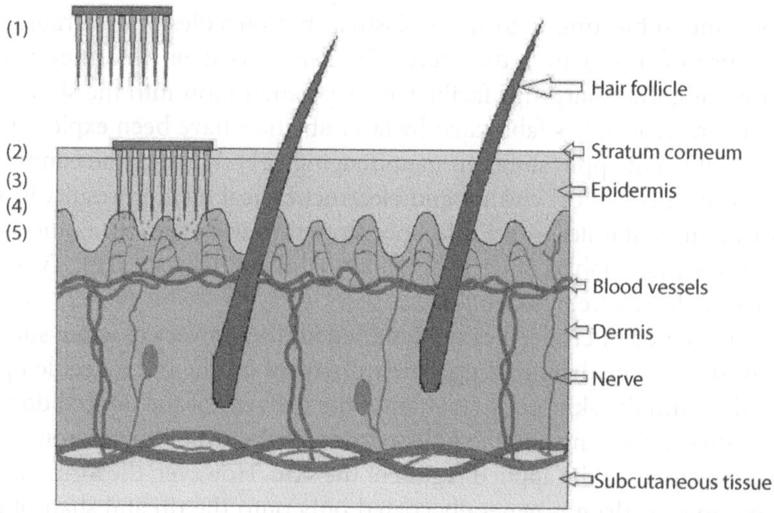

Figure 10.1 Mechanism of drug delivery by microneedle device [16].

remain even after its removal from the skin. These pores enable slow diffusion of drug molecules, from a formulation that is subsequently applied to the skin. Solid microneedles have demonstrated an increase in skin permeability of both small molecules as well as macromolecules such as insulin, bovine serum albumin and calcein by several orders of magnitude [17]. This is a significant achievement, as the stratum corneum of the skin is generally considered highly impermeable to drugs and macromolecules [18]. The needle height and the density of microneedle arrays have been found to have a considerable effect on the drug flux. Yan *et al.* investigated the degree of enhancement of drug penetration, by solid microneedle arrays with varying heights ranging from 100 to 1100 μm and needle densities in the range of 400 to 11,900 needles/cm^2. It was found that microneedle arrays with needles that are longer in height (> 600 μm) and with lower densities (<2000 needles/cm^2) exhibited a significantly higher flux of the drug acyclovir in comparison to those with shorter needle lengths [19]. The material of choice for the fabrication of solid microneedles must provide sufficient mechanical strength required to insert the microneedles into the dermis of the skin. Various materials and techniques have been investigated widely for its fabrication. Henry *et al.* utilized a deep reactive ion-etching technique to etch arrays of microneedles into silicon. The silicon wafers were coated by a chromium masking material, in the pattern of dots possessing a diameter approximately equal to the desired dimension of the microneedle base. On loading the wafers onto a reactive-ion etcher and subjecting it to deep plasma, microneedles with a radius of curvature of less than 1 μm were fabricated based on fluorine-oxygen chemistries. The sharp tips facilitated easy penetration into the skin [20]. Metallic microneedles fabricated by laser ablation have been explored for its cost-efficient applicability in nanoimaging [21]. Further, the combined process of mechanical cutting and electrochemical etching results in the fabrication of stainless-steel microneedle arrays with smooth bodies and sharp tips that enables efficient transdermal drug delivery [22]. Types of microneedles are depicted in Figure 10.2.

Drug-coated microneedles are designed for the delivery of water-soluble drugs. In this, the drug coating slowly dissolves off the microneedle upon insertion into the skin, thus facilitating the delivery of the desired dose to the tissues. Following complete dissolution of the drug, the microneedles are removed from the applied region of the skin. However, the formulation containing the drug is generally coated only onto the tip and shaft of the microneedle, which limits the dose that can be administered. Microneedles can be dip-coated using various aqueous or organic solvent-based formulations to obtain a uniform and thick coating. The choice of excipients is

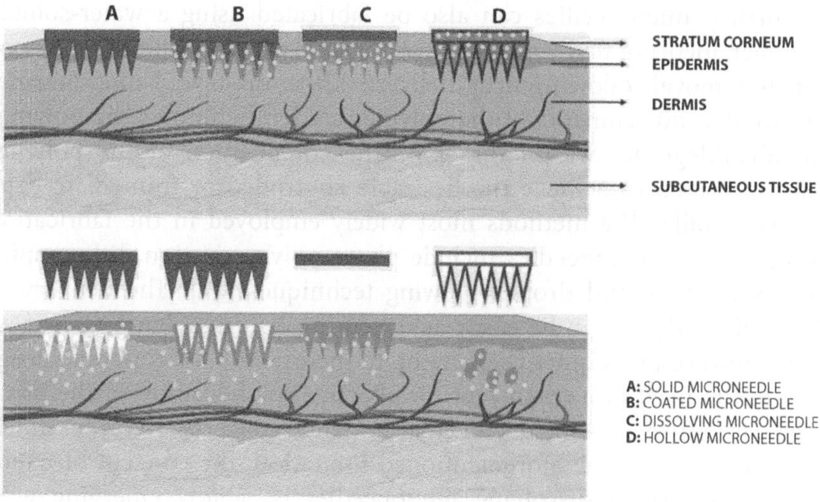

Figure 10.2 Types of microneedles.

crucial in the design of an appropriate coating solution. It has been investigated that the coating uniformity can be improved by incorporating excipients that reduce the surface tension of the coating solution [23]. Spraying onto the microneedles is also one of the widely adopted techniques for coating. The solutions used for coating are often formulated with excipients that tend to yield an increased viscosity so that more of the formulation can be retained during the subsequent drying process [24]. Various small molecules, proteins, deoxyribonucleic acids, and peptides have been explored for coating onto microneedles. Peters *et al.* demonstrated the successful coating of recombinant human erythropoietin on titanium microneedle arrays. A high delivery efficiency of around 75% to 80% of the erythropoietin-coated dose was found to be delivered across the skin. Further, *in vivo* studies suggested a linear dose response and a PK-profile comparable to that of the commercial subcutaneous injectable product of erythropoietin [25]. Drug-coated microneedles owe the potential to serve as a novel-device for the localized intratumoral delivery of anti-cancer drugs. Ma *et al.* formulated microneedles coated with doxorubicin-encapsulating poly(lactic-co-glycolic) acid nanoparticles for the treatment of oral carcinomas. The dose of doxycycline sufficient enough to exert cytotoxicity against cancer cells was found to be delivered effectively. Further, the coated microneedles were able to overcome the limitation of fluid leakage that is associated with intratumoral injections, thus preventing rapid escape of drug from the desired site into the systemic circulation [26].

Further, microneedles can also be fabricated using a water-soluble polymer, incorporated with a drug or vaccine, which eliminates the need for its removal following drug delivery. These dissolving microneedles ensure that no remnants of the polymer are left behind as the polymer dissolves/degrades within the skin. The composition of the polymer can be altered to achieve the desirable controlled or immediate drug release profile. The methods most widely employed in the fabrication of dissolving microneedles include photo-polymerization, lithography, micro-molding, and droplet-blowing techniques [27]. The delivery of hydrophilic drugs can be easily achieved through dissolving microneedles. However, it is quiet challenging to incorporate hydrophobic drugs due to the need of a co-solvent for effective solubilization of the drug, which is in turn associated with environmental concerns. In recent times, in order to avoid the aforementioned limitation, the concept of nanosuspension-loaded dissolving microneedles is rapidly emerging with numerous investigations being carried out by various researchers. Vora *et al.* explored the transdermal delivery potential of the lipophilic molecule cholecalciferol by formulating it in the form of a nanosuspension embedded into dissolving microneedles composed of hydrophilic polymers [28]. Further, transdermal delivery of the anti-retroviral drug etravirine has been explored, by formulating it as nanosuspension-loaded dissolving microneedle arrays, which enables self-administration and thereby improves patient compliance [29].

Lastly, hollow microneedles are intended for the injection of liquid formulations into the skin [24]. It comprises an empty cavity inside each needle and a bore on the needle tip that facilitates the injection of microvolumes of drug solution into the skin. Hollow microneedles possess a higher drug loading capacity in comparison to the other types of microneedles discussed above. They also possess high stiffness that can be attributed to the materials used in its fabrication such as ceramics, silicon, glass and metal [30]. Hollow microneedles have been investigated for the local delivery of immune checkpoint blockade (ICB) therapeutics that inhibit the immune suppressor components, thereby eliciting antitumor response. Hollow-structured microneedle patches loaded with an anti-programmed death-ligand 1 antibody (aPD-L1) in combination with cold atmospheric plasma therapy is also reported [31]. Further, poly(methyl methacrylate) hollow microneedles have been fabricated by Chen *et al.* for the transplantation of follicle dermal papilla cells, proving its potential application in cell delivery/transplantation [32].

10.3.1 Pulling Pipettes

This method is used for the preparation of hollow glass microneedles. Microneedles developed using this method demonstrated the successful delivery of 6-aminoquinolone and rose Bengal to the eye. This method proved that it is a less invasive method compared to the macroscale hypodermic needles [33]. The microneedles were created using a process involving the utilization of high-temperature-exposed fire-polished borosilicate glass pipettes, which were pulled and beveled using a micropipette puller. This particular technique proved to be highly efficient in administering a concentrated dose of insulin to individuals diagnosed with type 1 diabetes [34].

10.3.2 Droplet-Born Air Blowing Method

Droplet-born air blowing method have been proposed by Kim *et al.* [35]. In this method, polymer droplets are shaped into microneedle with the use of air blowing without the use of any radiation or heat [36]. In Figure 10.3, the method of preparation of microneedles is using Droplet-Born air blowing method. In this method, process initiates by dispensing the polymeric solution on two plates. Additionally, the upper plate is positioned in a downward orientation to facilitate the touching of droplets. As the upper plate moves upward, the viscous solution undergoes elongation. Moving on to the subsequent stage of the preparation process, the application of air blowing serves to eliminate any remaining water and solidify the droplets.

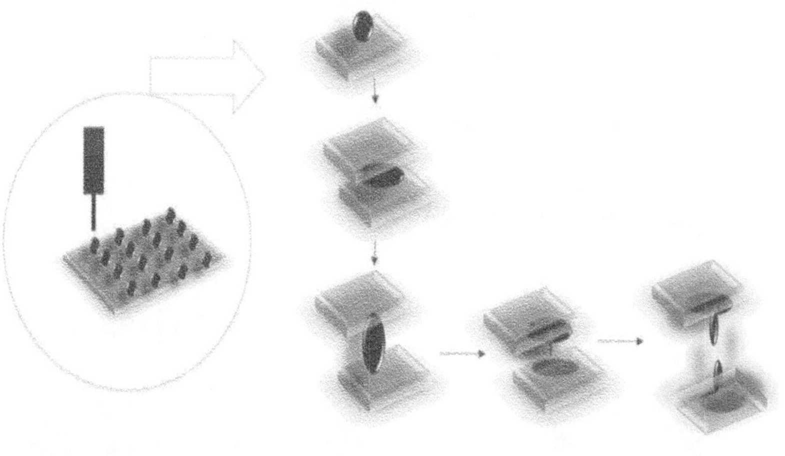

Figure 10.3 Representation of droplet born air blowing method for microneedle formation.

This solidification process involves pulling the droplets away from a substrate, ultimately giving them the desired shape.

This method has been used to develop insulin loaded dissolving microneedles to lower the blood glucose level in diabetic mice [37]. Interestingly, this method uses one drop of polymer for one microneedle formation.

10.3.3 Solvent Casting/Micromolding Method

Dissolving microneedles can be prepared by filling prepared microneedle moulds with liquid formulations. Moulds for microneedles generally made from silicon-based materials as a starting material [10, 38]. Polymeric solution is poured into moulds and then air voids are removed using centrifuge or vacuum [39]. Then, moulds are dried into the oven to dry the material. Prepared microneedles are removed after the cooling of mould and polymeric material. It is a simple, cost effective, and preparation method at ambient temperature. One report, investigators have prepared the ceramic based microneedles using this method [40]. Figure 10.4 depicts the method of preparation of microneedles using solvent casting method.

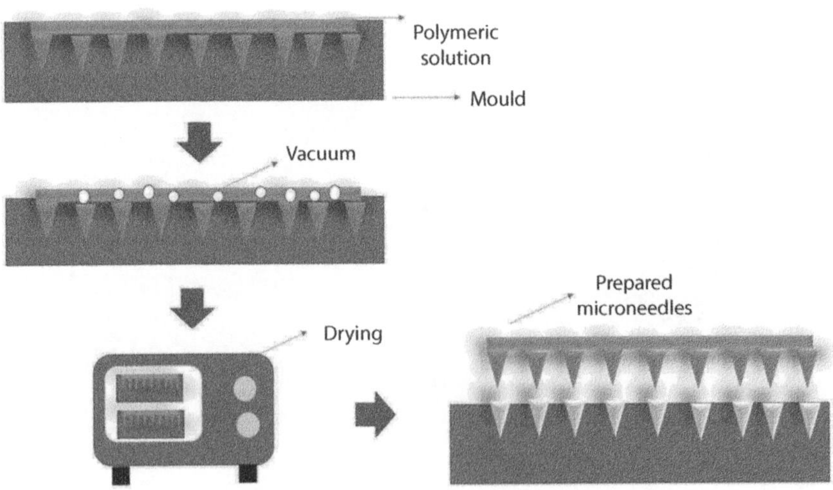

Figure 10.4 Solvent casting/micromolding method for microneedles.

10.3.4 Atomized Spraying Method

Atomised spraying instrument consist of nozzle connected to an air source and solution (formulation liquid) produces an atomized spray. This method reduces the limitations with the limited capacity of large production of dissolving microneedle with desired geometry and physical characteristics. Additionally, it reduces the limitations occurs due to liquid surface tension and viscosity while filling the microneedle molds. Figure 10.5 represents the processing of atomised spraying method for microneedle development This method able to fabricate the laminate-layered and horizontally layered dissolving microneedle. Materials like sugars and polymers can be used to produce microneedles using this method [41].

10.3.5 Laser Cutting

3D lased cutting method can be used for the preparation of metal microneedles [42, 43]. The desired geometry of microneedles is created by using computer aided design software. Array of solid microneedles are fabricated by cutting the stainless steel or titanium sheets into microneedles shape using infrared laser. The laser beam follows the predetermined geometry of needle. Then, microneedles are washed with hot water. This technique can be used to develop single row of microneedles of different geometry and metallic microneedles with two dimensional rows [42, 44].

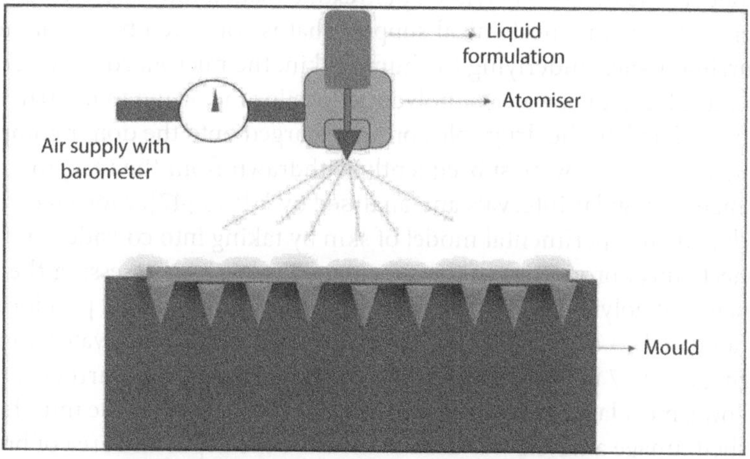

Figure 10.5 The processing of atomized spraying method for microneedle development.

10.3.6 Laser Ablation

Laser ablation method is top-down method used for the preparation of metal-based microneedles. Light pulses give the bulge of the desired shape on metal plate to form the solid metal arrays [45]. However, fabrication of structured material is not suitable due to high intensity laser pulses. Therefore, novel, and cost-effective fabrication method has been introduced to formulate metal-based microneedles [21]. Evens *et al.* developed new method for the development of solid polymer microneedles using laser ablated steel moulds.

10.4 *In Vitro* and *In Vivo* Evaluation of Microneedles

In vitro release testing guides a novel formulation toward further development and also helps in achieving control over the batch-to-batch variability, thus enabling consistent drug product quality and performance. The *in vitro* release testing method should be developed in a way such that it is able to test the key performance indicators of the formulation. Though dissolution testing was initially developed for solid oral dosage forms, it has now been extended widely to a number of novel dosage forms including microneedles. In the methods intended to evaluate the diffusion of drug from microneedles, pH of the medium is adjusted to around 5 to 6, whereas the temperature is maintained at 32°C to mimic the physiological skin conditions [46]. A flow-through diffusion cell system is generally used for *in vitro* diffusion studies. Minipig skin was employed by Milewski *et al.* to evaluate the flux of PEGylated naltrexone loaded into microneedles. In order to mimic the mechanical support that is conferred by the naturally occurring tissues underlying the human skin, the microneedle-treated pig skin was placed on a wafer of polydimethylsiloxane owing to its structural elasticity. Further, the drug solution was charged onto the donor compartment, and samples were subsequently withdrawn from the receptor compartment at regular intervals and analysed by HPLC [47]. Moronkeji *et al.* developed an experimental model of skin by taking into consideration the biomechanical properties of the subcutaneous layer for assessing the performance of polymethylmethacrylate microneedles. Neonatal porcine skin was attached to a second layer of gelatine gels of varying water content ranging from 67 to 96% to mimic the subcutaneous tissue. Further, a third and lowermost layer composed of Perma-Gel® that mimics the muscle was attached. Image analysis was used to measure the depth and area of breach caused by microneedle penetration. Larger perforations were observed on increasing the water-content of the subcutaneous layer, thereby demonstrating its role in the performance of microneedles [48]. However, the use of biological tissues for *in vitro* diffusion tests, leads to inherent variability in the results making it unfit for routine use in quality control tests or in industrial formulation development. In a study, Larraneta *et al.* [49] proposed a novel system comprising of Parafilm M® to study the release of the drug from microneedle arrays. Dissolving microneedle arrays, upon insertion into a single layer of parafilm results in the formation of a hermetic pouch that can in turn be placed in a dissolution bath to evaluate the release of drug from the microneedle arrays. This method can successfully overcome the difficulties associated with the use of biological tissues [49].

In vivo studies are carried out to estimate the pharmacokinetic parameters of the drug that ensures the attainment of therapeutic plasma drug concentrations. Further, it provides information regarding the safety and efficacy of a novel biological molecule of interest. Tort *et al.* developed microneedles coated with insulin-loaded nanoparticles by the electrospraying technique and estimated its *in vivo* efficacy in diabetic rats. The insulin-coated microneedles were rolled over the skin of rats for a period of 1 minute and it was observed that blood glucose levels were maintained at steady state up to 6 hours [50]. Similarly, 3D-printed insulin-loaded microneedle patches developed by Economidou *et al.* demonstrated a steady state hypoglycaemic effect up to 4 hours suggesting the release of insulin via passive diffusion [51]. A combination device of electro-modulated hydrogel and a microneedle array developed by Indermun *et al.* exhibited consistent release of the drug sodium indomethacin in a rat model, upon electro-stimulation of the device. Although an initial burst release of drug was observed due to hydration of the hydrogel matrix, upon subsequent electro-stimulations, the drug release was characterized by minimal fluctuations. Thus, the developed system, as supported by *in vivo* studies, serves as a good non-invasive delivery device for the management of chronic pain [52]. Transcutaneous vaccine delivery has been explored in female BALB/c mice by the administration of zein microneedles coated with the model antigen ovalbumin. A time-dependent enhancement (days 0, 21, 42, and 64) in the antibody titers corresponding to IgG, IgG1, and IgG2a levels was observed post immunization. Though the induction of antibody response was slower in the case of OVA-coated microneedles when compared to the intradermal injection, the responses obtained post day 42 and 64 were found to be significantly higher in both the cases. Further, on increasing the amount of ovalbumin coated onto the microneedles, the antibody response obtained was similar to intradermal administration. The results obtained imply the potential of zein-microneedles to deliver antigen for effective immunization against infectious agents such as the influenza virus [53]. *In vivo* safety assessment is crucial for dissolving microneedles in order to estimate the lack of occurrence of any untoward skin-related complications or toxicity that may occur due to the polymeric matrix that gradually dissolves in the skin. Zhang *et al.* evaluated the safety of polyvinyl alcohol-based microneedles after daily insertion in a 160-day uninterrupted trial on healthy mice. It was observed that the concentration of polyvinyl alcohol gradually decreased in the insertion sites and went completely undetectable after 6 days of administration [54]. Evaluation of the safety and toxicology of the matrix material for biomedical applications by *in vivo* studies is thus essential to support its translation from the preclinical stage to clinical trials.

10.5 Patents

Polymeric microneedles (MNs) are becoming increasingly significant due to their diverse attributes, including biocompatibility, biodegradability, strength, toughness, and optical transparency [55, 56]. Moreover, the production and consistency of polymeric MNs are considerably more economical.

Table 10.1 Recent patents on MNs [10].

Patent no.	Title of patent	Aim	Description
US8708966B2	Microneedle devices and methods of manufacture and use thereof	A method for delivering the active agent across a biological barrier	There are devices equipped with molecular nanosystems (MNs) that can be utilized to facilitate the transportation of substances through tissue barriers and act as miniature flame-holding mechanisms.
US10898703B2	Microneedle template and microneedle prepared using the same	Preparation of MNs template; Preparation of MNs, using a prepared template and MNs preparation method.	A template for microelectromechanical systems (MNs) consists of a base material that supports the formation of MNs structures. These structures are projected onto the template, and a layer of diamond is then deposited onto the surface of the MNs.

(*Continued*)

Table 10.1 Recent patents on MNs [10]. (*Continued*)

Patent no.	Title of patent	Aim	Description
US 2021/0008360 A1	Adhesion membrane and MNs patch	To provide a new patch with excellent flexibility to skin, which carries an MNs array patch.	MNs offer exceptional capabilities in terms of creating skin punctures and remaining adhered to the skin. Offers diverse puncture characteristics based on the skin's elasticity.
US10973757B2	Biodegardable microneedle device	To provide one or more biodegradable MNs capable of drug administration to the skin.	The newly created device takes the form of miniature nanosensors designed specifically for use on the skin. At least one biodegradable micro-needle (MNs) extends outward from the cap, positioned in such a way that its lower surface is in contact. The MN projection is created using a mixture of polymerized blend containing preserved factors derived from stem cells.
US20210106520A1	Conductive polymer microneedle arrays for electronically-controlled drug release	A method that delivers a therapeutic agent, provided with an MNs array, also includes a plurality of MNs, including conductive coating disposal.	A controlled release of therapeutic agents can be achieved by applying a conductive coating to an array of micro-needles, which are then implanted into the dura mater and penetrate into the central nervous system (CNS) of the individual.
US 10,987, 503 B2	Dissolvable MNs for skin treatment	A skin treatment includes MNs application on the skin and penetration to the stratum corneum.	Polymeric MNs and their methods as a skin treatment device

In this context, several patents are summarized below, highlighting the range of applications for MNs. Recent patents pertaining to MNs for various purposes are presented in Table 10.1 [10].

10.6 Conclusion

Transdermal administration offers advantages, such as bypassing the first-pass effect and enabling continuous drug release. However, the presence of the stratum corneum poses a challenge to drug delivery. Microneedles offer a painless, minimally invasive, and user-friendly transdermal delivery system while also ensuring high drug bioavailability. Microneedles are gaining popularity as they enhance patient access to medication, replacing other administration methods. There are four main types of microneedles: solid, coating, dissolving, and hydrogel formulations. Ongoing research in this field aims to achieve effective delivery of therapeutics, as there is an urgent need for innovative transdermal drug delivery approaches. The development of transdermal delivery methods is crucial to expand the market for hydrophilic molecules, protein macromolecules, and conventional drugs in new therapeutic areas. While clinical trials have demonstrated the efficacy of microneedles, further preclinical studies are still needed. Collaboration among academia, industry, and regulatory agencies is essential to address the drawbacks associated with these products and facilitate the safe and effective clinical use of microneedles.

References

1. Gowda, B.H.J., Ahmed, M.G., Sahebkar, A. *et al.*, Stimuli-Responsive Microneedles as a Transdermal Drug Delivery System: A Demand-Supply Strategy. *Biomacromolecules*, 23, 1519–1544, 2022, https://doi.org/10.1021/acs. biomac.1c01691.
2. Avcil, M. and Çelik, A., Microneedles in Drug Delivery: Progress and Challenges. *Micromachines*, 12, 1321, 2021, https://doi.org/10.3390/mi12111321.
3. Damiri, F., Gowda, B.H.J., Andra, S. *et al.*, Chitosan Nanocomposites as Scaffolds for Bone Tissue Regeneration, pp. 377–394, Springer, Singapore, 2023.
4. Gowda, B.H.J., Ahmed, M.G., Sanjana, A., Can Microneedles Replace Hypodermic Needles? *Resonance*, 27, 63–85, 2022, https://doi.org/10.1007/s12045-022-1294-5.
5. Bachra, Y., Damiri, F., Berrada, M. *et al.*, Introduction of Superabsorbent Polymers, in: *Properties and Applications of Superabsorbent Polymers*, pp. 1–18, Springer Nature, Singapore, 2023.

6. Damiri, F., Bachra, Y., Berrada, M. *et al.*, Future Challenges and Opportunities in the Field of Superabsorbent Polymers, in: *Properties and Applications of Superabsorbent Polymers*, pp. 231–242, Springer Nature, Singapore, 2023.
7. Ijaola, A.O., Akamo, D.O., Damiri, F. *et al.*, Polymeric Biomaterials for Wound Healing Applications: A Comprehensive Review. *J. Biomater. Sci. Polym. Ed.*, 33, 1–55, 2022, https://doi.org/10.1080/09205063.2022.2088528.
8. Idumah, C.I., Odera, R.S., Ezeani, E.O. *et al.*, Construction, characterization, properties and multifunctional applications of stimuli-responsive shape memory polymeric nanoarchitectures: a review. *Polym. Technol. Mater.*, 62, 1–26, 2023, https://doi.org/10.1080/25740881.2023.2204936.
9. Damiri, F., Bachra, Y., Berrada, M., Synthesis and characterization of 4-formylphenylboronic acid cross-linked chitosan hydrogel with dual action: Glucose-sensitivity and controlled insulin release. *Chin. J. Anal. Chem.*, 50, 100092, 2022, https://doi.org/10.1016/j.cjac.2022.100092.
10. Kulkarni, D., Damiri, F., Rojekar, S. *et al.*, Recent Advancements in Microneedle Technology for Multifaceted Biomedical Applications. *Pharmaceutics*, 14, 1097, 2022, https://doi.org/10.3390/pharmaceutics14051097.
11. Damiri, F., Kommineni, N., Ebhodaghe, S.O. *et al.*, Microneedle-Based Natural Polysaccharide for Drug Delivery Systems (DDS): Progress and Challenges. *Pharmaceuticals*, 15, 190, 2022, https://doi.org/10.3390/ph15020190.
12. Patel, P.J., Vala, R.M., Patel, S.G. *et al.*, Catalyst-free synthesis of imidazo [5,1-b]quinazolines and their antimicrobial activity. *J. Mol. Struct.*, 1285, 135467, 2023, https://doi.org/10.1016/j.molstruc.2023.135467.
13. Damiri, F., Bachra, Y., Berrada, M., Biomedical Applications of Quantum Dot–Polymer Composites, in: *Quantum Dots and Polymer Nanocomposites*, pp. 207–226, CRC Press, Boca Raton, 2022.
14. Damiri, F., Rojekar, S., Bachra, Y. *et al.*, Polysaccharide-based nanogels for biomedical applications: A comprehensive review. *J. Drug Deliv. Sci. Technol.*, 14, 104447, 2023, https://doi.org/10.1016/j.jddst.2023.104447.
15. Pardeshi, S., Damiri, F., Zehravi, M. *et al.*, Functional Thermoresponsive Hydrogel Molecule to Material Design for Biomedical Applications. *Polymers (Basel)*, 14, 3126, 2022, https://doi.org/10.3390/polym14153126.
16. Waghule, T., Singhvi, G., Dubey, S.K. *et al.*, Microneedles: A smart approach and increasing potential for transdermal drug delivery system. *Biomed. Pharmacother.*, 109, 1249–1258, 2019, https://doi.org/10.1016/j.biopha.2018.10.078.
17. Aldawood, F.K., Andar, A., Desai, S., A Comprehensive Review of Microneedles: Types, Materials, Processes, Characterizations and Applications. *Polym. (Basel)*, 13, 2815, 2021, https://doi.org/10.3390/polym13162815.
18. McAllister, D.V., Wang, P.M., Davis, S.P. *et al.*, Microfabricated needles for transdermal delivery of macromolecules and nanoparticles: Fabrication methods and transport studies. *Proc. Natl. Acad. Sci.*, 100, 13755–13760, 2003, https://doi.org/10.1073/pnas.2331316100.

19. Yan, G., Warner, K.S., Zhang, J. *et al.*, Evaluation needle length and density of microneedle arrays in the pretreatment of skin for transdermal drug delivery. *Int. J. Pharm.*, 391, 7–12, 2010, https://doi.org/10.1016/j.ijpharm.2010.02.007.
20. Henry, S., McAllister, D.V., Allen, M.G., Prausnitz, M.R., Microfabricated microneedles: A novel approach to transdermal drug delivery. *J. Pharm. Sci.*, 87, 922–925, 1998, https://doi.org/10.1021/js980042+.
21. Omatsu, T., Chujo, K., Miyamoto, K. *et al.*, Metal microneedle fabrication using twisted light with spin. *Opt. Express*, 18, 17967, 2010, https://doi.org/10.1364/oe.18.017967.
22. Zhibiao, L., Xiaoxiao, Y., Min, H. *et al.*, The Study on Fabrication of Solid Metal Microneedles Based on Optimized Process of Electrochemical Etching and Cutting. *2018 IEEE Int. Conf. Intell. Robot. Control Eng.*, pp. 56–59, 2018.
23. Gill, H.S. and Prausnitz, M.R., Coating Formulations for Microneedles. *Pharm. Res.*, 24, 1369–1380, 2007, https://doi.org/10.1007/s11095-007-9286-4.
24. Kim, Y.C., Park, J.H., Prausnitz, M.R., Microneedles for drug and vaccine delivery. *Adv. Drug Deliv. Rev.*, 64, 1547–1568, 2012, https://doi.org/10.1016/j.addr.2012.04.005.
25. Peters, E.E., Ameri, M., Wang, X. *et al.*, Erythropoietin-Coated ZP-Microneedle Transdermal System: Preclinical Formulation, Stability, and Delivery. *Biomed. Microdevices,* 17, 44, 1618–1626, 2012, https://doi.org/10.1007/s11095-012-0674-z.
26. Ma, Y., Boese, S.E., Luo, Z. *et al.*, Drug coated microneedles for minimally-invasive treatment of oral carcinomas: development and *in vitro* evaluation. *Biomed. Microdevices,* 17, 44, 2015, https://doi.org/10.1007/s10544-015-9944-y.
27. Dalvi, M., Kharat, P., Thakor, P. *et al.*, Panorama of dissolving microneedles for transdermal drug delivery. *Life Sci.*, 284, 119877, 2021, https://doi.org/10.1016/j.lfs.2021.119877.
28. Vora, L.K., Vavia, P.R., Larrañeta, E. *et al.*, Novel nanosuspension-based dissolving microneedle arrays for transdermal delivery of a hydrophobic drug. *Eur. J. Pharm. Biopharm.,* 3, 89–101, 2018. https://doi.org/10.1002/jin2.41.
29. Rojekar, S., Vora, L.K., Tekko, I.A. *et al.*, Etravirine-loaded dissolving microneedle arrays for long-acting delivery. *Eur. J. Pharm. Biopharm.*, 165, 41– 51, 2021, https://doi.org/10.1016/j.ejpb.2021.04.024.
30. Domínguez-robles, J., Vora, L.K., Anjani, Q.K., Donnelly, R.F., Hollow microneedles: A perspective in biomedical applications. *Acc. Chem. Res.*, 599, 2021, https://doi. org/10.1016/j.ijpharm.2021.120455.
31. Han, X., Li, H., Zhou, D. *et al.*, Local and Targeted Delivery of Immune Checkpoint Blockade Therapeutics. *Acc. Chem. Res.*, 53, 2521–2533, 2020, https://doi.org/10.1021/acs.accounts.0c00339.
32. Chen, Y.H., Wang, F.Y., Chan, Y.S. *et al.*, Biofabricating hollow micronee- dle array with controllable microstructure for cell transplantation. *J. Biomed. Mater. Res. B Appl. Biomater.*, 110, 1997–2005, 2022, https://doi.org/10.1002/JBM.B.35054.

33. Mahadevan, G., Sheardown, H., Selvaganapathy, P., PDMS embedded microneedles as a controlled release system for the eye. *J. Biomater. Appl.*, 28, 20–27, 2013, https://doi.org/10.1177/0885328211433778.
34. Gupta, J., Felner, E.I., Prausnitz, M.R., Minimally Invasive Insulin Delivery in Subjects with Type 1 Diabetes Using Hollow Microneedles. *Diabetes Technol. Ther.*, 11, 329, 2009, https://doi.org/10.1089/DIA.2008.0103.
35. Kim, J.D., Kim, M., Yang, H., Lee, K., Jung, H., Droplet-born air blowing: novel dissolving microneedle fabrication. *J. Control. Release,* 170, 430–436, 2013, doi:10.1016/J.JCONREL.2013.05.026.
36. Park, J.H., Allen, M.G., Prausnitz, M.R., Biodegradable polymer microneedles: Fabrication, mechanics and transdermal drug delivery. *Annu. Int. Conf. IEEE Eng. Med. Biol. - Proc. 26*, vol. IV, pp. 2654–2657, 2004, https://doi.org/10.1109/IEMBS.2004.1403761.
37. Kim, J.D., Kim, M., Yang, H. *et al.*, Droplet-born air blowing: novel dissolving microneedle fabrication. *J. Control. Release*, 170, 430–436, 2013, https://doi. org/10.1016/J.JCONREL.2013.05.026.
38. Park, J.-H., Allen, M.G., Prausnitz, M.R., Biodegradable polymer microneedles: Fabrication, mechanics and transdermal drug delivery. *J. Control. Release,* 104, 51–66, 2005. https://doi.org/10.1016/j.jconrel.2005.02.002
39. McCrudden MTC, Alkilani AZ, McCrudden CM, et al (2014) Design and physicochemical characterisation of novel dissolving polymeric microneedle arrays for transdermal delivery of high dose, low molecular weight drugs. *J. Control. Release,* 180, 71–80. https://doi.org/10.1016/J.JCONREL.2014.02.007
40. Indermun, S., Luttge, R., Choonara, Y.E. *et al.*, Current advances in the fabrication of microneedles for transdermal delivery. *J. Control. Release*, 185, 130–138, 2014, https://doi.org/10.1016/j.jconrel.2014.04.052.
41. McGrath, M.G., Vucen, S., Vrdoljak, A. *et al.*, Production of dissolvable microneedles using an atomised spray process: Effect of microneedle composition on skin penetration. *Eur. J. Pharm. Biopharm.*, 86, 200–211, 2013, https://doi. org/10.1016/J.EJPB.2013.04.023.
42. Gupta, J., Gill, H.S., Andrews, S.N., Prausnitz, M.R., Kinetics of skin resealing after insertion of microneedles in human subjects. *J. Control. Release*, 154, 148–155, 2011, https://doi.org/10.1016/J.JCONREL.2011.05.021.
43. Gill, H.S., Denson, D.D., Burris, B.A., Prausnitz, M.R., Effect of microneedle design on pain in human volunteers. *Clin. J. Pain*, 24, 585–594, 2008, https:// doi. org/10.1097/AJP.0B013E31816778F9.
44. Gill, H.S. and Prausnitz, M.R., Coated microneedles for transdermal delivery. *J. Control. Release*, 117, 227–237, 2007, https://doi.org/10.1016/j. jconrel.2006.10.017.
45. Nagarkar, R., Singh, M., Nguyen, H.X., Jonnalagadda, S., A review of recent advances in microneedle technology for transdermal drug delivery. *J. Drug Deliv. Sci. Technol.*, 59, 101923, 2020, https://doi.org/10.1016/J. JDDST.2020.101923.

46. Siewert, M., Dressman, J., Brown, C.K., Shah, V.P., FIP/AAPS guidelines for dissolution/*in vitro* release testing of novel/special dosage forms. *Pharm. Ind.*, 65, 129–134, 2003.
47. Milewski, M., Yerramreddy, T.R., Ghosh, P. *et al.*, *In vitro* permeation of a pegylated naltrexone prodrug across microneedle-treated skin. *J. Control. Release*, 146, 37–44, 2010, https://doi.org/10.1016/j.jconrel.2010.05.034.
48. Moronkeji, K., Todd, S., Dawidowska, I. *et al.*, The role of subcutaneous tissue stiffness on microneedle performance in a representative *in vitro* model of skin. *J. Control. Release*, 265, 102–112, 2017, https://doi.org/10.1016/j.jconrel.2016.11.004.
49. Larrañeta, E., Stewart, S., Fallows, S.J. *et al.*, A facile system to evaluate *in vitro* drug release from dissolving microneedle arrays. *Int. J. Pharm.*, 497, 62–69, 2016, https://doi.org/10.1016/j.ijpharm.2015.11.038.
50. Tort, S., Mutlu Agardan, N.B., Han, D., Steckl, A.J., *In vitro* and *in vivo* evaluation of microneedles coated with electrosprayed micro/nanoparticles for medical skin treatments. *J. Microencapsulation*, 37, 517–527, 2020, https://doi.org/10.10 80/02652048.2020.1809725.
51. Economidou, S.N., Pere, C.P.P., Reid, A. *et al.*, 3D printed microneedle patches using stereolithography (SLA)for intradermal insulin delivery. *Mater. Sci. Eng. C*, 102, 743–755, 2019, https://doi.org/10.1016/j.msec.2019.04.063.
52. Indermun, S., Choonara, Y.E., Kumar, P. *et al.*, *In Vitro* and *In Vivo* Evaluation of a Hydrogel-Based Microneedle Device for Transdermal Electro-Modulated Analgesia. *J. Pharm. Sci.*, 106, 1111–1116, 2017, https://doi.org/10.1016/j.xphs.2016.12.022.
53. Bhatnagar, S., Chawla, S.R., Kulkarni, O.P., Venuganti, V.V.K., Zein Microneedles for Transcutaneous Vaccine Delivery: Fabrication, Characterization, and *in Vivo* Evaluation Using Ovalbumin as the Model Antigen. *ACS Omega*, 2, 1321–1332, 2017, https://doi.org/10.1021/acsomega.7b00343.
55. Zhang, X.P., Wang, B.B., Li, W.X. *et al.*, In vivo safety assessment, biodistribution and toxicology of polyvinyl alcohol microneedles with 160-day uninterruptedly applications in mice. *Eur. J. Pharm. Biopharm.*, 160, 1–8, 2021, https://doi. org/10.1016/j.ejpb.2021.01.005.
55. Aoyagi, S., Izumi, H., Isono, Y. *et al.*, Laser fabrication of high aspect ratio thin holes on biodegradable polymer and its application to a microneedle. *Sens. Actuators A Phys.*, 139, 293–302, 2007, https://doi.org/10.1016/j.sna.2006.11.022.
56. Donnelly, R.F., Singh, T.R.R., Woolfson, A.D., Microneedle-based drug delivery systems: Microfabrication, drug delivery, and safety. *Drug Deliv.*, 17, 187–207, 2010, https://doi.org/10.3109/10717541003667798.

11

Smart Nanocarriers in Drug Delivery Systems

Alfredo Amaury Bautista-Solano, Emilia Ramos-Zambrano, Nadia Romero-Martínez, Alma Chu-Martínez and Alma Leticia Martínez-Ayala*

Centro de Desarrollo de Productos Bióticos, Instituto Politecnico Nacional, Carretera Yautepec-Jojutla, Col. San Isidro, Km. 6, calle CEPROBI No. 8, Yautepec, Morelos, Mexico

Abstract

Nanocarriers have been created with the purpose of enhancing the specificity of active compounds. Consequently, these nanoassemblies need to possess a high degree of permeability. Nanoassemblies are designed to contain and somehow protect the active compounds within a matrix or support, which also controls internal interactions, by conjugating with specific ligands capable of binding to antigens or receptors on target cells, nanocarriers offer protection and enable targeted delivery. Materials can be constituted by inorganic and organic elements, like combinations that present unique physicochemical properties on which its application and functionality will depend, being the object of study for the design of nanocarriers to increase the therapeutic index, improve pharmacokinetics, and decrease adverse effects and toxicity; to develop safer therapeutics to overcome the problems of physicochemical stability. The efficacy of these has been proven, especially in cancer because high specificity is required, by the which a more efficient and selective systems are required, through the response to these specific stimuli in response to different microenvironment. Therefore, this review discusses the development of smart drug nanocarrier systems and advances in materials used, as well as its physicochemical properties and applications.

Keywords: Nanomedicine, nanoassemblies, diseases treatment, drug release, nano-encapsulation, stimuli responsive system, target release, nanotechnology

**Corresponding author*: alayala@ipn.mx

Sathish-Kumar Kamaraj, Arun Thirumurugan, Muthuchamy Maruthupandy, Mercedes Guadalupe López Pérez and Shanmuga Sundar Dhanabalan (eds.) Nanoscience and Nanotechnology for Smart Prevention, Diagnostics and Therapeutics: Fundamentals to Applications, (261–286) © 2024 Scrivener Publishing LLC

11.1 Introduction

The development and implementation of nanotechnology as a concept of manipulation of matter at the atomic level at the beginning of the 21st century gave rise to its use in the medical field as nanomedicines. Nanosystems are also utilized in the diagnosis and therapy of diseases through controlled drug release. The nanocarriers, defined as a colloidal system with a size less than 100 nanometers, have proven to be successful because they improve the bioactivity of drugs, pharmacokinetics, and biodistribution, while reducing toxicity and increasing their stability and solubility, thus controlling the specific release of these drugs [1].

A smart drug delivery system comprises intelligent nanocarriers, targeting mechanisms, and stimulation techniques. These systems have demonstrated their ability to reduce the adverse effects associated with nonspecific distribution and uncontrolled drug release observed in traditional drug delivery systems [2]. So, research on the development of these systems has increased exponentially every year. The design of these systems primarily relies on utilizing nanocarriers such as liposomes, micelles, dendrimers, mesoporous silica nanoparticles, albumin nanoparticles, gold nanoparticles, superparamagnetic iron oxide nanoparticles, carbon nanotubes, and quantum dots. Each of these nanocarriers presents distinct advantages and disadvantages concerning toxicity and effectiveness. These systems can also be classified based on the stimulus to which they will respond for drug release. Response stimuli can be endogenous or exogenous. Endogenous stimuli encompass factors like pH, enzymatic transformation, temperature, and redox reactions. Conversely, external factors such as magnetic fields, ultrasonic waves, light, and electric fields have been employed as exogenous stimuli [2].

On the other hand, each of these materials used for the design of intelligent nanocarriers has physicochemical properties that determines its activity and gives it characteristics related to its responses to different stimuli, which make these delivery systems more effective and specific [3] Drug delivery systems utilizing nanocarriers have been extensively tested in a wide range of conditions and diseases that hold significant medical importance, for example: cancer, infections by different bacteria and fungi, hypertension, glaucoma, pulmonary embolism, inflammation, sexually transmitted diseases, human immunodeficiency virus (HIV), and arteriosclerosis, as well as other neurodegenerative, eye, and skin diseases [1]. Therefore, in this review, we will address aspects related to the composition of nanocarriers and their evolution, these intelligent systems have proven to be increasingly effective alternatives for the treatment of diverse diseases.

11.2 Progress in Materials Chemistry and Drug Delivery in Smart Nanocarriers

The need to explore new drug delivery systems that are more effective has led to increased study of nanoparticles and their components [4]. An important aspect in the synthesis of nanocarriers is the choice of the assembly material, and its choice depends on the diagnosis, the target tissue, the route of administration used, and the physicochemical characteristics of the drug, among others [5]. There are many types of materials used to form nanoparticles, such as polymers (cellulose, chitosan, alginate, collagen, starch), lipids (phospholipids, cholesterol), metals (copper, zinc, gold, silver), and others. It should be noted that smart materials are characterized by a larger surface area for drug delivery in the case of tumor environment smart nanocarriers can release a drug in a control manner after stimulation [6]. The following are some examples of intelligent materials.

11.2.1 Silica Nanoparticles

Mesoporous materials have the honeycomb-like porous structure of silica, with high surface area, tunable and large pore volume, high stability and compatibility, and allow the release of loaded drugs in response to changes in temperature, pH, redox reaction, light, magnetic field, or enzyme-mediated transformations [6]. Silica nanoparticles can be prepared with different sizes (50–300 nm) and can be loaded with anticancer drugs, such as camptothecin and doxorubicin, and modified with quantum dots and polyethylene glycol (PEG). Drug release can be controlled as they can be designed to release the drug in response to various stimuli reducing toxicity in healthy tissues; therefore, they have been heavily tested in the treatment and diagnosis of cancer [2].

11.2.2 Chitosan

Chitosan is a natural cationic polysaccharide polymer. The characteristics that make it eligible are its biodegradability and biocompatibility. Chitosan-based nanoparticles can be combined with mesoporous silica forming multi-stimuli-responsive surfaces that release drugs in response to pH changes [6]. Chitosan nanocomplexes can be modified with folic acid and carboxylate groups to form pH-sensitive gene nanocarriers. In addition, by modifying chitosan nanocarriers with PEG and stearic acid copolymers, micelles sensitive to various stimuli are obtained for drug release [7].

11.2.3 Metal-Based Nanoparticles

Zinc and gold are increasingly used for the design of metallic nanoparticles. Zinc-based nanoparticles can be designed for controlled release of anti-cancer drugs by ultrasound, microwave irradiation and pH changes with therapeutic efficacy, while reducing unwanted side effects, since zinc ion is biocompatible *in vivo*, so, due to its safety, it is used in the formation of anticancer nanocomposites, for example, zinc and quercetin nanoparticles have been effective against several types of cancer [6]. Gold nanoparticles have unique characteristics that lead to drug delivery of large biomolecules; however, gold nanoparticles need to be modified with PEG to avoid the uptake by cells of the reticuloendothelial system (RES) in the blood [8]. It can be functionalized with ligands for targeted drug delivery, and these nanoparticles can be stimulated to release the drug [2].

11.2.4 Quantum Dots

Quantum dots (QDs) are fluorescent semiconducting nanocarriers. Their structure is composed of 3 parts: a core, a shell, and a capping material. Quantum dots could be water-soluble with sizes of 2 to 4 nm [8]. To improve their effectiveness, they are often functionalized with PEG or antibodies. Due to their unique properties, QDs are applied in bioimaging, drug delivery, and biological labeling [2].

11.2.5 Liposomes

Conventional liposomes are phospholipid-based nanovesicles and can encapsulate hydrophilic or lipophilic drugs. As drug delivery systems present problems such as instability, faster drug release, and insufficient drug loading, they are functionalized in order to be considered smart [4]. Smart liposomes are capable of responding to exogenous or endogenous stimuli. Liposomes modified with fusogenic peptides and PEG were designed to be cleaved by tumor-expressed matrix metalloproteinase-2 for intracellular delivery of small interfering ribonucleic acid (RNA). Within this broad variety of nanoformulations, polymeric nanoparticles (NPs) and liposomes are the most studied at the clinic level.

11.2.6 Micelles

Micelles are amphiphilic molecules made of polylactide, PEG, and copolymers of 2-(methoxyethoxy) ethyl methacrylate. Also, they can be functionalized with different ligands and designed for drug release when a stimulus is applied, especially in cancer. Polymeric micelles could be thermoresponsive nanocarriers for the controlled delivery of drugs and genes [7].

11.2.7 Dendrimers

Dendrimers are a class of synthetic polymers with many branches that form a tree-like structure [4]. The modification of conventional dendrimers with linear polymers prevents them from being rapidly eliminated by the immune system, and they can be modified with aptamers, antibodies, proteins, and carbohydrates to target cancerous tissue. On their surface, they can be modified with different molecules, such as hydrazone, acid-activable ligands, peptides, collagen, anticancer drugs conjugated to dendrimers via ortho-nitrobenzyl, and others, in order to obtain multi-stimuli-responsive dendrimers [2].

There are diseases, such as cancer, that due to their complexity must be treated with unconventional treatments; therefore, new forms of drug delivery have been tested using therapeutic nanoparticles that carry poorly soluble or easily degradable drugs [5]. It is important to discuss the properties of smart drug delivery systems that make them necessary for the treatment of difficult-to-fight diseases, such as cancer. Smart nanocarriers can release the drug when an external or internal stimulus is applied to them, which minimizes adverse effects and increases the efficiency of cellular uptake because of the prolonged systemic circulation lifetime [4]. Due to the pathophysiological differences that exist between healthy and cancer cells, smart drug delivery systems can be designed to act on tumor tissue, referred to as passive targeting, when the nanocarriers accumulate in the interstitial tissue fluid due to enhanced permeability and retention effect (EPR), keeping the drug in the tumor tissue for a longer time at higher concentrations [9]. Another approach is active targeting, which is accomplished by functionalizing the surface of the nanoparticles with some ligands (antibodies, epidermal growth receptor, fibroblast growth receptor, peptides, aptamers, transferrin, carbohydrates) to target tumor tissues that are overexpressing some receptors [8].

11.3 Physicochemical Properties of Smart Nanocarriers

Nanoscience facilitates the convergence of various scientific disciplines, including biology, chemistry, and physics. It possesses a remarkable capability to explore and manipulate phenomena at atomic, molecular, and macromolecular levels. Furthermore, nanoscience holds tremendous potential in diverse areas, encompassing the design, synthesis, characterization, and implementation of intelligent nanocarriers [10]. When materials are examined on a nanoscale, they exhibit distinctive properties, with alterations occurring in their chemical reactivity, magnetism, optics, and conductivity. Consequently, this scale serves as a foundation for a multitude of materials with wide-ranging applications. A comprehensive grasp of the physicochemistry of nanomaterials is imperative to comprehend these phenomena (Figure 11.1) [3].

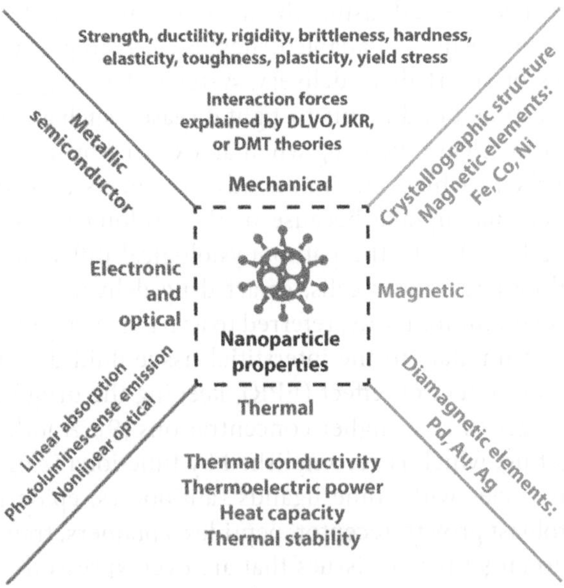

Figure 11.1 Physicochemical properties of nanoparticles. DLVO, Derjaguin, Landau, Verwey, and Overbeek; JKR, Johnson-Kendall-Roberts; DMT, Derjaguin, Muller, and Toporov; Pd, palladium; Au, gold; Ag, silver.

11.3.1 Mechanical Properties

Mechanical properties play a vital role in determining the response of a material under mechanical stress. Their manifestation depends on environmental conditions, external forces, and the composition of the material, be it organic, inorganic, or a hybrid with a specific atomic arrangement. At the nanoscale, mechanical properties undergo significant changes, resulting in enhanced mechanical strength compared to bulk materials. This phenomenon arises from increased surface area and quantum effects, primarily attributed to a variety of interactions occurring between nanocarriers and the surrounding surfaces. These interactions involve multiple forces such as Van der Waals forces, double-layer electrostatic forces, lateral capillary forces, solvation effects, structural factors, and hydration effects [11]. The explanation for these changes is provided by several theories. The Johnson-Kendall-Roberts (JKR) theory elucidates the adhesion of elastic spheres, considering both the sphere and the plane within the JKR boundaries as deformable entities. It involves an elastic response accompanied by energetic interactions between the two surfaces [12]. On the other hand, the Derjaguin, Landau, Verwey, and Overbeek (DLVO) theory describes how Van der Waals forces and double-layer electrostatic forces are independent of each other, allowing for their overlapping interaction at a specific distance between two particles [13]. Lastly, the Derjaguin, Muller, and Toporov (DMT) theory discusses elliptical and circular geometry in a Hertzian contact relationship. While it disregards surface forces, it introduces the concepts of surface energy and adhesive force in energy evaluations. This theory applies to the adhesion of rigid elastic solids with adhesive interactions, utilizing the law of specific attractive force, and finds applications in various fields [14].

11.3.2 Thermal Properties

Thermal properties refer to the characteristics of matter that relate to its interaction with thermal energy or heat. These properties describe the behavior and response of materials when subjected to thermal loads or changes in temperature. Thermal properties encompass a wide range of phenomena associated with the transfer, storage, and dissipation of heat within a material. Understanding these properties is crucial for studying heat transfer, thermal conductivity, expansion coefficients, specific heat capacity, and other thermal phenomena. The main components of thermal properties are thermoelectric energy, heat capacity, stability, and thermal conductivity. In the case of nanocarriers, their dimensions and size influence network vibrations. In bulk materials the composition of the elements that make it up largely determines their thermal properties, however, in nanocarriers size is the determining factor in thermal properties. The size has a guiding effect on the process of absorption of thermal energy and conductivity due to anharmonic vibrations that affects the volume and temperature of the object. Similarly, electron conduction is a primary mechanism through which heat is transferred. This phenomenon is influenced by the surface-to-volume ratio of a material, indicating that a larger surface area relative to its volume allows for a greater number of electrons involved in heat transfer. This relationship suggests that materials with higher surface areas are more efficient at conducting heat due to the increased availability of electrons for thermal energy transport [11]. The vibrations of the lattice influence all the thermal properties of materials, since the temperature will be a function of the degrees of freedom of this type of vibrations, particularly of the vibrational energy that arises from the oscillating atoms. As we know, nanocarriers tend to change their thermal properties due to their size. In general, this behavior is characterized by a notable reduction in the melting temperature when compared to bulk materials. This phenomenon can be explained by the fact that the energy associated with the liquid/vapor interface is lower than that of the solid/vapor interface. As the particle size decreases, the surface volume ratio increases. Consequently, a larger proportion of atoms or molecules are located near the surface, leading to a more pronounced influence of surface effects on the material's thermal properties, including the lowering of the melting temperature.

11.3.3 Magnetic Properties

Magnetism is known as a type of force, either attraction or repulsion. Certain elements within the d-block, such as iron (Fe), cobalt (Co), and nickel (Ni), as well as certain rare earth elements, exhibit magnetic phenomena, specifically ferromagnetism. Ferromagnetism is characterized by the alignment of electron spins within a material, resulting in a permanent magnetization even in the absence of an external magnetic field. These elements possess unpaired electrons in their atomic or ionic configurations, which contribute to their magnetic behavior of materials at the nanometric level, we need to know the can be influenced by several factors, such as composition, crystallographic structure, magnetic anisotropy, and the presence of nanoscale vacancy defects. Composition refers to the elemental composition of a material can significantly impact its magnetic properties. Different elements have varying magnetic characteristics, such as their magnetic moments and interactions with neighboring atoms or ions. The crystallographic arrangement of atoms or ions in a material can affect its magnetic behavior. Certain crystal structures, such as those with a close-packed arrangement, may exhibit enhanced magnetic properties due to favorable atomic interactions and ordering. The phenomenon of Magnetic anisotropy refers to the directional dependence of a material's magnetic properties. It arises from the preferred alignment of magnetic moments along specific crystallographic directions. The presence of magnetic anisotropy can influence factors such as the material's coercivity (resistance to demagnetization) and the stability of its magnetic domains. The Nanoscale vacancy defects, such as missing atoms or ion vacancies within a material's lattice, can affect its magnetic properties. These defects can disrupt the regular atomic arrangement and introduce localized magnetic moments or alter the magnetic interactions within the material, leading to changes in its overall magnetic behavior. These phenomena occurs at a characteristic size for each type of nanoparticle, where the anisotropy energy is equal to the thermal energy, giving rise to the random change of the magnetic moment. Coercivity, saturation magnetization, locking temperature, and relaxation time are other factors that determine the properties of nanocarriers. Coercivity refers to the strength of an external magnetic field to reinforce magnetization to zero. Moreover, saturation magnetization is said to be the maximum point of magnetization of a nanocarrier, which is possible when all magnetic dipoles are aligned in an external magnetic field. The blocking temperature arises from thermal fluctuations that are comparable or greater than an energy barrier in the inversion of magnetic moments, allowing the random change of magnetic moments [15].

Nanoparticles demonstrate interesting magnetic behavior, including a zero magnetic moment in the absence of an external magnetic field, as well as a rapid increase in the magnetic moment when subjected to such a field. This behavior is closely associated with the phenomenon of magnetic viscosity, which is observed in ferromagnetic and ferrimagnetic nanoparticles possessing a single magnetic domain because of superparamagnetism. The Néel relaxation time provides an explanation for this magnetic viscosity phenomenon. It refers to the characteristic time required for the magnetization of a nanoparticle to relax or respond to changes in the external magnetic field. In the case of superparamagnetic nanoparticles, which consist of a single magnetic domain, the magnetization can rapidly fluctuate between different orientations due to thermal energy. These fluctuations cause the magnetic moment to change direction, resulting in a characteristic time scale for relaxation. When an external magnetic field is applied, it can align the magnetic moments of the nanoparticles, causing an increase in the overall magnetization. However, once the external field is removed, the thermal energy can rapidly disrupt the alignment, returning the nanoparticles to a state of zero magnetic moment. This behavior is responsible for the magnetic viscosity observed in these nanoparticles [15].

Superparamagnetism is a magnetic phenomenon that shares characteristics with ferromagnetism and paramagnetism. In nanocarriers, this phenomenon will only occur with ferromagnetic or ferrimagnetic materials because nanocarriers have effects on magnetic coercivity. Nanocarriers can indeed possess single magnetic domain structures below a certain critical radius, where all magnetic spins align unidirectionally. However, instead of generating a superparamagnetism effect, this alignment leads to a phenomenon called single-domain ferromagnetism. In single-domain ferromagnetic nanocarriers, the magnetic moments of all atoms or ions within the nanocarrier align in a uniform direction due to strong magnetic interactions. This alignment results in a net magnetic moment and makes the nanocarrier behave as a single magnetic domain. Single-domain ferromagnetic nanocarriers exhibit enhanced magnetic properties, such as a high saturation magnetization and strong response to external magnetic fields. if the nanocarrier radius is less than the threshold radius. The relationship between superparamagnetism, critical radius and size of the nanocarrier influences the magnetic coercivity since increasing the size of the nanocarrier up to the maximum critical radius generates a high magnetic coercivity in the nanocarriers. Finally, we can say that another point to consider its geometry due to the change of magnetic effects, example of this are the nanoparticles of COFe2O4 in its cubic or spherical geometry. Cubic structures possess less coercivity than spherical ones, which can be explained due to surface fixation. Overall, a combination of factors, including composition, crystallographic structure, magnetic anisotropy, and the presence of nanoscale vacancy defects, collectively contribute to the magnetic properties exhibited by a material. Understanding and manipulating these factors can enable the design and engineering of materials with tailored magnetic behavior for specific application [11].

11.3.4 Electronic and Optical Properties

The optical and electronic properties of nanoparticles and nanocarriers depends on the size and chemical element of the nanometer structure. Electronic properties are those that describe the state and behavior of electrons in a material. At the nanometric level, electrons are confined to a space of certain dimensions according to the structure, such as nanowires and nanoparticles, producing a quantum confinement effect. Quantum confinement basically indicates a restriction of electronic movement in a certain spatial region, this effect is associated with a reduction in size in one or more spatial directions, which produces a change in electronic density, and this in turn leads to changes in certain features, such as the formation of an ensemble of electronic states, in discrete energy states that depend on confined spatial directions, and expanding these discrete energy states by increasing quantum confinement. This directly affects the electronic properties of the material going from conductor to semiconductor or semiconductor to conductor modifying the insulating properties. Examples of this are metals like bismuth and semimetals like silicon. Therefore, we can say that the type of chemical element is of equal importance to the size. Metal and metalloid nanoparticles have linear absorption, optical properties, photoluminescence emission, to name a few; due to quantum confinement effects and surface-localized plasmon resonance (collective excitations of conducting electrons in metals); this type of resonance arises when the frequency of the incident photon is constant with the collective excitation of electrons. Optical properties, such as refractive index, damping constant, and absorbance, are those that relate the structure of a material and electromagnetic radiation in the form of waves or particles. Like electronic properties, optical properties depend on the size and chemical element of the material, where effects such as the quantum confinement of spatial directions intervene. In addition to this, it is important to mention that the shape of the nanocarrier influences these properties. Studies indicate that the resonance wavelength of the plasmon makes a shift in the visible wavelength, contrary to the plasmons of bulk materials which cannot be excited by visible light because nanoparticles tend to have a flat surface, so the resonance plasmon is highly dependent on the size. Another point to note is that the electric di-environment also directly influences the wavelength of the plasmon, being linearly dependent on the refractive index of the solvent in which nanocarriers are found [15].

11.4 Stimuli-Responsive Nanosystems in Smart Nanocarriers

Nanocarriers have different properties depending on the type of element, geometry, and size. These properties can be changed and manipulated if exposed to endogenous and exogenous stimuli. Due to this, a broad array of applications is offered, including the release at desired sites of bioactive

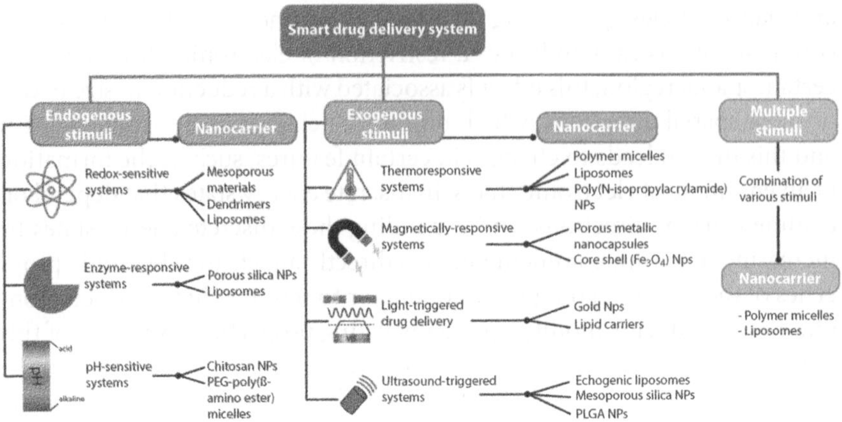

Figure 11.2 Schematic presentation of smart nanomaterials employed as nanocarriers. NPs, nanoparticles; PLGA, poly(lactic-co-glycolide).

compound. Therefore, we can classify these nanostructures as stimuli-sensitive nanocarriers (Figure 11.2). These types of nanocarriers are normally formed by amphiphilic molecules using lipids or surfactants to create a hydrophilic outer shell and a hydrophobic inner core. The main endogenous and exogenous stimuli that are used in the administration of drugs through these nanocarriers are shown below [16].

11.4.1 Exogenous Stimuli-Responsive Drug Delivery

Exogenous factors are those that are outside a defined system and can be of different nature or composition. Due to this, these nanocarriers are susceptible to different stimuli, so they are used to perform releases based on temperature, light, and magnetism, among others.

11.4.1.1 Heat-Sensitive Nanocarriers

The nanocarriers affected by heat can be classified as thermosensitive systems for which reason materials are used that keep the active compound encapsulated at body temperature and at the same time are susceptible to temperature changes around 310.15 K, generating a response by the nanocarrier, that is, a thermoresponse. These nanocarriers are used in tissues with inflammation, injury, infection, and carcinogenic damage due to the temperature increases that occur in these sites. For example, in cancer treatments, where the tumor site has a temperature around 314.15 K, the drug is delivered as a response to this slight change in temperature in addition to counteracting the passage of blood. On the other hand, there are also affected areas where the temperature decreases, causing an increase in the porosity of the nanotransporters, which leads to the diffusion of the encapsulated drug [11].

11.4.1.2 Light-Sensitive Nanocarriers

This type of nanocarrier has a spatiotemporal response to light. They are used for drug delivery in noninvasive systems, utilizing wavelengths within the ultraviolet or infrared range of the electromagnetic spectrum to trigger the release of the encapsulated drug by reversible reactions or irreversible structural modifications. An example of this is the use of azobenzene and its silica-based nanocarriers, which create a system sensitive to ultraviolet light that causes a structural modification, releasing the compound [17].

11.4.1.3 Magnetism-Sensitive Nanocarriers

This type of nanocarriers use magnetic elements to form a structure with a magnetic core and a surface coating, causing the bioactive molecules are charged on the surface, producing modifications in it. This type of transporter is used for the delivery of drugs by focusing methods, either passive or active. The mechanism of these transporters is based on the exposure of magnetic fields that generate thermal and mechanical effects, causing a reversal of the magnetization, which is determined by two factors the amplitude and frequency of the magnetic field ensures that the drug penetrates the right place, and an increase in the effect of the drug [18].

11.4.1.4 Ultrasound-Sensitive Nanocarriers

Ultrasound is defined as longitudinal waves with a frequency and direction of propagation parallel to the direction of displacement of the molecules of a system as a means of propagation. The nature of a specific intensity and frequency is because ultrasound occurs in a cross section, where more energy is transported than in the rest of the sound waves. In general, there are two types of ultrasounds: those of low intensity with a frequency less than 20 kHz and those of high intensity with a frequency greater than 20 kHz. High intensity ultrasound is useful in the administration of certain dosage forms that generate sonoporation mechanisms that induces other phenomena in the cells, such as the entry of calcium into the cytoplasm and the exit of potassium, which produces hyperpolarization of the cell.

On the other hand, low intensity ultrasound is used in sonodynamics, which induces cavitation processes, a busy strategy for the treatment of deep tumors. This type of exogenous factor is correlated with thermal and mechanical effects and radiation forces, offering a controlled release of drugs, safe penetration into tissues, noninvasiveness, and better spatiotemporal control. Examples of this are mesoporous silica nanostructures PEGylated by a thermosensitive linker, poly (lactic-co-glycolic acid) (PLGA) nanoparticles, and liposomes [11].

11.4.2 Endogenous Stimuli-Responsive Drug Delivery

Endogenous stimuli, also called internal stimuli, are those where the nanocarriers will be in contact with these internal stimuli when administered in a system. Some of these factors, such as pH, redox potential, and enzyme levels, are described below.

11.4.2.1 pH-Sensitive Nanocarriers

pH is used as a controllable switch for the controlled and specific release of drugs from nanocarriers. There are several mechanisms of pH-induced release, such as nanocarrier dissolution, nanocarrier swelling, or a combination of both; the above mechanisms have the same principle based on the use of a specific pH using various chemical species. Different types of nanocarriers exhibit pH sensitivity, such as polymer particles, liposomes, micelles, and dendrimers. The chemical structure of these nanocarriers will influence their sensitivity to changes in pH depending on the functional groups, the length of the carbon chain and the intervention of the monomers that make up the nanocarriers. These nanocarriers are generally designed based on bonded surface materials that are sensitive to acids, generating a dissociation when undergoing a pH change, which releases the drug at the specific site. They are generally used in pathologies like cancer or inflammation [19].

11.4.2.2 Nanocarriers Sensitive to Redox Reactions

Drug release from nanocarriers can occur from oxidation-reduction reactions by formulating transporters based on redox-sensitive bonds. The release mechanism is based on the dissociation of bonds, which release of the drug. An example of this is the reduced glutathione disulfide bonds, as they occur in various intracellular compartments. The disulfide bonds of the nanocarriers are reduced to a thiol due to the process of endocytosis and high levels of reduced glutathione, resulting in the dissociation of the nanocarriers and subsequent release of the encapsulated drug [20].

11.4.2.3 Enzyme-Sensitive Nanocarriers

Furthermore, enzymes play a vital role in the metabolism and regulation of organism, additionally they are useful in the diagnosis of various diseases because they function as indicators in certain diseases, since when enzyme concentrations are abnormal, they indicate a deregulation in their expression. Example of these enzymes are proteases, phospholipases, and glucosidases, which are found in tumors and inflammatory tissues; the substrate is used as a component of the delivery system of the nanocarrier interacting through enzymatic reactions, which achieves a controlled release in the target tissue [19].

11.4.3 Multiple Stimuli-Responsive Drug Delivery

Nanocarriers are highly specific according to the site where they act and the factors involved; therefore, they have designed transporters that combine various stimuli with the aim of increasing specificity. In addition to this, multi-response nanocarriers possess higher drug loading efficiency and longer release times. The combination of these stimuli can be either endogenous, exogenous or a combination of both and react simultaneously or at separate times. One of the applications for these transporters is in the use of genes and drugs that have a synergistic effect on alterations, such as cancer, as a dual response therapy design [20].

11.5 Clinical Status of Stimuli-Responsive Nanocarriers

Research on nanocarrier-based therapies against various diseases has improved in recent years. Most of the smart drug delivery systems are focused on cancer treatment, due to the physiochemical differences between cancer and healthy cells [21, 24]. The effect of nanoparticles has also been evaluated against autoimmune or infectious diseases such as pneumonia, hepatitis, HIV and Acquired Immune Deficiency Syndrome (AIDS) [24].

In the clinic, nanomedicines show low effectiveness due to the pharmaceutical problems with the nano-drug delivery systems and the lack of efficient models to emulate the human body [22]. Also, the reasons for not producing the desired effects of nanoparticles in clinical applications have to do with several events, including that the particle delivery systems that are in the blood must enter the surrounding tissue when the local area is activated by some stimulus and that the nanoparticles must release their contents into the target tissue quickly [21]. However, implementing endogenous stimuli in clinical practice has been very challenging, especially with changes in pH and enzyme concentration; therefore, stimuli-responsive nanoparticles also need to be formulated and meet reproducibility, scalability and effectiveness characteristics [8].

Animal models have been useful in the evaluation of the effect of nanomedicines; however, they are ineffective for analyzing specific physiological or molecular mechanisms due to the complexity of animal physiology, and these models fail to accurately reproduce human diseases. However, *in vitro* cultures and animal models have been very useful for evaluating the effectiveness and toxicity of the nanocarriers in the short and long term [9]. Advances in the efficacy of nanoparticle formulations have been observed in small animal models, leading to expectations of improved drug delivery efficacy in human patients [7].

In order to overcome pharmaceutical problems, many nanoformulations using liposomes, polymeric nanoparticles, polymeric micelles, dendrimers, and QDs have been studied at the clinical level, it should be noted that the characteristics mentioned in Table 11.1 are taken into account

Table 11.1 Desirable characteristics of smart nanocarriers to be considered at the clinical phase.

Pharmaceutical characteristics	**Safe and easy application, low cost, upscaling, easy to store, easy to sterilize, adequate containers**
Blood circulation	Hemocompatibility, prolonged circulation time, controlled drug release, no aggregation, no uptake by cells of the reticuloendothelial system, no lung clearance, does not cause embolism
Target tissue	Tissue/cell specific targeting, penetration of biological barriers, uptake into cells/organs, biodegradability, biocompatibility, no cytotoxicity or immunogenicity, low molecular weight
Metabolism	Nontoxic degradation products, no accumulation in organs, excretion by liver and kidneys

when designing a nanoformulation [9, 11]. A promising strategy has been the development of stimuli-responsive nanodevices to improve drug specificity and effectiveness with a smaller dose [8]. The internal or external stimuli systems can be classified as smart if they display reversible structural changes, respond proportionally to the stimulus intensity, and can be considered as an adaptive medicine [4]. If the stimulus interferes with the normal structure of the nanoparticle, disassembly leads to drug release [2].

The search for advanced excipients that can give rise to responsive formulations without losing sight of the main functions of excipients, which are to facilitate drug preparation and maintain drug stability, has prompted an astounding attempt for finding suitable stimuli-sensitive excipients. In this sense, the behavior of the excipients is very important because the materials also interact with the surrounding environment and biological functions. To prevent nanocarriers from causing toxicity, they should be administered when they are stable to avoid leakage of drugs from nanocarriers before reaching the target site [7]. The formulation of nanocarriers still needs to be improved, as they are known to show some degree of toxicity, although their effectiveness is greater than that of conventional drugs. Some disadvantages of drugs that use drug efflux transporters to release the bioactive molecule can be countered by using nanotransporters to achieve better solubility, bioavailability and pharmacokinetics [4].

For smart nanoparticles to exert the desired effect, it must be ensured that external or internal stimuli can be reliably activated [7]. It has been reported that in the preclinical setting, nanomedicines are often more effective than in clinical practice, reports indicate that patients receiving nanocarrier treatments benefit primarily from decreased adverse effects. As a result, compared to the numerous reports of clinical trials using nanocarriers, it can be considered that nanomedicine-based products that are approved in the clinical area are few in number [22]. Multiple biological and pharmaceutical barriers contribute to this imbalance [7]. Currently, research should focus on decreasing the problems associated with nanoformulation and the production of smart nanocarriers so that the effectiveness of smart nanocarriers is improved; however, they are also more affordable since their manufacturing cost is high. Therefore, for a nanomedicine to be effective, a key point is to select the right drug and to administer it in combination with other drugs [8]. Nowadays, multiple smart stimuli-responsive nanostructures are under clinical trials; however, few of them are approved, such as Opaxio®, Guardian® Real-time Continuous Glucose Monitoring System, Visudyne®, Cornell Dots®, and Nanotherm® [9].

11.5.1 Nanocarriers in Clinical Cancer Care and Other Diseases

The advances in nanomedicine have been attributed to improvements in efficiency, toxicity, drug stability, and diffusion [23]. Smart nanocarriers are encapsulated in a nanoparticle that can be delivered to target cells. Once inside the cells, the drugs are released from the nanoparticle carrier [24]. Furthermore, nanocarriers with multi-stimulus responses may be able to achieve a more precise dosage control [25].

11.5.1.1 Nanocarriers in Cancer

The efficacy of cancer treatments is constrained by inherent limitations. Chemotherapeutic agents based on conventional pharmaceutical preparations are generally distributed throughout the body, and their concentration in the tumor(s) is low, resulting in severe side effects and a diminished therapeutic efficacy. Despite the progress of treatment modalities, they often damage healthy tissues [24, 27]. Instead, nanomedicine is used to distribute medicines by causing the cancerous cell(s) to ingest the nanocarrier. Nanoparticles possess a small size and surface characteristics that facilitate their penetration through blood vessel walls and into tissues [27]. Considering that tumors have leaky blood vessels and poor lymphatic drainage, nanoparticles may accumulate in them. Several classes of nanocarriers have been used for anticancer drug delivery, including liposomes, organic, inorganic, or hybrid nanoparticles, polymeric micelles, polymer drug conjugates, and nanogels. Besides solubilizing poorly soluble drugs and protecting them from degradation in vivo, drug-loaded nanocarriers could prolong the circulation time of free drugs and selectively deliver them to the target tissue [26].

Smart drug carrier vehicles are capable of delivering medications to a patient in a manner that enhances the concentration of therapeutic agents in a specified target site [25]. The literature reported that Doxil® was the first approved nanodrug by the Food and Drug Administration (FDA) in 1995. Due to EPR, Doxil® is "passively targeted" to tumors. As doxorubicin is released and becomes available to tumor cells, it extravasates and accumulates as intact liposomes in tumors [28]. The nanodrug is administered intravenously at 1 mg/min in order to minimize the risk of adverse reactions. It is indicated for the treatment of ovarian cancer, AIDS-related Kaposi sarcoma, and multiple myeloma [29]. Abraxane® is another nanomedicine that acts as a microtubule inhibitor and is indicated for the treatment of metastatic breast cancer, which is applied after chemotherapy treatments for metastatic disease. In other types of metastatic cancer, such as pancreatic adenocarcinoma, it is the first-line treatment in combination with gemcitabine, whose mechanism of action is the inhibition of DNA synthesis and the inhibition of the enzyme ribonucleotide reductase and RNA [30, 31]. Vincristine sulfate is used either alone or in conjunction with other oncolytic drugs for the treatment of leukemias, including acute lymphocytic leukemia, chronic lymphocytic leukemia, acute myelogenous leukemia, and the blast crisis of chronic myelogenous leukemia. The medication is administered intravenously on a weekly basis [32].

11.5.1.2 Nanocarriers in HIV/AIDS

There are several nanomedicine methods for the oral distribution of antiretroviral medication that are being investigated and tested. The current treatment approach employs solid medication nanoparticle formulations, which have the potential to alter tissue dissemination and provide a potential cure for HIV. Nanocarrier devices that exploit the mannose receptor have been suggested for HIV therapy [27].

11.5.1.3 Nanocarriers in Tuberculosis

Tuberculosis can be treated with nanocarriers through different routes, such as intravenous, intraperitoneal, oral, topical, and intranasal routes. They aim to improve drug permeability in the bacterial microbiome, reduce drug-related side effects, increase antimicrobial potency, prevent the emergence of drug resistance, reduce dose size, dosing frequency, and the duration of tuberculosis [33]. Liposomes, which are more easily absorbed by cells than larger molecules, made antitubercular drugs more effective against mycobacterial infections. Liposomes are a good way to transport and distribute drugs. Because medications are durable and release continuously from nanoparticles, oral administration is possible [27].

11.5.1.4 Nanomedicines in Other Diseases

There are approved treatments for other diseases and conditions, such as multiple myeloma, AIDS-related Kaposi sarcoma, acute lymphoblastic leukemia, metastatic pancreatic cancer, soft tissue sarcoma, epithelial cancer, solid tumors, and gastric cancer [9]. Additionally, numerous studies have demonstrated efficacy in the treatment of severe acute respiratory syndrome coronavirus 2 [34] and as therapeutic strategies for diabetes [35].

11.6 The Role of Bioinformatics and *In Silico* Analysis in the Design of Smart Nanocarriers

Bioinformatics and *in silico* analysis as tools for the design of intelligent nanocarriers are currently the most used resources, since they help predict, albeit partially, how the different materials used for the design and production of nanocarriers can be assembled in smart nanocarriers and allow us to study the effect of various characteristics of the nanoassemblies. For example, the size, the modification of its surface, the shape, the chemistry of its surface, the elasticity, its toxicity, its hydrophobicity, and even the effect of the concentration of these nanosystems can help predict the possible behavior in *in vivo* systems. In addition, these studies allow us to elucidate how these nanoassemblies penetrate the tissues or, more internally, the cell membrane [36]. In order for nanocarriers to be effective, they need to be able to overcome or cross these barriers but also be able to travel through the vasculature and avoid being captured by the endothelial reticulum, so they can reach their place of action, such as a tumor tissue. Then, by endocytosis, they reach the interior of the cells to release the bioactive compound or drug. The recent computational advances for *in silico* analysis and other simulations at different scales allow the ability to improve the design of nanocarriers by evaluating parameters that are closer to the reality of biological scenarios [37].

Likewise, analyses *in silico* influence the decisions for the choice of ligands that serve to recognize certain targets of therapeutic action and favor its specificity and efficacy, as well as help save time and money in the development of expensive experiments and biological tests to determine the most efficient or most favorable ligands. *In silico* analysis also allows the study of gene expression and metabolic pathways by the action of nanocarriers [38, 39].

Some of the most recent work includes the publication of Haider and Soni [40], where, with the help of *in silico* analyses, they were able to design nanocarriers directed at phosphatidylserine for the release of methotrexate in a targeted manner in tumors in conjugation with annexin, which presented a good affinity with phosphatidylserine found on the outside of the membrane plasma, achieving a specific release. In other works, Rahdar *et al.* [41] designed nanomicelles from the Pluronic F127 polymer in water and oil with the aim of encapsulating and releasing the antineoplastic drug carfilzomib (CFZ) in a controlled manner, which showed high encapsulation efficiency and a sustained release pattern, in addition to anticancer activity. They used *in silico* analysis to learn about the interactions between the components of this nanosystem.

Another use of *in silico* analysis is the identification of new components or drugs that serve to treat some diseases to be included in the design of innovative smart carriers. For example, in a study carried out by Chuffa *et al.* [42], it was possible to identify the phytochemicals through *in silico* analysis. Berberine, curcumin, epigallocatechin, luteolin, and quercetin were determined to be compounds with high potential due to their regulatory activity of different microRNAs and their antitumor effects, which was corroborated both *in vitro* and *in vivo* in the treatment of hepatocellular carcinoma.

11.7 Conclusions

For the design of intelligent nanocarriers, three main points are considered: the matrix that will serve as a support to contain the bioactive compound drugs of interest, the mechanisms of specificity, and stimulation techniques. The matrix can be composed of inorganic elements (gold, magnetic nanocarriers, quantum dots, mesoporous silica, etc.), organic elements (liposomes, dendrimers, micelles, and polymeric and viral nanocarriers), or a combination of inorganic and organic components to form hybrid nanocarriers with the aim of improved efficiency and diffusion, less toxicity, and increased drug stability. These materials will have unique properties at the nanometric level compared to a larger size, because, at this scale, physicochemical properties as: magnetism, chemical reactivity, conductivity, as well as optics properties, etc. are modified, as already discussed in this review, and its application, mechanisms of specificity, and stimulation techniques that can be carried out will also depend on this. The mechanisms of specificity will also depend on the materials used, as well as on the conjugation with targeted ligands that can bind to specific antigens or receptors on the target cells. Finally, the stimulation techniques will depend on the nanocarrier and, therefore, on the components and materials used for its design, as well as the objective or target of the nanocarrier. The design of these complex systems is the subject of research with an exponential increase per year due to the various benefits that can be obtained from them, as well as for the treatment of diseases of global importance, such as cancer and other diseases that require high specificity for treatment. In this review, the successful cases that have been developed in relation to nanocarriers of drugs and the treatment of various diseases are disclosed while emphasizing the challenges that still must be overcome.

These challenges include the lack of sufficient models to emulate the human body that allow glimpses of the responses that these drug delivery systems may have in living systems, as well as other details in terms of toxicity, efficacy, and costs. In this aspect, bioinformatics and *in silico* analyses are of great help for the design of intelligent nanocarriers that are increasingly more efficient and specific.

References

1. Abed, S.N. *et al.*, Nanocarriers in different preclinical and clinical stages, in: *Basic Fundamentals of Drug Delivery*, pp. 685–731, Elsevier, Amsterdam, 2018, doi: 10.1016/B978-0-12-817909-3.00017-0.
2. Hossen, S., Hossain, M.K., Basher, M.K., Mia, M.N.H., Rahman, M.T., Uddin, M.J., Smart nanocarrier-based drug delivery systems for cancer therapy and toxicity studies: A review. *J. Adv. Res.*, 15, Elsevier B.V., 1–18, Jan. 01, 2019, doi: 10.1016/j.jare.2018.06.005.
3. Sangeetha, J., Thangadurai, D., Hospet, R., Purushotham, P., Karekalammanavar, G., Mundaragi, A.A., Harish, E.R., Agricultural Nanotechnology: Concepts, Benefits, and Risks. In: Prasad, R., Kumar, M., Kumar, V. (eds) *Nanotechnology*, pp. 10-11, Springer, Singapore, 2017. doi: 10.1007/978-981-10-4573-8_1
4. Huda, S., Alam, M.A., Sharma, P.K., Smart nanocarriers-based drug delivery for cancer therapy: An innovative and developing strategy. *J. Drug Deliv. Sci. Technol.*, 60, 1–20, Editions de Sante, Dec. 01, 2020, doi: 10.1016/j.jddst.2020.102018.
5. Zhang, H., Fan, T., Chen, W., Li, Y., Wang, B., Recent advances of two-dimensional materials in smart drug delivery nano-systems. *Bioact. Mater.*, 5, 4, 1071–1086, Dec. 2020, doi: 10.1016/J.BIOACTMAT.2020.06.012.
6. Chauhan, N., Saxena, K., Jain, U., Smart Nanomaterials Employed Recently for Drug Delivery in Cancer Therapy: an Intelligent Approach. *Bionanoscience*, 1, 1–10, Aug. 2022, doi: 10.1007/S12668-022-01022-9/ TABLES/1.
7. Waheed, S., Li, Z., Zhang, F., Chiarini, A., Armato, U., Wu, J., Engineering nano-drug biointerface to overcome biological barriers toward precision drug delivery. *J. Nanobiotechnology*, 20, 1, 1–25, 2022, doi: 10.1186/s12951-022-01605-4.
8. Kenchegowda, M. *et al.*, Smart Nanocarriers as an Emerging Platform for Cancer Therapy: A Review. *Molecules*, *27*, 1, 146, 2022Dec. 2021, doi: 10.3390/MOLECULES27010146.
9. Liao, Z., Wong, S.W., Yeo, H.L., Zhao, Y., Smart nanocarriers for cancer treatment: Clinical impact and safety. *NanoImpact*, 20, 1–10, Elsevier B.V., Oct. 01, 2020, doi: 10.1016/j.impact.2020.100253.

10. Bharat, B., Encyclopedia of Nanotechnology, in: *Encyclopedia of Nanotechnology*, 2012.
11. Joudeh, N. and Linke, D., Nanoparticle classification, physicochemical properties, characterization, and applications: a comprehensive review for biologists. *J. Nanobiotechnology*, 20, 1, 1–29, BioMed Central Ltd, Dec. 01, 2022, doi: 10.1186/s12951-022-01477-8.
12. Johnson, K.L., and Greenwood, J.A., *Encyclopedia of Tribology*, Springer, US, 2013, doi: 10.1007/978-0-387-92897-5.
13. Ogwu, A.A., Bouquerel, E., Ademosu, O., Moh, S., Crossan, E., Placido, F., An Extended Derjaguin-Landau-Verwey-Overbeek Theory Approach to Determining the Surface Energy of Copper Oxide Thin Films Prepared by Reactive Magnetron Sputtering. *Metall. Mater. Trans. A*, 36, 2435–2439, 2005, doi:10.1007/s11661-005-0116-1.
14. Greenwood, J. A. "Elastohydrodynamic Lubrication," In: Wang, Q.J., Chung, YW. (eds) *Encyclopedia of Tribology*. Springer, Boston, MA, 2020, doi:10.1007/978-0-387-92897-5_100010
15. Kolhatkar, A.G., Jamison, A.C., Litvinov, D., Willson, R.C., Lee, T.R., Tuning the magnetic properties of nanoparticles. *Int. J. Mol. Sci.*, 14, 8, 15977–16009, 2013, doi: 10.3390/ ijms140815977.
16. Varalakshmi, B., Karpagam, T., Anand, A.V., Balamuralikrishnan, B., Nanoscale Smart Drug Delivery Systems and Techniques of Drug Loading to Nanoarchitectures, in: *Nanotechnology in the Life Sciences*, pp. 29–82, 2022.
17. Thakuria, A., Kataria, B., Gupta, D., Nanoparticle-based methodologies for targeted drug delivery—an insight. *J. Nanopart. Res.*, 23, 1–30, Springer Science and Business Media B.V., Apr. 01, 2020, doi: 10.1007/ s11051-021-05190-9.
18. Majumder, J. and Minko, T., Multifunctional and stimuli-responsive nanocarriers for targeted therapeutic delivery. *Expert Opin. Drug Deliv.*, 18, 2, 205–227, 2021, doi: 10.1080/17425247.2021.1828339.
19. Fang, Z., Shen, Y., Gao, D., Stimulus-responsive nanocarriers for targeted drug delivery. *New J. Chem.*, 45, 10, Royal Society of Chemistry, 4534–4544, Mar. 14, 2021, doi: 10.1039/d0nj05169a.
20. Karimi, M., Zangabad, P.S., Ghasemi, A., Hamblin, M.R., *Smart internal stimulus-responsive nanocarriers for drug and gene delivery*, Morgan and Claypool Publishers, United States, 2015, doi: 10.1088/978-1-6817-4257-1.
21. Sabir, F. *et al.*, DNA Based and Stimuli-Responsive Smart Nanocarrier for Diagnosis and Treatment of Cancer: Applications and Challenges. *Cancers (Basel)*, 13, 1–39, Jul. 2021, doi: 10.3390/CANCERS13143396.
22. van der Meel, R., Sulheim, E., Shi, Y., Kiessling, F., Mulder, W.J.M., Lammers, T., Smart cancer nanomedicine. *Nat. Nanotechnol.*, *14*, 11, 1007–1017, 2019Nov. 2019, doi: 10.1038/ s41565-019-0567-y.
23. Vankayala, R. and Hwang, K.C., Near-Infrared-Light-Activatable Nanomaterial- Mediated Phototheranostic Nanomedicines: An Emerging

Paradigm for Cancer Treatment. *Adv. Mater.*, 30, 1–27, Wiley-VCH Verlag, Jun. 06, 2018, doi: 10.1002/adma.201706320.
24. Ullah, R. *et al.*, Factors Influencing the Delivery Efficiency of Cancer Nanomedicines. *AAPS PharmSciTech*, 21, 4, 1–16, Springer, May 01, 2020, doi: 10.1208/s12249-020-01691-3.
25. Doroudian, M., O' Neill, A., mac Loughlin, R., Prina-Mello, A., Volkov, Y., Donnelly, S.C., Nanotechnology in pulmonary medicine. *Curr. Opin. Pharmacol.*, 56, Elsevier Ltd, 85–92, Feb. 01, 2021, doi: 10.1016/j.coph.2020.11.002.
26. Dai, W. *et al.*, Combination antitumor therapy with targeted dual-nanomedicines. *Adv. Drug Deliv. Rev.*, 115, Elsevier B.V., 23–45, Jun. 01, 2017, doi: 10.1016/j.addr.2017.03.001.
27. Malviya, R. *et al.*, Commercial utilities and future perspective of nanomedicines. *PeerJ*, 9, 1–30, 2021, doi: 10.7717/peerj.12392.
28. Barenholz, Y., Doxil® - The first FDA-approved nano-drug: Lessons learned. *J. Control. Release*, 160, 2, 117–134, Jun. 10, 2012, doi: 10.1016/j.jconrel.2012.03.020.
29. Janssen Products. (1995). Doxil: Highlights of Prescribing Information. Retrieved from www.accessdata.fda.gov/drugsatfda_docs/label/2019/050718s055lbl.pdf
30. Celgene Corporation. (2005). Abraxane: Highlights of Prescribing Information. Retrieved from www.accessdata.fda.gov/drugsatfda_docs/label/2012/021660s031lbl.pdf
31. Gerson, R., Serrano, A., Villalobos, A., Chavarri, Y., Gemcitabina en cáncer de páncreas. *An. Med. Asoc. Med. Hosp. ABC*, 47, 4, 211–216, México, D.F., Sep. 2002.
32. Talon Therapeutics. (2012). Vincristine Sulphate: Highlights of Prescribing Information. Retrieved from www.accessdata.fda.gov/drugsatfda_docs/label/2012/202497s000lbl.pdf
33. Sarkar, K., Kumar, M., Jha, A., Bharti, K., Das, M., Mishra, B., Nanocarriers for tuberculosis therapy: Design of safe and effective drug delivery strategies to overcome the therapeutic challenges. *J. Drug Deliv. Sci. Technol.*, 67, 1–13, Editions de Sante, Jan. 01, 2022, doi: 10.1016/j.jddst.2021.102850.
34. Vahedifard, F. and Chakravarthy, K., Nanomedicine for COVID-19: the role of nanotechnology in the treatment and diagnosis of COVID-19. *Emergent Mater.*, 4, 1, 75–99, 2021, doi: 10.1007/s42247-021-00168-8/Published.
35. Lemmerman, L.R., Das, D., Higuita-Castro, N., Mirmira, R.G., Gallego-Perez, D., Nanomedicine-Based Strategies for Diabetes: Diagnostics, Monitoring, and Treatment. *Trends Endocrinol. Metab.*, 31, 6, Elsevier Inc., 448–458, Jun. 01, 2020, doi: 10.1016/j. tem.2020.02.001.
36. Jayasinghe, M.K. *et al.*, The Role of *in silico* Research in Developing Nanoparticle-Based Therapeutics. *Front. Digit. Health*, 4, 35, Mar. 2022, doi: 10.3389/FDGTH.2022.838590/BIBTEX.

37. Stillman, N.R., Kovacevic, M., Balaz, I., Hauert, S., *In silico* modelling of cancer nanomedicine, across scales and transport barriers. *NPJ Comput. Mater.*, 6, 1, 1–10, 2020Jul. 2020, doi: 10.1038/s41524-020-00366-8.
37. Shimizu, H. and Toya, Y., Recent advances in metabolic engineering–integration of *in silico* design and experimental analysis of metabolic pathways. *J. Biosci. Bioeng.*, 132, 5, 429–436, Nov. 2021, doi: 10.1016/J. JBIOSC.2021.08.002.
39. Murray, D., Doran, P., MacMathuna, P., Moss, A.C., *In silico* gene expression analysis - An overview. *Mol. Cancer*, 6, 1–10, Aug. 2007, doi: 10.1186/1476-4598-6-50.
40. Haider, T. and Soni, V., Response surface methodology and artificial neural network-based modeling and optimization of phosphatidylserine targeted nanocarriers for effective treatment of cancer: *In vitro* and *in silico* studies. *J. Drug Deliv. Sci. Technol.*, 75, 103663, Sep. 2022, doi: 10.1016/J. JDDST.2022.103663.
41. Rahdar, A. *et al.*, Pluronic F127/carfilzomib-based nanomicelles as promising nanocarriers: synthesis, characterization, biological, and *in silico* evaluations. *J. Mol. Liq.*, 346, 118271, Jan. 2022, doi: 10.1016/J. MOLLIQ.2021.118271.
42. de A. Chuffa, L.G., de Souza, M.C., Cruz, E.M.S., Ferreira, F.B., de Morais, J.M.B., Seiva, F.R.F., Hepatocellular carcinoma and miRNAs: An *in silico* approach revealing potential therapeutic targets for polyphenols. *Phytomed. Plus*, 2, 2, 100259, May 2022, doi: 10.1016/J. PHYPLU.2022.100259.

12

Recent Era of Smart Nanocarrier–Based Drug Delivery System for Inhibition of Azole-Resistant Biofilm Forming *Candida albicans*

Rajivgandhi Govindan[1], Mohankumar Narayanan[2], Franck Quero[1*], Ramachandran Govindan[2], Chackaravarthy Gnanasekaran[2], Balamurugan Palanisamy[2], Chenthis Kanisha Chelliah[3] and Manoharan Natesan[2]

[1]*Laboratorio de Nanocelulosa y Biomateriales, Departamento de Ingeniería Química, Biotecnología y Materiales, Facultad de Ciencias Físicas y Matemáticas, Universidad de Chile, Avenida Beauchef, Santiago, Chile*
[2]*Marine Pharmacology and Toxicology laboratory, Department of Marine Science, Bharathidasan University, Tiruchirappalli, Tamil Nadu, India*
[3]*Department of Nanotechnology, Noorul Islam Centre for Higher Education, Kumaracoil, Kanyakumari, Tamil Nadu, India*

Abstract
Recently, the WHO announced the priority list of antibiotic-resistant bacteria and fungi based on the healthcare, transmissibility, preventability, hospital dissemination, resistance behavior, formulation of community diseases, high mortality, and morbidity. Based on their observation, the azole-resistant *Candida albicans* is the major drug-resistant fungi due to the production of plenty of extracellular antigenic matrix material and variety of enzymes production. In particular, the β-glucan and extracellular DNA components are the major factors to promote the biofilm formation with high resistance to multiple antifungal agents. Due to this defect, the scientists are promoted future R&D strategies should focus on the discovery and development of new kind of drug delivery method for inhibiting drug-resistant fungi, especially azole-resistant *Candida albicans*. Last few decades, nanotechnology is an efficient platform to design novel drug delivery approaches

**Corresponding author*: fquero@uchile.cl

Sathish-Kumar Kamaraj, Arun Thirumurugan, Muthuchamy Maruthupandy, Mercedes Guadalupe López Pérez and Shanmuga Sundar Dhanabalan (eds.) Nanoscience and Nanotechnology for Smart Prevention, Diagnostics and Therapeutics: Fundamentals to Applications, (287–310) © 2024 Scrivener Publishing LLC

against various kind of fungal pathogens with biocompatible, biodegradable and safer treatment in biomedical field. Interestingly, the smart nanocarrier systems are promising numerous benefits for drug delivery, and it is the highly demanding research field in biomedical application without making disturbance. The complement nature of high biocompatibility, excellent penetration and entrapment capacity are helped to control release of drugs in targeted sites, and also delivered specific amount of engineered therapeutic drug on the targeted places. More number of successive nanocarriers is frequently used in the biomedical treatment process, including liposomes, micelles, dendrimers, carbon nanotubes, quantum dots, aptamers, small molecules, peptides, and polymeric materials. Therefore, the current study research is focused on advantages of drugs, antibiotics, or bioactive compound intended smart nanocarriers as an efficient drug delivery molecule in target sites of biofilm forming azole-resistant *Candida albicans*.

***Keywords*:** Azole-resistant fungi, biofilm formation, alternative route, nanocarriers, target sites, drug delivery

12.1 Introduction

Nanotechnology has made significant progress over the last decade, and it has become an essential tool in various fields such as medicine, electronics, and energy. The exceptional properties of nanoparticles, such as their small size, high reactivity, and large surface area, enable them to interact differently with biological systems, contributing to their broad range of applications in the medical field. Nanomedicine, which uses nanotechnology for medical purposes, has emerged as an exciting field for targeted drug delivery, imaging, and therapy. Nanoparticles can be used as drug carriers, which can target specific sites in the body, such as cancer cells, while leaving healthy cells untouched, reducing the side effects of chemotherapy.

12.2 Role of Nanocarriers in Drug Delivery System

Nanocarriers can be modified to have specific properties like drug release profiles, solubility, immunogenicity, and bioavailability, making them a versatile and efficient drug delivery system. These modifications can be tailored according to the specific requirements of the drug being delivered, making it a highly targeted and efficient method of drug delivery. There are two types of drug targeting—active and passive. In active targeting, peptides and antibodies are coupled with the carrier to anchor them to the targeted site. In contrast, in passive targeting, the carrier complex circulates over the bloodstream, driven by affinity influenced by temperature, pH, and molecular site. These targeting methods enable highly specific delivery of drugs to the target site, leading to better therapeutic outcomes and reduced side effects. Moreover, nanocarriers have enormous potential as a novel drug delivery system, offering highly targeted and efficient delivery of drugs with minimal side effects. The continued advancement of nanotechnology will undoubtedly bring more exciting developments and applications in the medical field in the coming years [1].

12.3 Role of Nanomaterial-Based Nanocarriers

Nanomaterials have become increasingly popular due to their exceptional structural and physical properties, which differ significantly from those of atoms, molecules, and bulk materials made of the same elements. Nanoparticles, ranging in size from 1-100 nm, have transformed the fields of pharmaceuticals and diagnostics. One of the key benefits of nanoparticles is their small size, which enables them to be readily absorbed and transported by cells and tissues in the body, making them supreme for use in drug delivery and other medical applications. Furthermore, nanoparticles can be easily modified with a variety of molecules, such as drugs, antibodies, or imaging agents, allowing for a high degree of specificity and targeting. The use of nanotechnology and nanomaterials has expanded the possibilities for pharmaceutical and diagnostic applications, offering new and more effective ways to treat and diagnose a wide range of diseases and conditions [2].

The unique properties of nanoparticles, such as improved electrical conductivity, unique fluorescence properties, high surface-to-volume ratio and superparamagnetic behavior, are due to their small size and the number of constituents they possess. Nanomedicine, which uses nanomaterials for medical purposes such as drug delivery and diagnostics, has gained significant attention. Nanocarriers, which are nanomaterials that can transport drugs or other agents to specific parts of the body, have become popular as an alternative to traditional drug-delivery methods. They can improve the solubility of hydrophobic drugs and have minimal impact on the appearance of final food products. Studies have demonstrated that nanocarriers can be effective in treating diseases such as infectious diseases, cancer and neurodegenerative diseases.

12.4 Various Nanocarriers and Their Uses in Drug Delivery System

There are more types of nanocarriers are used in the drug delivery system, and their uses are very important in clinical filed. Dendrimers, carbon nanotubes, nanoemulsions, and liposomes are major nanomaterials designed for use in the pharmaceutical industry. Nanoparticles are excellent candidates for carrying drugs to targeted areas in organs, specific tissues, or cells. Various types of nanocarriers have gained significant attention in current years due to their applications in diagnostics, therapeutics, regenerative medicine and drug delivery systems. These small particles, which are typically less than 100 nm in size, can improve drug delivery to specific regions of the body and penetrate barriers such as the blood-brain barrier. To construct a nanocarrier, three essential ingredients are required: an organic or inorganic material, a photothermal material, and a drug. Optional ingredients include targeting ligands, which help the nanocarrier to bind to specific cells or tissues, and drug binding frames, which help to control the release of the drug. Nanocarriers have been extensively studied as a potential therapeutic and diagnostic tool for an extensive range of diseases, including infectious diseases, cancer, and neurogenerative diseases. They are typically classified into three main categories: lipid-based nanocarriers, inorganic nanocarriers, and polymeric nanocarriers. Each type has its own unique properties and advantages, depending on the specific application [3, 4].

12.4.1 Liposomes

Liposomes are highly versatile supramolecular assemblies that have been extensively researched since their discovery in 1965. They have numerous applications in various fields, including drug and gene delivery. In analytical sciences, liposomes aid two purposes: as analytes in quality assessment procedures and as functional components in various analytical systems. Encapsulated markers that offer amplification are useful in liposome immunoassays, and networks of liposomes connected by nanotubes have become ultrasmall-scale analytical tools. Because of their capacity to integrate different compounds and their shape, which is comparable to that of biological membranes, liposomes are regarded as almost ideal drug-carrier systems. They are the best delivery vehicles for biologically active compounds both *in vitro* and *in vivo* due to their biological and technological benefits [13, 16, 17].

They are also the best medicine delivery method currently available [18]. Significant advancements have been made in the industry over the past 20 years, and liposomes are now either in clinical trials or are going to be commercialized for a variety of biomedical uses. There are already a few liposomal medicine delivery technologies that can be used by the general public. Liposomes are a type of drug delivery system that share two important characteristics: they are biocompatible and have the capability to encapsulate both hydrophobic and hydrophilic drugs. The lipid bilayer that forms the structure of liposomes offers protection to the drugs against degradation by enzymes, chemical inactivation, and the immune system. This protective feature allows the drugs to reach their intended target tissues without being metabolized or deactivated along the way. Furthermore, liposomes help to minimize acquaintance of healthy tissue to the drug while it circulates in the bloodstream. The effectiveness of drug encapsulation in a liposome is influenced by the drug's polarity and partition coefficient.

These factors determine the drug's location within the liposomal membrane and its encapsulation efficiency. Hydrophobic drugs are incorporated into the acyl hydrocarbon chain, which means that the drug's encapsulation is dependent on the properties of the acyl chains, including their length and packing density. The efficiency of encapsulation for hydrophobic drugs is also influenced by the drug-to-lipid ratio. In summary, liposomes offer a good method of drug delivery, providing protection against natural degradation and minimizing exposure of healthy tissue to drugs. The ability of liposomes to encapsulate both hydrophobic and hydrophilic drugs makes them a versatile drug delivery system, with their effectiveness influenced by factors, such as drug polarity, partition coefficient, acyl chain properties, and drug-to-lipid ratio [19, 20]. The study focused on testing the effectiveness of thymoquinone liposomes (Lip-TQ) against *Candida albicans*, a type of fungus that can cause infections in humans.

The researchers prepared Lip-TQ and tested it against both fluconazole-resistant and fluconazole-susceptible isolates of *C. albicans*. The researchers found that Lip-TQ was able to target the *C. albicans* present intracellular locations of the macrophages, which are a type of immune cell. This means that Lip-TQ was able to reach and destroy the fungal cells inside the macrophages. The researchers also tested the effectiveness of Lip-TQ in mice infected with either fluconazole-susceptible or -resistant *C. albicans*. The results showed that mice treated with Lip-TQ were almost completely cured of the infection, and had the comparatively minimal fungal load in their nephrons compared to other treatments. This suggests that Lip-TQ may be a promising treatment option for both drug-susceptible

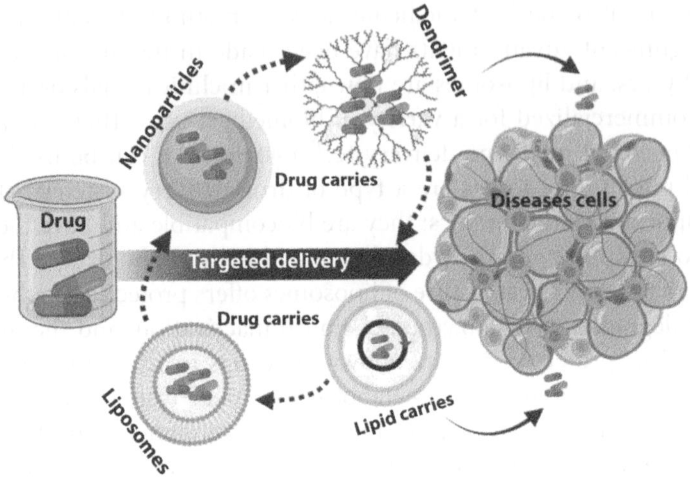

Figure 12.1 Various nanocarrier based targeted delivery system against for infected cells.

and -resistant *C. albicans* infections. This study provides evidence that Lip-TQ is an effective treatment option for *C. albicans* infections, particularly in cases where the fungus is resistant to commonly used antifungal drugs like fluconazole. The study highlights the potential of liposomal formulations of antifungal drugs to improve their efficacy and target fungal infections more effectively [21]. A recent study has revealed that the use of liposomal formulation of voriconazole (VCZ) can significantly condense the expression of azole-resistant genes compared to VCZ alone. The liposomal form also showed a higher rate of growth inhibition due to its superior ability to penetrate cell membranes. Currently available antifungal agents have limitations, including toxicity and drug resistance. The results of this study suggest that the synthesized liposomal VCZ can be an effective antifungal agent against *C. albicans*. Furthermore, the formulation has the potential to partially overcome resistance to azole drugs. However, further *in vivo* research is needed to confirm these findings [22]. Various nanocarriers based drug delivery approaches in target site against various infections details are available in Figure 12.1.

12.4.2 Dendrimers

Dendrimers are multifunctional monomeric components that are repeatedly added to produce complex polymeric structures with a central core that resemble trees. Any functional monomer with at least two functional groups can be used as the core for attaching further layers or "generations" of dendrimers, giving this particular structure a tremendous amount of structural freedom. Either divergent synthesis or convergent synthesis techniques can be used to assemble the dendrimer structure. Unconjugated dendrimers frequently have primary amine groups on their outside surfaces, which contribute to their polycationic charge. Unconjugated dendrimers are effective at binding to and disrupting the tissue membrane because of this property, which enables strong ionic bonding with anionic charges on tissue membranes and cells. This property creates pores through which dendrimers can be absorbed [23]. Their hyperbranched and three-dimensional structure, along with multiple surface functional groups, enables conjugation of targeting moieties to specifically attract to receptors overexpressed on various cancer cells. This makes dendrimers a versatile carrier for bioactive with substantial potential for delivering therapeutic agents.

However, it should be noted that unconjugated dendrimers may exhibit increased cytotoxicity in *in vitro* assays due to their strong tissue binding capability. Nonetheless, with further research, dendrimers hold significant promise as a nanocarrier for various biomedical applications. It has highly versatile structures that offer tremendous potential for drug delivery and gene therapy applications. Their highly branched architecture, combined with their easily modifiable surfaces, enables functionalization and coupling with drugs and DNA/RNA. Dendrimers can be synthesized using different processes to control their size, shape, charge, and solubility, making them highly customizable. Dendrimers can improve the solubility and bioavailability of hydrophobic medicines, which is a significant advantage. The dendrimer's intramolecular cavity can be used to enclose hydrophobic medicines, or they can be conjugated to the functional groups on the surface.

Additionally, cationic dendrimers' positively charged surface and nucleic acids' propensity to form complexes, making them an attractive delivery vehicle for gene therapy. Dendrimers' exterior functional groups also offer the potential for active targeting of specific diseases by adding targeting moieties like folate and antibodies. These targeting strategies can improve delivery to the intended site, especially for tumors. These are highly versatile structures that can be tailored to specific biomedical applications. Their ability to enhance drug solubility and bioavailability, as well as their compatibility with gene therapy, make them a promising nanocarrier platform. The addition of targeting moieties to their surface offers the potential for more effective and specific delivery to diseased tissue [24, 25]. Researchers have discovered a highly effective dendritic molecule, BDSQ024, which possesses cationic carbosilane properties, making it potent in damaging preexisting biofilms and eliminating *C. albicans* at low concentrations.

During their research, they found that it was essential to plate the well suspension on agar plates in order to precisely calculate the minimal biofilm eradication concentration (MBEC) values. Additionally, they discovered that in order to account for the existence of metabolically inactive cells, which have the potential to renew the biofilm, biocide experiments must be carried out in conjunction with colorimetric viability studies. Finally, when BDSQ024 was mixed with commercial antifungal medications, there was a marked reduction in effective concentrations and cytotoxicity, as well as an increase in Candida cells' susceptibility to antifungals. With lower dosages of commercial antifungals, this combination may prove to be a potent therapeutic strategy for preventing the formation of biofilms, leading to the creation of antiseptic topical treatments [26].

12.4.3 Carbon Nanotubes

Carbon nanotubes **(CNTs)** are cylindrical structures made up of rolled-up sheets of carbon atoms arranged in a hexagonal lattice pattern. They have diameters that are only a few nanometers wide, while their lengths range from micrometers to millimeters. CNTs can exist in two forms: single-walled and multi-walled. Their electrical conductivity or semiconducting properties depend on the alignment of their carbon lattice relative to the tube axis. CNTs possess a unique set of properties that make them an exciting prospect in many fields, including biotechnology. Their high aspect ratio makes them well suited for cell penetration. Furthermore, their anisotropic conductivity or semi conductivity along the tube axis makes them ideal for integrating with nervous and muscular tissue. The ultra-high surface area of CNTs maximizes their ability to interact with biological matter. Additionally, their hollow interior provides a large cargo-carrying capacity for drug delivery [27].

The exteriors of CNTs can also be modified or functionalized, which allows for custom solubility and biological recognition. It possesses a range of properties that make them a promising candidate for use in biotechnology. Carbon nanotubes (CNTs) possess several unique properties, including their anisotropic conductivity, high aspect ratio, customizable surface modifications, ultra-high surface area, and large cargo-carrying capacity. As a result, CNTs have garnered significant attention from researchers in recent years, particularly in the field of biomedical applications [28]. They are being explored for drug delivery in cancer treatment, as scaffolds for stem cell culture, and for biosensing applications. Additionally, the integration of CNTs with proteins has led to the development of functional nanocomposites with antimicrobial properties. One of the main reasons why CNTs are ideal for biomedical applications is due to their exceptional mechanical, electrical, and optical properties. Furthermore, techniques for facile functionalization of CNTs have expanded the potential range of applications. This means that CNTs are being utilized not only to develop functional nanocomposites but also as efficient carriers for delivering various drugs *in vivo*.

Over the past decade, extensive research has been conducted on the use of nanomaterials, especially CNTs, as potent drug delivery vehicles. The enhanced permeability and retention (EPR) effect of CNTs has been harnessed for the successful delivery of various anti-cancer drugs. Therefore, carbon nanotubes are currently being investigated for a wide range of biomedical applications due to their unique properties. They have already demonstrated potential as drug carriers for cancer treatment, scaffolds for stem cell culture, and biosensors, and their use is expanding as techniques for functionalization improve. The ability of CNTs to enter tumors passively and effectively deliver anti-cancer drugs has been extensively studied and holds great promise for future cancer therapies [29]. Carbon nanotubes have exclusive properties that make them striking for drug delivery applications. They can penetrate living cells without causing damage, and Incorporating medicines into these structures is simple. One example of this is amphotericin B, which was found to be effective against amphotericin B–resistant *C. albicans* when incorporated into carbon nanotubes. Even after the drug's wash-off period from the skin, topical distribution of these particles demonstrated drug restoration in the skin. This implies that carbon nanotubes might be useful for long-term drug delivery, which might enhance the efficacy of therapies. Additionally, it has been demonstrated that adding medications to multi-walled carbon nanotubes increases their effectiveness against fungi. In comparison to conventional formulations, it was discovered that the minimum concentration of nystatin needed to achieve fungicidal action was lower when it was integrated into multi-walled carbon nanotubes. In general, carbon nanotubes show promise as a method of drug administration that can increase the effectiveness of medications and lessen their negative effects [30].

12.4.4 Polymeric Nanomicelles

Polymeric nanomicelles are spherical structures made up of amphiphilic polymers, which are stable in aqueous solutions and have both hydrophilic and hydrophobic regions. They have a unique structure that makes them useful for drug delivery and imaging applications. It can be encapsulate hydrophobic drugs in their hydrophobic cores, while their hydrophilic coronas prevent them from aggregating in aqueous solutions. This allows them to be used for targeted drug delivery to specific organs or tissues. Polymeric nanomicelles are also being studied for their potential use in imaging applications, as they can be easily modified with imaging agents to visualize specific areas of the body and aid in diagnosis and treatment. Researchers reported that polymeric micelles made from P(PEGMA-b-DEAEMA) have been shown to be effective in inhibiting the formation of *C. albicans* biofilms and removing pregrown single- and dual-species fungal biofilms, consisting of both *C. tropicalis* and *C. albicans*, by preventing the growth of hyphae. The micelles alone demonstrated superior antibiofilm activity against both single- and dual-species biofilms compared to the antifungals Fluconazole and Amphotericin B. In fact, when co-delivered with these antifungals, the micelles interacted synergistically and enhanced their efficacy against pregrown biofilms. This suggests that agents with anti-morphologic transition activity may provide a new strategy for defeating biofilms of polymorphic resistance strains and could be potentially useful in counteracting Candida virulence. Polymeric micelles composed of a pH-responsive diblock copolymer P(PEGMA-b-DEAEMA) were successfully developed for efficient delivery and potentiation of the action of itraconazole against *C. albicans* biofilms. The polymers in the micelles were protonated at low pH, resulting in strong electrostatic interactions with the biofilm structural components and improved local drug release. As a result, the micelles had improved affinity towards Candida biofilms and enhanced itraconazole activity against biofilm-embedded microbes. These pH-responsive micelles have the potential to control Candida infections and can also be used for delivering other antimicrobials to overcome the challenges posed by biofilms. The study highlights the use of agents with anti-morphologic transition activity as a novel strategy for defeating biofilms of polymorphic resistance strains and offers a potential approach to offsetting Candida virulence [31].

In this study, amphotericin B (AmB) was formulated into AmB/MPEG-PCL-g-PEI micelles to improve its aqueous solubility. The micelles were spherical, nano-sized, and had high drug loading and encapsulation efficiency. Encapsulation of AmB was confirmed by Differential scanning calorimetry (DSC) analysis. The micelles exhibited sustained release of AmB under normal and *C. albicans* infection conditions, without pH dependence. Buccal tablets containing AmB/MPEG-PCL-g-PEI micelles were prepared and evaluated for weight variation, hardness, and friability. *In vitro* drug release of the tablets was measured using USP dissolution apparatus and a dissolution rate test apparatus that simulated the oral cavity. The tablets exhibited sustained release for 8 hours. *In vivo* safety evaluation of the micelles showed no irritation to rat buccal mucosa, and their cytotoxicity was lower than that of AmB alone in MTT experiments. The micelles had nonsystemic toxicity when delivered through the gastrointestinal route. Although the antifungal effect against planktonic *C. albicans* was not significantly improved, the micelles enhanced the antifungal activity against biofilm *C. albicans*. Thus, AmB/MPEG-PCL-g-PEI micelles have the potential to be developed as a more convenient and effective oral buccal delivery system for AmB [33, 35]. Previously available reports of various nanocarriers and their importance against biofilm forming *C. albicans* are available in Table 12.1.

12.4.5 Nanocapsules

Nanocapsules are hollow and spherical structures made up of an outer polymer shell and an inner hollow space. They can be produced through a process called nanoprecipitation or mini emulsion polymerization and have several advantages in drug delivery, including reducing toxicity, preventing drug degradation, enabling controlled release, and masking unpleasant taste. Nanocapsules can also guard the drug from the peripheral environment and increase the bioavailability of peptides and proteins, making them a promising tool for drug delivery [36]. Polymeric nanocapsules (NC) are a promising system for improving antimicrobial therapy. Researchers developed NC/ZnS4Pc + AMB, which consists of zinc phenyl-thio-phthalocyanine associated with amphotericin B, using the preformed polymer interfacial deposition method with poly(lactide-co-glycolide) (PLGA) 50:50.

The theoretical concentration of NC was 0.05 mg mL^{-1}. The average particle diameter was 253.8 ± 17.3, with a polydispersity index of 0.36 ± 0.01, and a negative zeta potential average of −31.03 ± 5.54 for 158 days. UV-vis absorption and emission spectroscopy analyses showed no changes in photophysical properties compared to free ZnS4Pc. NIH/3T3 ATCC® CRL-1658 ™ cell viability assays exhibited no cytotoxicity for the tested concentrations. The photodynamic activity assay using NC/ZnS4Pc + AMB diluted showed fungal toxicity against *Candida albicans* yeast with energetic fluences of 12 J cm^{-2} and 25 J cm^{-2}, leading to a decrease in cell viability. The minimum fungicidal concentration assay demonstrated a fungistatic activity for the conditions employed in the photodynamic therapy assay [37]. The aim of this research was to evaluate the effectiveness of a suspension of poly(ε-caprolactone) nanocapsules containing diphenyl diselenide (NC-1) as an antifungal agent, both *in vitro* and *in vivo*. The *in vitro* antifungal activity was assessed using the broth microdilution

Table 12.1 Nanocarriers and their applications against azole-resistant *Candida albicans.*

Nanocarrier System	Therapeutic agents	Applications	Reference
Lipid nanocapsules	Fluconazole	Reduction in the effective dose of fluconazole	[14]
Chitosan-PLGA nanoparticles	Fluconazole	pH modulated drug release, localization and	[15]
Liposomes	Thymoquinone	Targeting *C. albicans* inside macrophages	[20]
Liposomes	Voriconazole	Effective antifungal agent against *C. albicans*	[21]
Dendritic molecule, BDSQ024	Antifungal drugs	Preventing biofilm formation of *C. albicans*	[26]
Carbon nanotubes	Amphotericin B	Improved drug delivery for treatment	[30]
Polymeric micelles P(PEGMA-b-DEAEMA)	Itraconazole	Strong interaction with biofilm and pH mediated drug release	[31]
Micelles dispersed in ureasil-polyether	Curcumin	Improved adherence rate and effective antifungal property	[32]
AmB/MPEG-PCL-g-PEI micelles	Amphotericin B	Enhanced antifungal activity against biofilm forming *C. albicans*	[33]
AgNPs	Iturin	Increased damage of membrane integrity and antifungal activity	[34]

method, and it was found that NC-1 had similar or lower minimum inhibitory concentration (MIC) values than the free compound against fifteen strains of Candida.

A hydrogel containing the nanocapsules was prepared using gellan gum, and its pH, compound content, particle size distribution, stability, and bioadhesive properties were evaluated. The animal model of vulvovaginal candidiasis was induced in female Swiss mice, and the hydrogels were topically applied for seven days. After euthanasia, the total fungal burden was assessed, and it was found that the nano-based formulation was more effective than the hydrogel containing nonencapsulated diphenyl diselenide. The study demonstrated that the nanoencapsulation of diphenyl diselenide maintained its antifungal activity and that the nano-based hydrogel formulation may be a promising approach against vulvovaginal candidiasis [38]. This study compared two types of lipid-based drug delivery systems, namely lipid nanocapsules (LNC) and nanostructured lipid carriers (NLC), for their ability to deliver the antifungal drug itraconazole (ITC) topically.

The results showed that ITC-LNC had better *in vitro* properties, including size, uniformity, drug release profile, and antifungal activity against *Candida albicans* compared to ITC-NLC. In addition, when applied topically in a gel form, ITC-LNC showed better skin retention compared to conventional ITC gel and ITC-NLC gel. The researchers conducted a preclinical study on rats with induced cutaneous candidiasis to assess the therapeutic efficacy of ITC-LNC gel and conventional ITC gel therapy. The results showed that both formulations were effective in treating the infection without causing any histopathological changes in the skin after 14 days of consecutive application. Overall, the study suggests that ITC can be an effective topical antifungal agent, and nanocapsulation can further improve its efficacy [39].

12.4.6 Metallic Nanoparticles

Metallic nanoparticles have captivated scientists for more than a century, and today they are extensively used in biomedical sciences and engineering. These tiny particles are of great interest due to their vast potential in the field of nanotechnology. Modern technology has made it possible to synthesize and modify these particles with various chemical functional groups. This ability enables them to be conjugated with antibodies, ligands, and drugs of interest, thereby opening a wide range of potential applications in biotechnology. One of the main advantages of metallic nanoparticles is their ability to be used in various applications such as magnetic separation and preconcentration of target analytes, targeted drug delivery, and vehicles for gene and drug delivery [40, 41]. They are also useful in diagnostic imaging, which is of great importance in medicine. These tiny particles can be easily functionalized with specific molecules, such as antibodies, ligands, and drugs, which bind specifically to certain types of cells or tissues. This allows for targeted drug delivery, where drugs can be delivered directly to the site of interest, reducing potential side effects. Moreover, metallic nanoparticles have unique optical and magnetic properties that make them useful in diagnostic imaging. For example, they can be utilized as contrast agents in magnetic resonance imaging (MRI) and computed tomography (CT) scans, allowing for clearer and more accurate images. Metallic nanoparticles have become an indispensable tool in biomedical sciences and engineering. Their unique properties and versatility have made them an attractive option for many applications, ranging from targeted drug delivery to diagnostic imaging. As technology continues to advance, we can expect even more exciting developments in the field of metallic nanoparticles [42].

Iturin-AgNPs have demonstrated outstanding antifungal activity with low minimum inhibitory concentration (MIC), without inducing drug resistance in *Candida albicans*. The mechanisms of action of Iturin-AgNPs involve an increase in cell membrane permeability, damage to membrane integrity, increased levels of reactive oxygen species (ROS) in cells, and leakage of nucleic acids and cellular proteins. Iturin-AgNPs have exhibited no toxicity to HaCaT cells at concentrations below 10 μg/mL. Furthermore, a chitosan dressing based on Iturin-AgNPs has shown the potential to promote wound healing. These results suggest that Iturin-AgNPs may be a promising antifungal agent for promoting wound healing, due to their potent antifungal activity against *C. albicans* [34].

12.5 Importance of Fungal Infections (*C. albicans*)

Fungi, including yeast species of the genus Candida, are a significant source of human infections. The majority of pathogenic fungal infections in humans are caused by *Candida* species, namely. Although many Candida species exist in the environment, only about a dozen of them are linked to colonization and infection in humans. Candida species are generally found as commensals in the oral cavity, intestinal tract, and vagina, and are frequently acquired by newborns soon after birth. Although these species are typically harmless in most individuals, they can become opportunistic and cause various diseases under certain circumstances [5]. A crucial stage in the infection process is the adhesion of *Candida albicans* to host epithelial cells, which is mediated by different surface structures known as adhesins. Adhesins are proteins or glycoproteins found on the fungal cell surface that engage with certain receptors on the surface of the host cell to facilitate attachment and colonization. In order to adhere to epithelial cells, *C. albicans* expresses a variety of adhesins, such as the Als proteins, Hwp1, Eap1, and others. These adhesins mediate adhesion through several mechanisms and are variably expressed on yeast opposed to hyphae. For example, Als proteins are expressed on both yeast and hyphal forms of *C. albicans* and interact with multiple host cell surface receptors, while Hwp1 is mainly expressed on hyphae and binds to specific receptors on the host cell surface. The expression of adhesins is also influenced by environmental factors such as pH, temperature, and nutrient availability, which can affect the morphology of the fungal cell and its ability to adhere to host cells [6].

12.6 Biofilm-Forming *C. albicans*

The recent studies have suggested that the microbiot may play a major role in the host's health and disease, including immune regulation, nutrient acquisition, and pathogen exclusion. One potential interaction between *C. albicans* and the gut microbiota is through competition for nutrients and space. *C. albicans* can use various carbon sources, including simple sugars and complex polysaccharides, and may compete with other fungi and bacteria in the gut for these nutrients. Another potential interaction is through the modulation of the host immune response. *C. albicans* is known to activate immune responses in the gut, including the production of cytokines and chemokines, which can affect the overall immune balance in the gut. *C. albicans* and other fungi may interact with the gut bacterial community directly or indirectly. For example, fungi can produce metabolites that inhibit or promote bacterial growth, or they can physically interact with bacterial cells through biofilm formation or other mechanisms. The interactions between *C. albicans* and the gut microbiota are complex and poorly understood, but they may have significant implications for human health and disease [7, 8].

Recently in some cases, *Candida albicans* is responsible for causing life-threatening infections, with a death rate of up to 40% even with treatment. This fungus has three different biological phases, including yeast, pseudo hyphae, and hyphae. The hyphae phase is particularly important in

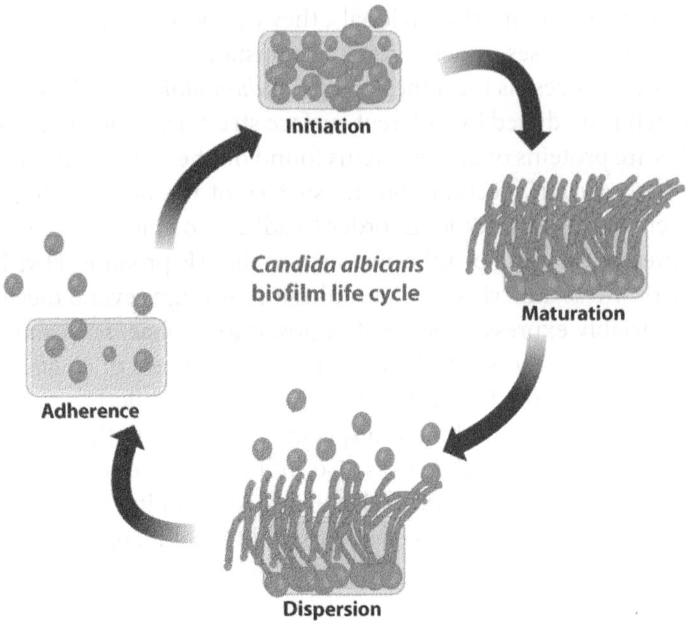

Figure 12.2 Development, regulation, and molecular mechanisms of *Candida albicans* biofilms.

the disease process because it can invade mucosal epithelial cells and cause tissue damage that leads to blood infection. Recent research in fungal cell biology has provided valuable insights into how hyphal growth is regulated and maintained. Researchers have been studying the differences between the three phases of *C. albicans*, as well as the mechanisms that control hyphal growth, extension, and polarity. They have also been investigating how hyphal development is regulated and how virulence factors contribute to serious infections. These infections have become increasingly common in recent years, particularly among immunocompromised patients. The mechanism of *C. albicans* biofilm formation is effectively depicted in Figure 12.2.

12.7 Virulence Factors of *C. albicans* in Biofilm Formation

This is due to a various factor, including the widespread use of antibiotics, which can disrupt the normal balance of microorganisms in the body, as well as the use of immunosuppressant drugs, which can impair the immune system's ability to fight off infections. Among all of the Candida species, *Candida albicans* is the most common cause of candidiasis. In fact, studies have shown that the detection rate of *C. albicans* is between 70% and 90% among all candidiasis-causing fungi. This is partly because *C. albicans* is a usual part of the human microbiota and can be found on the skin, mucous membranes, and in the gastrointestinal tract of healthy persons. However, when the immune system is compromised or when the normal microbial balance is disrupted, *C. albicans* can cause infections. These infections can range from superficial infections such as vaginal yeast infections and thrush, to more serious systemic infections such as candidemia, where the fungus enters the bloodstream and can spread throughout the body. Candidemia is particularly dangerous, as it can lead to sepsis and organ failure. In fact, the mortality rate among patients with candidemia is up to 43.6%. This is partly because *C. albicans* is able to form biofilms, which can protect it from the immune system and make it more resistant to antifungal drugs. The treatment of Candida infections typically involves antifungal drugs such as fluconazole or amphotericin B. However, the increasing prevalence of drug-resistant strains of *C. albicans* is a growing concern, and researchers are working to develop new antifungal drugs to combat these infections. In addition to drug therapy, prevention is also important in the management of Candida infections [9].

12.8 Eradication of *C. albicans*

There are several groups of chemical compounds available to treat *Candida* infections, each with varying effectiveness depending on the type and location of the infection, as well as the sensitivity of the Candida species involved. These compounds include nucleoside analogs, polyenes, echinocandins, azoles, and allylamines. Fluconazole, an azole antifungal, is the most commonly prescribed medication for most candidal infections. Azoles work by inhibiting the activity of the 14-α-sterol demethylase enzyme, which is encoded by the ERG11 gene and plays a crucial role in the production of ergosterol, a fungal-specific membrane sterol. However, some non–*Albicans candida* (NAC) species are intrinsically resistant to azoles, and the overuse of these medications may contribute to the increased incidence of contaminations caused by NAC species and numerous studies have demonstrated that Candida can develop advanced resistance to azole antifungals over time, further reducing the effectiveness of these drugs [10].

12.9 Azole-Resistant *C. albicans* and Its Role in Biofilm Formation

The rise of fluconazole-resistant *Candida albicans* is becoming an increasingly complex issue. Years of indiscriminate medication prescribing and unnecessary drug exposure have led to a lack of alternative therapeutic options, which has led to the development of this issue. Unfortunately, azole-resistant Candida vaginitis patients have less effective treatment options due to the lack of new antifungal drug classes that have emerged while the azole class of medications has grown. The difficulty of treating azole-resistant non–*Albicans candida* spp. has only increased with the appearance of fluconazole-resistant *C. albicans*. The progress of drug resistance in Candida is a well-established phenomenon. However, despite this understanding, there has been a lack of useful therapeutic advantages for patients with azole-resistant Candida infections. This has resulted in limited treatment options for women experiencing refractory symptoms. Urgent attention is needed to develop new strategies and therapeutic options to combat the challenge of azole drug resistance. Overall, indiscriminate drug prescription and unnecessary fluconazole exposure have contributed significantly to the current challenges of treating Candida infections. While knowledge of azole resistance in yeast has been established, there is a need to translate this knowledge into useful therapeutic options to address the limited treatment options for patients with azole-resistant *Candida vaginitis* [11].

When physicians are faced with the difficult task of selecting the most appropriate antifungal agent to treat distributed Candidal infections, they may resort to conducting susceptibility testing to help determine the most effective therapy. The decline in the prevalence of *C. albicans* infections during the 1990s can be attributed to the victory of fluconazole therapy, rather than being a sign of an overwhelming problem of azole-resistant yeast. However, it is important to note that during the height of the pre-HAART AIDS epidemic, there was a rapid increase in azole resistance in oral Candida isolates. This serves as a cautionary tale about the serious threat that azole resistance can present in circumstances where there are enough yeast that are subject to the right selection pressure. This highlights the need for proactive measures to manage and prevent the spread of azole-resistant Candida infections, especially in high-risk populations such as those with compromised immune systems [12]. Accordingly, in this chapter we discuss about the types and role of different nanocarriers against azole-resistant biofilm forming *Candida albicans*. The virulence factors of azole resistant *C. albicans* and their biofilm mechanisms are available in Figure 12.3.

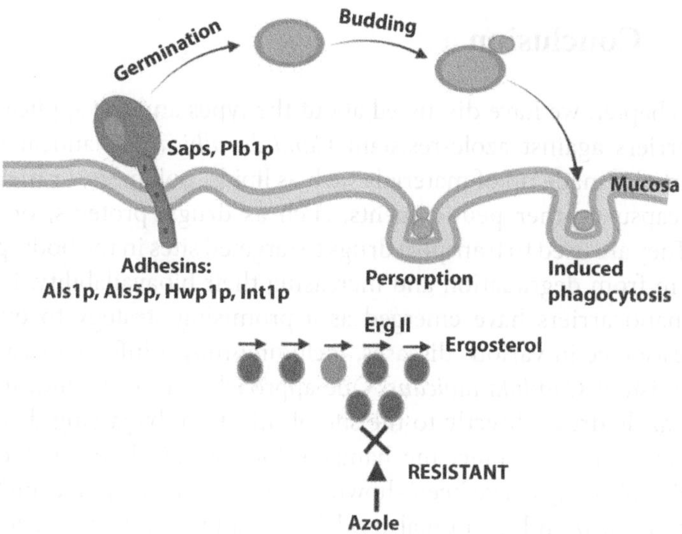

Figure 12.3 Schematic representation of azole-resistant *Candida albicans* and its virulence factors.

12.10 Inhibition of Azole-Resistant Biofilm Forming *C. albicans*

Here, the schematic representation of various drugs loaded nanoparticles and their importance against biofilm forming *C. albicans* are explained in details of Figure 12.4.

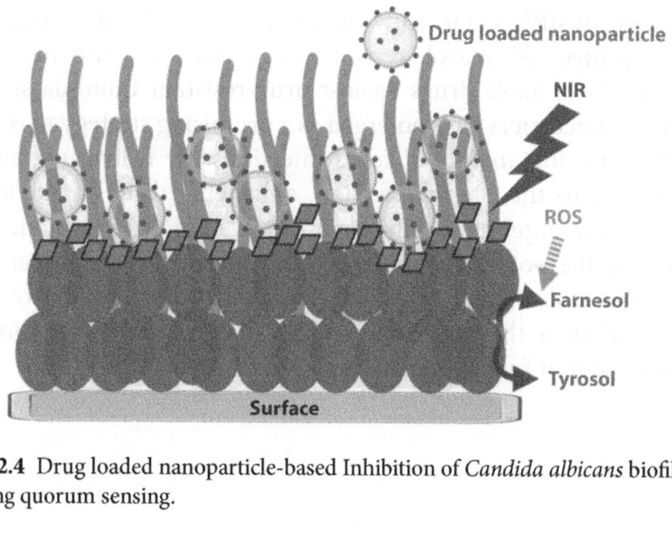

Figure 12.4 Drug loaded nanoparticle-based Inhibition of *Candida albicans* biofilms with interfering quorum sensing.

12.11 Conclusion

In this chapter, we have discussed about the types and the applications of nanocarriers against azole-resistant *Candida albicans.* Nanocarriers are tiny particles made up of materials such as lipids, polymers, or metals that can encapsulate therapeutic agents, such as drugs, proteins, or nucleic acids. They are used to transport drugs to targeted sites in the body, protecting them from degradation and increasing their bioavailability. In recent years, nanocarriers have emerged as a promising strategy to overcome drug resistance in various diseases, including fungal infections caused by azole-resistant *Candida albicans.* One approach is to use nanocarriers to deliver azole drugs directly to the site of infection, bypassing drug efflux pumps and other resistance mechanisms, for example, liposomal formulations of azole drugs have been shown to enhance drug uptake and reduce toxicity *in vitro* and in animal models of Candida infections. Similarly, polymeric nanoparticles have been used to encapsulate azole drugs and enhance their bioavailability and antifungal activity against drug-resistant Candida strains. Another approach is to use nanocarriers to deliver alternative antifungal agents that can overcome drug resistance mechanisms in *Candida albicans.* For example, researchers have developed nanocarriers that can deliver plant-derived antifungal compounds, such as curcumin or resveratrol, which have been shown to have synergistic effects with azole drugs against drug-resistant Candida strains. Finally, nanocarriers can also be used to deliver RNA-based therapeutics, such as siRNAs or antisense oligonucleotides, which can target specific genes involved in drug resistance in *Candida albicans.* For example, researchers have used liposomal formulations of siRNAs targeting the MDR1 gene, which encodes a drug efflux pump involved in azole resistance in Candida, to enhance the antifungal activity of azole drugs against drug-resistant Candida strains. In summary, nanocarriers have emerged as a promising strategy to overcome drug resistance in *Candida albicans* infections. By delivering antifungal drugs directly to the site of infection, enhancing drug uptake and bioavailability, and targeting specific genes involved in drug resistance, nanocarriers have the potential to improve the efficacy of existing antifungal drugs against azole-resistant Candida strains. However, further research is needed to optimize the design and efficacy of nanocarriers for clinical use in the treatment of fungal infections.

Acknowledgement

Dr. G. Rajivgandhi and Dr. Franck Quero acknowledge the financial support from ANID-FONDECYT (Chile) under the Postdoctoral Fellowship No. 3220019.

References

1. Hsu, J.-F., Chu, S.-M., Liao, C.-C., Wang, C.-J., Wang, Y.-S., Lai, M.-Y., Wang, H.-C., Huang, H.-R., Tsai, M.-H. Nanotechnology and Nanocarrier-Based Drug Delivery as the Potential Therapeutic Strategy for Glioblastoma Multiforme: An Update, *Cancers*, 13, 1-22, 2021.
2. Surendiran, A., Sandhiya, S., Pradhan, S.C., Adithan, C., Novel Applications of Nanotechnology in Medicine. *Indian J. Med. Res.*, 130, 689-701, 2009.
3. Colone, M., Calcabrini, A., Stringaro, A. Drug Delivery Systems of Natural Products in Oncology. *Molecules*, 25, 1-23, 2020.
4. Avramović, N., Mandić, B., Savić-Radojević, A., Simić, T., Polymeric Nanocarriers of Drug Delivery Systems in Cancer Therapy. *Pharmaceutics*, 12, 1-17, 2020.
5. Moran, G.P., Coleman, D.C., Sullivan, D.J., Candida Albicans *versus* Candida Dubliniensis: *Why Is* C. Albicans *More Pathogenic?. Int. J. Microbiol.*, 2012, 205921, 2012.
6. Zhu, W. and Filler, S.G., Interactions of Candida Albicans with Epithelial Cells. *Cell. Microbiol.*, 12, 273, 2010.
7. Rakoff-Nahoum, S., Paglino, J., Eslami-Varzaneh, F., Edberg, S., Medzhitov, R., Recognition of Commensal Microflora by Toll-like Receptors Is Required for Intestinal Homeostasis. *Cell*, 118, 229, 2004.
8. Angebault, C., Djossou, F., Abélanet, S., Permal, E., Soltana, M.B., Diancourt, L., Bouchier, C., Woerther, P.-L., Catzeflis, F., Andremont, A., Candida Albicans Is Not Always the Preferential Yeast Colonizing Humans: A Study in Wayampi Amerindians. *J. Infect. Dis.*, 208, 1705, 2013.
9. Chen, H., Zhou, X., Ren, B., Cheng, L., The Regulation of Hyphae Growth in Candida Albicans. *Virulence*, 11, 337, 2020.
10. Pfaller, M.A., Diekema, D.J., Gibbs, D.L., Newell, V.A., Ellis, D., Tullio, V., Rodloff, A., Fu, W., Ling, T.A., Results from the ARTEMIS DISK Global Antifungal Surveillance Study, 1997 to 2007: A 10.5-Year Analysis of Susceptibilities of Candida Species to Fluconazole and Voriconazole as Determined by CLSI Standardized Disk Diffusion. *J. Clin. Microbiol.*, 48, 1366, 2010.

11. Sobel, J.D. and Sobel, R., Current Treatment Options for Vulvovaginal Candidiasis Caused by Azole-Resistant Candida Species. *Expert Opin. Pharmacother.*, 19, 971, 2018.
12. Tobudic, S., Kratzer, C., Presterl, E., Azole-Resistant Candida Spp. - Emerging Pathogens? *Mycoses*, 55, 24, 2012.
13. Elsabahy, M. and Wooley, K.L., Design of Polymeric Nanoparticles for Biomedical Delivery Applications. *Chem. Soc. Rev.*, 41, 2545, 2012.
14. Crucho, C., II and Barros, M.T., Polymeric Nanoparticles: A Study on the Preparation Variables and Characterization Methods. *Mater. Sci. Eng. C*, 80, 771, 2017.
15. Bianchin, M.D., Borowicz, S.M., da Rosa Monte Machado, G., Pippi, B., Guterres, S.S., Pohlmann, A.R., Fuentefria, A.M., Külkamp-Guerreiro, I.C., Lipid Core Nanoparticles as a Broad Strategy to Reverse Fluconazole Resistance in Multiple Candida Species. *Colloids Surf. B: Biointerfaces*, 175, 523, 2019.
16. Chan, J.M., Valencia, P.M., Zhang, L., Langer, R., Farokhzad, O.C., Polymeric Nanoparticles for Drug Delivery. *Cancer Nanotechnol. Methods Protoc.*, 163–175, 163, 2010.
17. Kolge, H., Patil, G., Jadhav, S., Ghormade, V., A PH-Tuned Chitosan-PLGA Nanocarrier for Fluconazole Delivery Reduces Toxicity and Improves Efficacy against Resistant Candida. *Int. J. Biol. Macromol.*, 227, 453, 2023.
18. Johal, H.S., Garg, T., Rath, G., Goyal, A.K., Advanced Topical Drug Delivery System for the Management of Vaginal Candidiasis. *Drug Deliv.*, 23, 550, 2016.
19. Jesorka, A. and Orwar, O., Liposomes: Technologies and Analytical Applications. *Annu. Rev. Anal. Chem.*, 1, 801, 2008.
20. Bozzuto, G. and Molinari, A., Liposomes as Nanomedical Devices. *Int. J. Nanomed.*, 10, 975, 2015.
21. Khan, M.A., Aljarbou, A.N., Khan, A., Younus, H., Liposomal Thymoquinone Effectively Combats Fluconazole-Resistant Candida Albicans in a Murine Model. *Int. J. Biol. Macromol.*, 76, 203, 2015.
22. Hassanpour, P., Hamishehkar, H., Baroughi, B.B., Baradaran, B., Shotorbani, S.S., Mohammadi, M., Shomali, N., Aghebati-Maleki, L., Nami, S., Antifungal Effects of Voriconazole-Loaded Nano-Liposome on Fluconazole-Resistant Clinical Isolates of Candida Albicans, Biological Activity and ERG11, CDR1, and CDR2 Gene Expression. *Assay Drug Dev. Technol.*, 19, 453, 2021.
23. Mlynarczyk, D.T., Dlugaszewska, J., Kaluzna-Mlynarczyk, A., Goslinski, T., Dendrimers against Fungi–A State of the Art Review. *J. Control. Release*, 330, 599, 2021.
24. Palmerston Mendes L, Pan J, Torchilin VP. Dendrimers as Nanocarriers for Nucleic Acid and Drug Delivery in Cancer Therapy. *Molecules*, 22, 1-21, 2017.
25. Kaminskas, L.M., McLeod, V.M., Porter, C.J.H., Boyd, B.J., Association of Chemotherapeutic Drugs with Dendrimer Nanocarriers: An Assessment of

the Merits of Covalent Conjugation Compared to Noncovalent Encapsulation. *Mol. Pharm.*, 9, 355, 2012.

26. Heredero-Bermejo, I., Gómez-Casanova, N., Quintana, S., Soliveri, J., de la Mata, F.J., Pérez-Serrano, J., Sánchez-Nieves, J., Copa-Patiño, J.L., *In Vitro* Activity of Carbosilane Cationic Dendritic Molecules on Prevention and Treatment of Candida Albicans Biofilms. *Pharmaceutics*, 12, 1-17, 2020.
27. De Volder, M.F.L., Tawfick, S.H., Baughman, R.H., Hart, A.J., Carbon Nanotubes: Present and Future Commercial Applications. *Sci. (80-.)*, 339, 535, 2013.
28. Serpell, C.J., Kostarelos, K., Davis, B.G., Can Carbon Nanotubes Deliver on Their Promise in Biology? Harnessing Unique Properties for Unparalleled Applications. *ACS Cent. Sci.*, 2, 190, 2016.
29. Mundra, R.V., Wu, X., Sauer, J., Dordick, J.S., Kane, R.S., Nanotubes in Biological Applications. *Curr. Opin. Biotechnol.*, 28, 25, 2014.
30. Waghule, T., Sankar, S., Rapalli, V.K., Gorantla, S., Dubey, S.K., Chellappan, D.K., Dua, K., Singhvi, G., Emerging Role of Nanocarriers Based Topical Delivery of Anti-Fungal Agents in Combating Growing Fungal Infections. *Dermatol. Ther.*, 33, e13905, 2020.
31. Albayaty, Y.N., Thomas, N., Ramírez-García, P.D., Davis, T.P., Quinn, J.F., Whittaker, R., Prestidge, C.A., PH-Responsive Copolymer Micelles to Enhance Itraconazole Efficacy against Candida Albicans Biofilms. *J. Mater. Chem. B*, 8, 1672, 2020.
32. Nicolau Costa, K.M., Sato, M.R., Barbosa, T.L.A., Rodrigues, M.G.F., Medeiros, A.C.D., Damasceno, B.P.G.D.L., Oshiro-Júnior, J.A. Curcumin-Loaded Micelles Dispersed in Ureasil-Polyether Materials for a Novel Sustained-Release Formulation. *Pharmaceutics*, 13, 1-10, 2021.
33. Zhou, L., Zhang, P., Chen, Z., Cai, S., Jing, T., Fan, H., Mo, F., Zhang, J., Lin, R., Preparation, Characterization, and Evaluation of Amphotericin B-Loaded MPEG-PCL-g-PEI Micelles for Local Treatment of Oral Candida Albicans. *Int. J. Nanomed.*, 12, 4269, 2017.
34. Zhou, L., Zhao, X., Li, M., Lu, Y., Ai, C., Jiang, C., Liu, Y., Pan, Z., Shi, J., Antifungal Activity of Silver Nanoparticles Synthesized by Iturin against Candida Albicans *in Vitro* and *in Vivo*. *Appl. Microbiol. Biotechnol.*, 105, 3759, 2021.
35. Zhou, L., Li, A., Wang, H., Sun, W., Zuo, S., Li, C., Preparation and Characterization of Luteolin-Loaded MPEG-PCL-g-PEI Micelles for Oral Candida albicans Infection. *J. Drug Deliv. Sci. Technol.*, 63, 102454, 2021.
36. Mora-Huertas, C.E., Fessi, H., Elaissari, A., Polymer-Based Nanocapsules for Drug Delivery. *Int. J. Pharm.*, 385, 113, 2010.
37. Evangelista, R.P., Amantino, C.F., Pietro, R.C.L.R., Sorrechia, R., Piazza, R.D., Marques, R.F.C., Tedesco, A.C., Primo, F.L., Copolymer-Nanocapsules of Zinc Phenyl-Thio-Phthalocyanine and Amphotericin-B in Association with Antimicrobial Photodynamic Therapy (A-PDT) Applications against Candida albicans Yeasts. *Photodiagnosis Photodyn. Ther.*, 34, 102273, 2021.

38. Zimmermann, E.S., Ferreira, L.M., Denardi, L.B., Sari, M.H.M., Cervi, V.F., Nogueira, C.W., Alves, S.H., Cruz, L., Mucoadhesive Gellan Gum Hydrogel Containing Diphenyl Diselenide-Loaded Nanocapsules Presents Improved Anti-Candida Action in a Mouse Model of Vulvovaginal Candidiasis. *Eur. J. Pharm. Sci.*, 167, 106011, 2021.
39. El-Sheridy, N.A., Ramadan, A.A., Eid, A.A., El-Khordagui, L.K., Itraconazole Lipid Nanocapsules Gel for Dermatological Applications: *In Vitro* Characteristics and Treatment of Induced Cutaneous Candidiasis. *Colloids Surf. B: Biointerfaces.*, 181, 623, 2019.
40. Thakkar, K.N., Mhatre, S.S., Parikh, R.Y., Biological synthesis of metallic nanoparticles, *Nanomed. Nanotechnol. Biol. Med.*, 6, 257-262, 2010.
41. Kamli MR, Alzahrani EA, Albukhari SM, Ahmad A, Sabir JSM, Malik MA. Combination Effect of Novel Bimetallic Ag-Ni Nanoparticles with Fluconazole against Candida albicans. *J Fungi (Basel)*, 8, 1-19, 2022.
42. Mittal, A.K., Chisti, Y., Banerjee, U.C., Synthesis of Metallic Nanoparticles Using Plant Extracts. *Biotechnol. Adv.*, 31, 346, 2013.

13

Role of Polymer-Based Nanocarriers in Drug Delivery System

Mohankumar Narayanan[1], Chackaravarthy Gnanasekaran[1], Balamurugan P.[1], Ramachandran Govindan[1], Chenthis Kanisha Chelliah[2], Rajivgandhi Govindan[1]* and Manoharan Natesan[1†]

[1]*Marine Pharmacology and Toxicology laboratory, Department of Marine Science, Bharathidasan University, Tiruchirappalli, Tamil Nadu, India*
[2]*Department of Nanotechnology, Noorul Islam Centre for Higher Education, Kumaracoil, Kanyakumari, Tamil Nadu, India*

Abstract
Nanotechnology made an enormous growth progress over the last decade. This growth is facilitated by its unique properties and potential along with broad spectrum application like pharmaceutical, cosmetics, and medical application. Its unique physical and chemical properties enable them to interact differently with the biological systems and contribute to antimicrobial activity. Nanomedicines used science of nanotechnology and cure a lot of diseases using biocompatible nanoparticles and nanorobots for various application includes delivery, diagnosis, and sensory in living organisms. It becomes target-oriented and site-specific delivery of medicines. Now there are numerous advancements in possibility to modify properties like drug release profiles, solubility, immunogenicity and bioavailability. So it is nanocarriers which can be modified according to the requirements. Drug targeting is classified into active and passive. In active targeting, peptides and antibodies are coupled with the carrier to anchor them to target. But in the case of passive targeting, carrier complex circulates through the bloodstream driven by affinity influenced by temperature, pH and molecular site. So, with all these advantages, nanocarriers can be a novel efficient drug delivery system. One of the leading causes of death worldwide is cancer. Several conventional treatments, such as immunotherapy, radiation, chemotherapy, etc., are available after understanding

**Corresponding author*: rajivgandhimicro@yahoo.com
†*Corresponding author*: biomano21@gmail.com

Sathish-Kumar Kamaraj, Arun Thirumurugan, Muthuchamy Maruthupandy, Mercedes Guadalupe López Pérez and Shanmuga Sundar Dhanabalan (eds.) Nanoscience and Nanotechnology for Smart Prevention, Diagnostics and Therapeutics: Fundamentals to Applications, (311–338) © 2024 Scrivener Publishing LLC

the biology of cancer, but these are low-specific treatments. Scientist discovered many natural compounds having the potential to promote apoptosis, reduce the clonogenicity and restrict the migration of tumour cells without affecting the normal cells. During drug administration, whole dose will not reach the targeting site. Hence, loading the compound in a nanocarrier will deliver the drug to the tumour region. Accordingly, this study concentrated on the transfer of natural antitumor compound to the targeted site using polymer based nanocarriers.

Keywords: Polymers, tumor, nanocarriers, apoptosis, drug targeting, drug delivery

13.1 Introduction

Nanotechnology and nanomaterials have garnered significant attention due to their exceptional Structural and physical properties, which differ significantly from those of atoms, molecules, and bulk materials made of the same elements. Nanoparticles, which range in size from 1-100 nm, have revolutionized the fields of pharmaceuticals and diagnostics. One of the key advantages of nanoparticles is their small size, which allows them to be easily absorbed and transported by the body's cells and tissues. This makes them ideal for use in drug delivery and other medical applications. In addition, nanoparticles can be easily functionalized with a variety of molecules, such as drugs, antibodies, or imaging agents, allowing for a high degree of specificity and targeting. The use of nanotechnology and nanomaterials has greatly expanded the possibilities for pharmaceutical and diagnostic applications, offering new and more effective ways to treat and diagnose an extensive range of diseases and conditions [1]. As Schmid *et al.* have describes, "Nanotechnology is dealing with functional systems based on the use of sub-units with specific size-dependent properties of the individual sub-units or of a system of those". The only way to explain size dependent physical properties such as chemical, optical, magnetic, or thermodynamic properties are number of constituents makes the nanoparticles unique [2]. These materials have distinctive properties because of their small size, such as high surface-to-volume ratio, improved electrical conductivity, unique fluorescence properties and superparamagnetic behaviour. Nanomedicine is in use of nanomaterials for medical purposes, such as drug delivery and diagnostics. Nanocarriers, which are nanomaterials that can transport drugs or other agents to specific parts of the body, have gained interest as an alternative to traditional drug-delivery methods. They can enhance the solubility of hydrophobic drugs and have minimal impact on the appearance of final food products. Studies have shown that nanocarriers can be effective in treating diseases, such as cancer, infectious diseases, and neurodegenerative diseases. However, additional research is needed to fully understand the major risks and benefits of using nanomaterials in medicine. Dendrimers, carbon nanotubes, nano emulsions, and liposomes are major nanomaterials designed for its uses in the field of pharmaceuticals. The main reason for the scientifical spotlight in this field is nano-sized carriers can be a significant drug development and treatment by using nanoparticles. On the whole nanoparticles are possible candidate for carrying drugs into targeted area in organ, particular tissue or cell [3]. Various type of nanocarriers that have gained significant attention in recent years due to their applications in diagnostics, therapeutics, regenerative medicine, and drug delivery systems as represented in Figure 13.1. These small particles, which are typically less than 100 nm in size, have the ability to improve drug delivery to specific regions of the body and to penetrate barriers such as the blood-brain barrier. To construct a nanocarrier, there are three essential ingredients: an organic or inorganic material, a photothermal material and a drug. Optional ingredients include targeting ligands, which help the nanocarrier to bind to specific cells or tissues, and drug binding frames, which help to control the release of the drug. Nanocarriers have been studied extensively as a potential therapeutic and diagnostic tool for a wide range of diseases, including infectious diseases,

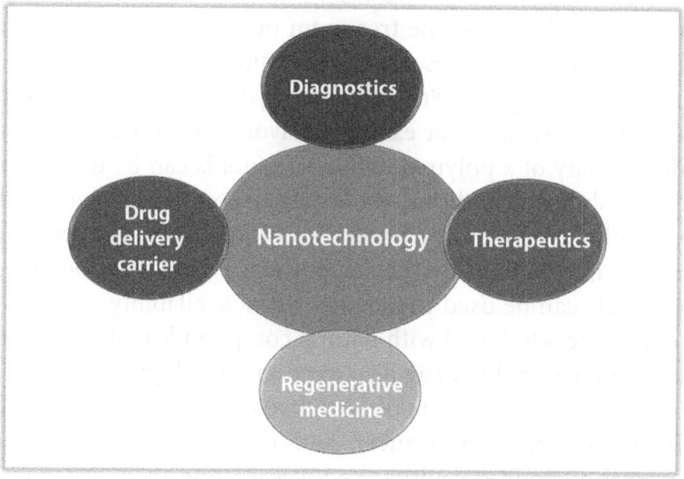

Figure 13.1 Applications of nanotechnology in various fields.

cancer and neurogenerative diseases. They are typically classified into three main categories: lipid-based nanocarriers, inorganic nanocarriers, and polymeric nanocarriers. Each type has its own unique properties and advantages, depending on the specific application [4].

Polymer based nanocarriers are easy to prepare and much more stable than lipid based nanocarriers. It has excellent properties such as controlled release and surface modification of polymer nanoparticle is convenient. It has already been reported that the polymeric nanocarriers having the potential in antibacterial, antifungal, antiproliferative, anticancer and antitumor properties. Even the synthetic drugs for tumor and cancer also been transferred to exact region on dose dependent manner. However, the potential toxic effects of nanoparticles on human health have raised concerns among researchers and regulators. Studies have shown that nanoparticles can cause inflammation, oxidative stress, and animals DNA damage after entering the body and interacting with cells. They can also trigger a cellular process called autophagy, which can lead to nanotoxicity and there is concern that biopersistent nanoparticles may be able to penetrate the skin and cause damage to live, viable cells or accumulate in secondary organs. Despite these concerns, there is growing evidence that certain nanoparticles, when loaded with bioactive substances, can be safe and effective for use in food and medicine. As a result, there is a growing interest in developing and evaluating the safety of these nanocarriers for use in various applications [5, 6]. These compounds have been shown to have various biological activities, such as anti-inflammatory, antioxidant and anti-tumor effects. They have also been found to be useful in the development of novel drugs for the treatment of various diseases. In terms of blending with polymers, these natural compounds can be used to modify the properties of the polymer, such as its hydrophilicity, biodegradability, and mechanical strength. For example, flavonoids can be used to improve the hydrophilicity of a polymer, while terpenoids can be used to improve its biodegradability. In addition, these natural compounds can also be used to improve the drug delivery properties of polymers. For example, fatty acids can be used to increase the solubility of a drug in a polymer matrix, while alkaloids can be used to enhance its bioavailability. Extensively, the ability of polymers to blend with natural compounds makes them a versatile and useful material in drug discovery and development.

Microbes also produce lots of bioactive compounds which is responsible for antimicrobial, anticancer, anti-analgesic, antitumor properties [7]. Cancer is a complex multifaceted disease, and there are many different approaches to treating it. One of the most common treatments worldwide for cancer is chemotherapy, which involves using drugs to eliminate cancer cells. However, chemotherapy can have many side effects, including damage to healthy cells and tissues. Nanomedicine is a promising approach to treating cancer that involves using nanotechnology to deliver drugs directly to cancer cells. This can help to reduce the number of drugs needed and improve their effectiveness, as well as increasing their specificity and reducing the side effects of treatment. One of the key advantages of using nanomedicine for cancer treatment is that it allows for targeted delivery of drugs to specific areas of the body. This can help to reduce the overall number of drugs needed, as well as increasing their effectiveness. Additionally, nanomedicine can help to improve the pharmacokinetics of drugs, which can further enhance their effectiveness and reduce the likelihood of side effects. It offers a promising approach to cancer treatment that can help to improve the effectiveness of chemotherapy and other treatments, while reducing the side effects and improving the overall quality of life for cancer patients. While there is still much research to be done in this field, the potential benefits of nanomedicine for cancer treatment are significant and offer hope for the future of cancer treatment [8, 9]. Presently in this chapter, we will discuss about the types of polymeric nanocarriers, advantages and disadvantages of polymeric nanocarriers with natural compounds in pharmaceutical field.

13.2 Polymeric Nanocarriers

Nanocarriers defined as the particles sizes below 100 nm and engaged in transporting substances like protein, drug, gene etc., polymers in designing nanocarriers received lot of attention because of vast medical application. It possesses significant advantage in hosting wide range of functional groups and the nanocarriers are macromolecules composed of repeating subunits. Most of the polymers are biodegradable it makes them the suitable candidate and there is no need to concern of clearance from the body. Surface modification of polymeric nanocarriers also considerably much convenient than others. Drug encapsulated nanocarriers increases the blood circulation time, extends bioavailability and enhancement of permeability. Other than that, carriers coupled with targeting ligands with activating targeting property, enhancing therapeutic property and decreasing toxicity. As the size of the carriers are small, it can pass through mucous membrane, tiny capillaries and even intracellular organelles [10, 11]. Basically, polymer based nanocarriers are divided into nano microcapsules and nanosphere category. Microcapsules are majorly "repository type" in which the drug molecules are present all over the matrix while in nanosphere drug molecules are present in the core covered with polymeric shell is "matrix type" [5]. Chitosan, pectin, dextran, hyaluronic acid, alginate, cellulose are some of the natural polymers derived from microbes, plants, animals and algal resources. Polycaprolactone (PCL), poly Lactic acid (PLA) and poly (lactic-co-glycolic acid) (PLGA), poly (ethylene glycol) (PEG), poly (ethyleneimine) (PEI), poly (vinyl alcohol) (PVA) and poly (acrylamide) (PAM) are the hydrophobic and hydrophilic synthetic polymers. Polymeric nanocarriers are in various structure based on the requirement like dendrimers, polymer micelles, nanogels, nanocomposites, nanospheres, polymersomes and nanocapsules (Figure 13.2). Bioimaging, tissue engineering, gene therapy, drug delivery and regenerative medicines are the major fields focused through polymeric nanocarriers [12]. Polysaccharides are complex sugars that are found in many plants and animals. They are known for their biocompatibility and biodegradability, making them an attractive option for use in food preservation. However, their natural properties may not always be sufficient for the desired application, so they are often chemically modified to improve their performance. One example of a modified polysaccharide is chitosan, which is derived from the exoskeleton of crustaceans. Chitosan is mucoadhesive, meaning that it can stick to

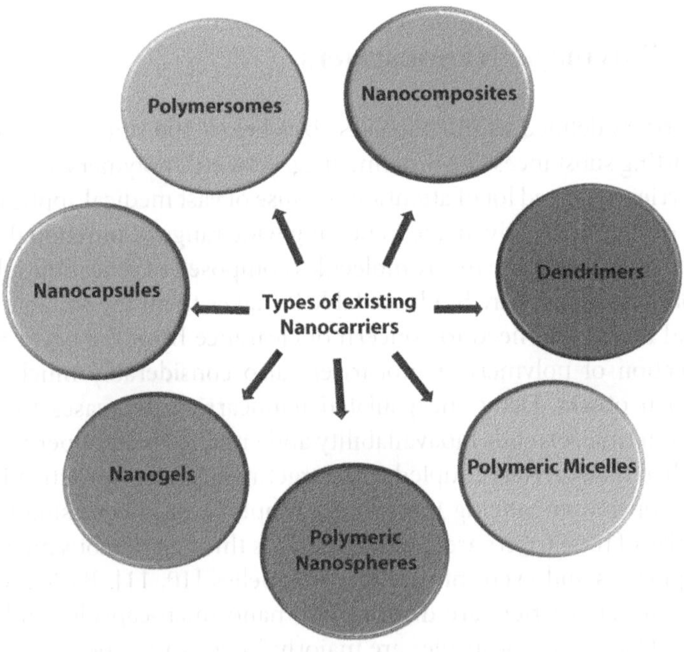

Figure 13.2 Types of existing nanocarriers.

mucous membranes, and is only soluble at acidic pH levels. These properties make it useful for certain food preservation applications, but they can also limit its usefulness in other applications. To overcome these limitations, researchers and scientists have developed chitosan derivatives, such as trimethyl chitosan (TMC) and thiolated chitosan, which have improved solubility and mucoadhesive properties. These modified polysaccharides are being studied for their potential use in food preservation and other applications.

There are numerous stimuli are initiating response in nanocarriers which is termed as smart drug delivery system. Examples for such stimuli are pH, oxidation-reduction in chemical stimuli, glucose, enzymes, inflammation in biological stimuli and temperature, photon, ultrasound in physical stimuli.

13.3 Structures of Polymeric Nanocarriers

13.3.1 Polymeric Micelles

Polymeric nanomicelles are spherical structures that are composed of amphiphilic polymers, which have both hydrophobic and hydrophilic regions. These structures are stable in aqueous media, and are commonly used in a various applications, including drug delivery and imaging (Figure 13.3). The unique structure of polymeric nanomicelles allows them to encapsulate hydrophobic drugs in their hydrophobic cores, while their

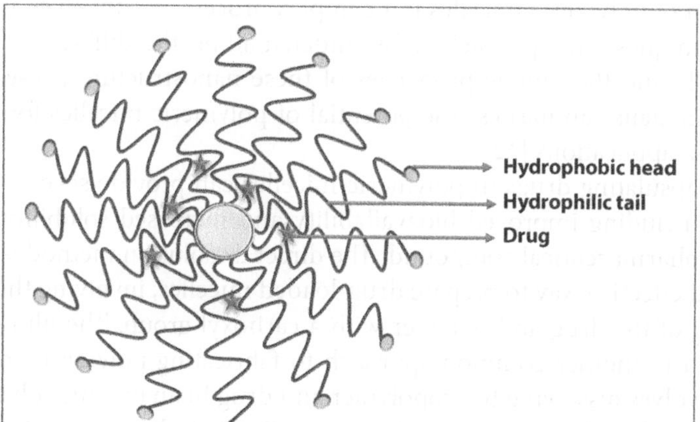

Figure 13.3 Polymeric micelle with drug materials.

hydrophilic coronas prevent them from aggregating in aqueous solutions. This makes them a promising platform for delivering drugs to specific site of the body, as they can be fabricated to target particular organ or tissues. In addition to their potential use in drug delivery, polymeric nanomicelles are also being studied for their potential use in imaging applications. Their unique structure allows them to be easily modified with imaging agents, which can help to visualize specific areas of the body and aid in diagnosis and treatment. Therefore, polymeric nanomicelles are a versatile and promising platform for a variety of applications, including diagnostic imaging and drug delivery. Their unique structure and stability in aqueous solutions make them a valuable tool for researchers and clinicians working in these fields [13]. There are several methods that can be used to fabricate polymeric micelles, including dialysis, film casting, direct dissolution, and oil-in-water emulsion. Every method has its own advantages and disadvantages, and the most appropriate method will depend on the specific application and requirements. In general, polymeric micelles are small, with sizes ranging from 20 to 200 nanometer. At concentrations below the critical micelle concentration (CMC), the amphiphilic polymers remain as individual molecules. At concentrations above the CMC, the polymers will spontaneously form micelles, with the hydrophobic parts of the polymers surrounded by the hydrophilic coronas. One of the key advantages of using polymeric micelles for drug delivery is their ability to encapsulate hydrophobic drugs in their hydrophobic cores, while their hydrophilic coronas prevent them from aggregating in aqueous solutions. This makes them a promising platform for delivering drugs to specific areas of the body, as they can be designed to target specific tissues or organs. Moreover, the fabrication of polymeric micelles is a complex process that involves a variety of techniques and approaches. By understanding the different methods available and the unique properties of these nanostructures, researchers and clinicians can harness the potential of polymeric micelles for a wide range of applications [12].

Encapsulating drugs in polymeric micelles can provide several advantages, including improved bioavailability and increased solubility of the active pharmaceutical compound. The direct dissolution method is a simple and effective way to prepare drug-loaded micelles, involving the direct mixing of the drug and polymer with a carboxyl group. The film casting method is another common approach to fabricating polymeric micelles, and involves dissolving the copolymer and drug in an organic solvent and then allowing the solution to dry down. The micelle preparation using this method involves hydrating the amphiphilic polymer at 60°C in a saline solution. Both of these methods have been shown to be effective for preparing drug-loaded micelles, and the appropriate method will depend on the specific application and requirements. By understanding the advantages and disadvantages of these methods, researchers and clinicians can choose the most appropriate approach for their specific needs. Dialysis method is used to make micelle from polymer having low soluble nature in water. In this scheme, copolymer is first dissolved well in solvent which is water miscible, then dialysed to remove solvent against the aqueous media. The polymer helps to stabilize the micelles and prevent them from breaking down or dissolving too quickly in the body. This can help to increase the amount of drug available in the bloodstream, and can also reduce the rate at which the drug is cleared from the body. As a result, polymer-stabilized micelles can be useful for delivering drugs that would otherwise be cleared too quickly to be effective, or for increasing the effectiveness of drugs that have a short half-life. This can broaden the range of pharmaceutical applications for micelles and may help to improve the treatment of certain conditions [14–16]. Oil-in-water emulsion method involves dissolving a copolymer and a drug in a non-water miscible and volatile organic solvent to form micelles. The micelles are composed of poly (ethylene glycol)-b-poly (2-diisopropyl methacrylate) and are activated only at a pH of 7.4, but remain inactive at lower pH levels such as 5-4. This type of micelles is considered to be externally stimuli mediated, meaning that they respond to external stimuli such as pH changes [17]. Poloxamer 407 is a type of micellar drug carrier that can be used in combination with oleanolic acid. It has been shown to be effective in enhancing skin hydration and reducing the appearance of wrinkles when used in dermal drug delivery. This makes it a promising potential treatment for skin-related conditions such as dehydration and wrinkles [18].

13.3.2 Nanocapsules

Nanocapsules are hollow and spherical structures that are composed of an outer polymer shell and an inner hollow space as shown in Figure 13.4 (a). They are produced through a process called nanoprecipitation or mini emulsion polymerization, which involves forming a shell around a sacrificial template, which is then removed to create an empty polymeric shell according to the Figure 13.4 (b). This process can help to reduce the toxicity of the drug on normal cells, prevent drug degradation, enable controlled release, and mask unpleasant taste. In addition, nanocapsules can protect the drug from the external environment and increase the bioavailability of peptides and proteins. These properties make nanocapsules a promising tool for drug delivery [19]. Chitosan nanocapsules have been shown

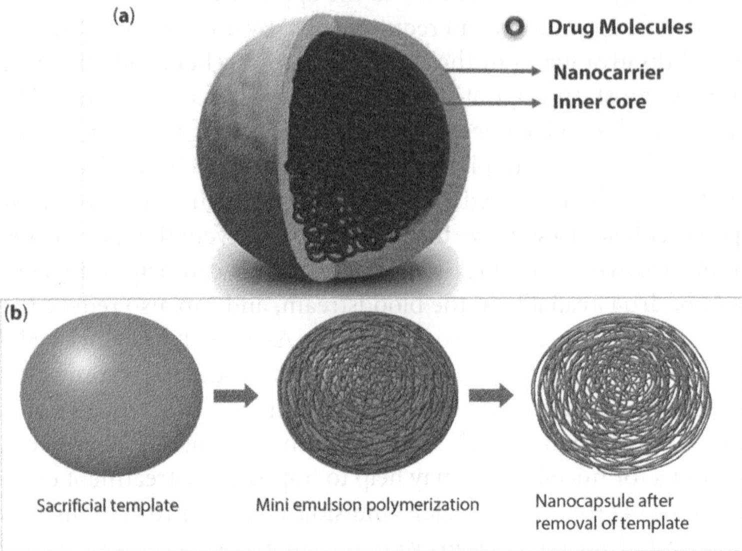

Figure 13.4 (a) Drug encapsulated polymeric nanocapsule, (b) nanocapsule formation.

to increase the absorption of cyclosporine A in mouse skin, resulting in increased hair follicles in the treated area. The mechanism by which this occurs is not well understood, but it is thought to be related to the unique properties of chitosan, which is a biocompatible and biodegradable polysaccharide that can be easily modified to create nanocapsules with desired properties. One common method for synthesizing chitosan nanocapsules is the layer-by-layer self-assembly method, which involves applying multiple layers of oppositely charged polyelectrolytes to form a multi-layered shell. This method allows for the precise control of the properties of the nanocapsules, such as size, shape, and release rate [20]. The self-templating method is a recent approach to synthesizing nanocapsules, in which both the polymer shell and the template core are assembled from the same polymeric material. This method can produce stable nanocapsules without significant issues during topical administration. The emulsion-solvent evaporation method is one common technique for synthesizing nanocapsules using the self-templating method. In addition to their potential use as drug delivery vehicles, nanocapsules can also be used in theragnostic applications by loading a magnet inside the shell and attaching fluorescence tags to the surface. This allows the nanocapsules to be tracked and monitored in the body. Furthermore, the use of nanocapsules can help to protect drugs from degradation by the harsh biological environment [21].

Targeted gene editing has been successfully accomplished *in vitro* without cytotoxicity by creating a thin, glutathione (GSH)-cleavable covalently crosslinked polymer coating, known as a nanocapsule (NC), around a preassembled ribonucleoprotein (RNP) complex between a Cas9 nuclease and a sgRNA. The NCs can be modified easily on their surface and have a hydrodynamic diameter of 25 nm. Additionally, they have been demonstrated to cause effective gene editing *in vivo* in skeletal muscle and mouse retinal pigment epithelium (RPE) tissue after local delivery. A promising method for effectively delivering CRISPR RNP complexes for both *in vitro* and *in vivo* somatic gene editing is this adaptable NC nanoplatform [22].

13.3.3 Polymersomes

Polymersomes, also known as nanovesicles or polymeric vesicles, are tiny hollow structures composed of a hydrophobic bilayer with an aqueous solution in the hollow core region. They typically range in size from 80-150 nm, and have a number of useful properties, such as an impermeable physical barrier, mechanical stability, and decreased permeability. The hydrophobic segments of the block copolymers tend to gather together in the interior of the vesicle, forming the bilayer membrane structure, while the hydrophilic segments interact with the surrounding water. This allows polymer vesicles to encapsulate hydrophobic substances within the membrane, while also sub-encapsulating hydrophilic substances in the water-based core. Polymer vesicles have several advantages over other types of vesicles, such as liposomes. For example, polymer vesicles are typically more stable and tougher than liposomes, which makes them better suited for applications such as drug delivery. Additionally, polymer vesicles have a lower permeability than liposomes, which makes them more effective at protecting their contents from the outside environment. These properties make polymer vesicles attractive for a variety of applications, such as drug delivery and biocatalysis. These characteristics make polymersomes an ideal vehicle for the co-delivery of hydrophobic and hydrophilic functional materials, such as proteins, genes, and anticancer compounds. One of the main advantages of polymersomes is their ability to protect these functional materials from degradation or damage during delivery. The impermeable nature of the polymersomes bilayer prevents the release of the functional material until it reaches its target site, allowing for more precise and controlled delivery. In addition, the mechanical stability of polymersomes allows them to withstand the rigors of the body's environment, such as changes in pH and temperature, ensuring that the functional material remains intact until it reaches its target. Overall, polymersomes offer an effective and versatile means of co-delivering hydrophobic and hydrophilic functional materials, making them a valuable tool in the field of drug delivery and other applications [23]. One example of the use of photo-responsive nanovesicles for codelivery is the self-assembly of hydrophobic and hydrophilic drugs. In this method, a hydrophobic model drug, such as Nile red, is mixed with a hydrophilic anticancer drug, such as doxorubicin hydrochloride, to form a solution. This solution is then exposed to light, which triggers the self-assembly of the drugs into nanovesicles, or polymersomes. There are two main methods for the preparation of polymersomes: solvent-free and organic solvent-based. In the solvent-free method, the drugs are mixed and self-assembled without the use of any additional solvents. This method is simpler and more environmentally friendly, but it can be less efficient in terms of drug loading and release. In the organic solvent-based method, the drugs are dissolved in an organic solvent and then hydrated to form the polymersomes. This method allows for better control over the size and properties of the nanovesicles, but it requires the use of potentially harmful organic solvent [24]. Polymersomes are vesicles, or tiny hollow structures, composed of a polymer material. They are produced by a process called solvent-switch, or phase inversion, which involves dissolving a copolymer in an organic solvent and then adding water slowly to the solution to form the polymersomes structures. Polymersomes have several advantages over other drug delivery systems because they are made of a polymer material. And they are biocompatible and can be easily modified to tissues or specific cells in the body. They can also be designed to release their payload in a controlled manner, allowing for more precise drug delivery. In addition, polymersomes can be used for diagnostic applications, such as in imaging technologies, making them a versatile tool in the field of medicine. Overall, the development of functionalized polymersomes has brought us closer to the vision of the "magic bullet" in drug delivery, providing a more targeted and effective way to deliver drugs to the body [25].

13.3.4 Dendrimers

Dendrimers are highly ordered polymeric macromolecules with a size that typically ranges between 5-10 nm. They have a number of desirable properties, including high biocompatibility, a hyperbranched composition, and good control over their molecular structure. However, the preparation of dendrimers is a time-consuming and tedious process. Dendrimers are a type of highly branched and highly symmetrical molecule that have been used for improving the solubility of drugs, including refractory anticancer drugs. Dendrimers have a high number of terminal groups, or "dendrons," which can interact with drugs through various mechanisms, including hydrogen bonding, ionic interactions and hydrophobic interactions as revealed in Figure 13.5. This may aid in enhancing the drug's solubility and bioavailability. Additionally, drug molecules can be covalently bonded to dendrimers to increase their stability and decrease their susceptibility to degradation. This is one of the advantages of dendrimers over other drug delivery systems, such as liposomes. There are two main methods for synthesizing dendrimers: the convergent method and the divergent method. The divergent method is more suitable for large-scale preparation and can produce high-generation dendrimers. This method involves starting with a central core and then repeatedly adding branches in a stepwise fashion until the desired number of generations is reached [26]. The "nanoperiodic" concept is a theoretical framework that proposes that the properties of dendrimers, a type of highly branched, tree-like molecule, can be predicted based on their size and shape. However, this approach has been criticized for failing to take into account the complex interactions between dendrimers and their environment, such as changes in their surface characteristics and aggregation state when they encounter other molecules. These interactions can result in significant variations in the pharmacokinetics, biodistribution, and safety profiles of dendrimer-based medicines, making it difficult to predict their behaviour using the "nanoperiodic" concept. The number of steps required to synthesize dendrimers is the same for both the convergent and divergent methods, but the convergent method typically requires more starting materials. Recent research has demonstrated that cisplatin encapsulated in Polyamidoamine (PAMAM) dendrimers is more

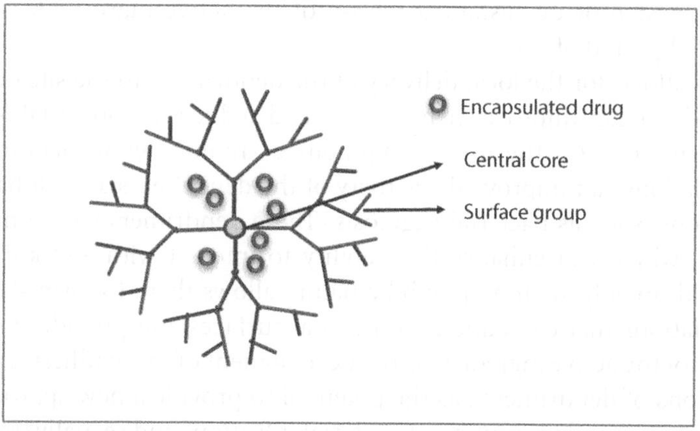

Figure 13.5 Polymeric dendrimers doped with drug molecules.

effective at targeting tumor tissue than free cisplatin, and has lower toxicity and a more gradual release profile.

This suggests that PAMAM dendrimers could be a promising drug delivery platform for cancer therapy. In general, dendrimers have a number of attractive properties for use in drug delivery, such as their small size, biocompatibility, and ability to control drug release. These properties make them a promising tool for delivering a wide range of therapeutic agents [27]. Higher generation dendrimers have a larger number of surface groups, which increases their capacity to entrap guest molecules. However, as the density of surface groups increases, the interior of the dendrimer becomes increasingly protected from the bulk solution by the "de Gennes dense packing" phenomenon. This can limit the ability of the dendrimer to entrap guest molecules, but it also provides a protective barrier for the interior of the dendrimer. The strength of the intramolecular interactions between adjacent surface groups, as well as the physicochemical conditions of the bulk solution, play a crucial role in the "de Gennes dense packing" phenomenon. This feature, known as the "dendritic box" approach, allows dendrimers to function as efficient containers for guest molecules. Overall, the "dendritic box" approach offers a versatile and efficient way to control the entrapment and release of guest molecules within dendrimers It is small enough to effectively cross the Blood–cerebrospinal fluid barrier and has a high permeability to allow for efficient drug delivery to the brain tumor. The biodegradation feature allows for controlled and targeted release of the dual drugs at the tumor site, ensuring maximum efficacy and minimizing side effects. This unique drug release mechanism allows for more precise and effective treatment of brain tumors [28]. Mucoadhesive gel formulations are designed to adhere to mucosal surfaces, such as the vagina, and provide sustained release of the active ingredients over an extended period of time.

This allows for the local delivery of the dendrimers to the site of infection, which can improve their efficacy and reduce the potential for systemic side effects. The use of amphiphilic anionic dendrimers in these formulations can improve the activity of the dendrimers against bacterial infections, such as bacterial vaginosis. These dendrimers have a negative charge, which can enhance their ability to interact with and kill bacteria. Additionally, their amphiphilic nature allows them to form stable gel formulations that can adhere to mucosal surfaces and provide sustained release of the active ingredients. The development of mucoadhesive gel formulations of dendrimers has the potential to provide a new approach for the treatment and prevention of bacterial infections and sexually transmitted diseases. These formulations are currently in clinical trials, and their effectiveness and safety are being evaluated. The PAMAM G4 hydroxyl dendrimer conjugated with methylprednisolone and spacers such as succinic acid and glutaric acid has shown to have a higher anti-inflammatory effect against epithelial lung cancer cell A549 when using succinic acid compared to glutaric acid. This is likely due to the higher stability and improved drug delivery provided by the succinic acid spacer, leading to a more efficient and effective treatment against lung cancer. Further research is needed to fully understand the mechanisms behind the increased efficacy of succinic acid-conjugated dendrimers [29].

13.3.5 Polymeric Nanogels

Polymeric nanogels are a promising type of nanocarrier for anticancer drug delivery due to their ability to respond to various stimuli and trigger drug release. These nanogels are constructed from amphiphilic or hydrophilic polymer chains that are crosslinked through physical or chemical means, resulting in a highly swellable and responsive nanosized hydrogel. The thermo-responsive, degradation-induced, and acid-triggered hydrolysis processes of these nanogels allow for precise and controlled drug release in the tumor microenvironment. This makes polymeric nanogels an attractive option for the delivery of anticancer drugs [30]. Polymeric nanogels can be prepared using various methods, including precipitation polymerization, emulsion polymerization, and one-step polymerization. However, the self-assembly method is the most widely used and extensively studied method for the production of nanogels. Through a variety of physiochemical interactions, including stereo complexation, electrostatic interactions, hydrogen bonds, and hydrophobic interactions, this approach achieves controlled aggregation of amphiphilic or hydrophilic polymers. Furthermore, temperature-sensitive self-assembly of double-hydrophilic copolymers in aqueous medium, followed by photo-cleavage and photo-dimerization of coumarin moieties, can be used to create thermal and photo-sensitive nanogels. *In vitro* experiments have shown that the low stiffness of polymeric nanogels allows them to effectively pass through the blood brain barrier, making them promising candidates for brain cancer treatment. These nanogels have demonstrated consistent results in the treatment of brain cancer, highlighting their potential as a valuable tool in the fight against this disease [31]. Lidocaine drug fabricated with PCEC nanoparticles with pluronic hydrogel can be act as a long-lasting infiltration anaesthesia for more than 300 mins of circulation. PEGylated nanogels are a type of drug delivery system that utilizes the unique properties of polyethylene glycol (PEG) to improve the stability and targeting capabilities of hydrogel nanocarriers. PEG is a water-soluble polymer with a number of useful properties, including high biocompatibility, low toxicity, and exceptional stability in extreme physiological conditions. In the synthesis of PEGylated nanogels, PEG molecules are attached to the surface of hydrogel cores through a cross-linking process. This results in a nanogel with a stable, self-assembled structure that is protected by a layer of PEG molecules. The PEG layer also imparts pH-dependent swelling behaviour to the nanogel, causing it to swell in acidic environments and shrink in basic environments. These properties make PEGylated nanogels an attractive drug delivery system for targeting drugs to tumor sites. The pH-dependent swelling behaviour allows the nanogels to selectively accumulate at tumor sites, where the acidic environment causes them to expand and release their payload. Additionally, the PEG layer protects the nanogel from degradation or clearance by the body's immune system, allowing it to remain in the body for extended time period and potentially improve drug efficacy. Nanogels are used in a wide variety of biomedical applications, including scaffolds for controlled drug and gene release systems, regenerative medicine, imaging diagnostics and *in vivo* biological sensors. The advantage of particle disintegration, which ensures the release of the constituent polymer network, is particularly advantageous for biodegradable nanogels. Nanogels' ability to entrap bioactive molecules and release them in a controlled and sustained manner improves the therapeutic efficacy of drugs and gene delivery. This makes nanogels a promising tool for improving drug and gene delivery in the medical field. Nanogel conjugated with polyamine and ethylene glycol di-glycidyl ether (EGDE) backbone to hold drugs. So, the drugs will be delivered when the EGDE crosslinking is broken by ultraviolet rays [32, 33].

13.3.6 Polymeric Nanosphere and Nanocomposite

Nanospheres are spherical nanostructures that are similar to nanocapsules but are smaller in size, typically less than one micron. They are made up of biodegradable or non-biodegradable polymers and have a hollow core that can entrap drug molecules. This allows for controlled and targeted drug delivery, leading to improved therapeutic efficacy and reduced side effects. Nanospheres have lots of applications in a variety of medical fields, including cancer therapy and drug delivery. One of the examples for the size of nanospheres can be controlled by regulating the ratio of citrate to gold in the preparation process. Increasing the amount of citrate can result in larger nanospheres. The two-phase method, also known as the Giersig-Mulvaney method, is another method for preparing nanospheres that uses phase transfer reagents to control the size of the nanospheres. The average size of the nanospheres produced using this method depends on the ratio of -SH to gold. A higher ratio of -SH to gold and a faster addition of the reductant can produce larger, monodisperse gold nanospheres, but at the cost of lower yield [34]. Polymeric nanocomposites are a class of nanocarriers that consist of nanoscale organic or inorganic fillers and a copolymer matrix. The biopolymer matrix can include proteins, polysaccharides, and aliphatic polymers, while the inorganic matrices can include gold, titanium oxide, silver, nano calcium carbonate, silica, nanocellulose crystal, and even clay. Several methods, including *in situ* polymerization, electrospinning, melt processing and solution casting, can be used to construct these nanocomposites. Green carbon nanocomposites are an important emerging technology in the field of materials science. They have a number of unique properties and potential applications, due to the combination of the high surface area and excellent conductivity of carbon nanoparticles with the biocompatibility and sustainability of agricultural waste materials. One of the key advantages of using agricultural waste materials as the carbon source for green carbon nanocomposites is their low environmental impact. These materials are typically abundant, renewable, and less toxic than many other carbon sources used in traditional carbon nanocomposite manufacturing processes. This makes them an attractive alternative to fossil fuel-based carbon sources, which are finite and contribute to greenhouse gas emissions.

The use of nanoparticles in green carbon nanocomposites also enhances the material's performance in several ways. The high surface area of nanoparticles increases the external surface area of the composite, which can improve its reactivity and catalytic activity. Additionally, the size of nanoparticles allows them to penetrate into the pores of the support material, increasing its strength and durability. Then green carbon nanocomposites offer a promising solution for sustainable and eco-friendly materials with a wide range of potential applications. These materials could be used in a variety of fields, including energy storage, environmental remediation, and biomedicine.

Polymeric nanocomposites have potential applications in drug delivery, tissue engineering, and other biomedical fields due to their capability to improve the physical and chemical properties of the copolymer matrix. The inclusion of nanoscale fillers can enhance the stability, biocompatibility, and drug release properties of the nanocomposites, making them a promising tool for the development of advanced nanocarriers [35, 36]. Polyethylene glycol-grafted graphene oxide (pGO) has been shown to successfully co-deliver the photosensitizer Ce6 and the anti-cancer drug doxorubicin. This co-delivery system has demonstrated an ability to promote accumulation on tumor tissue and increase cellular uptake of the photosensitizer. This makes pGO a promising nanocarrier for the treatment of cancer, as it can effectively deliver multiple drugs to the tumor site for improved therapeutic efficacy. Further research is needed to fully understand the mechanisms behind the enhanced drug delivery capabilities of pGO [37].

13.4 Applications of Polymeric Nanoparticles in Cancer Drug Delivery

Epigallocatechin-3-gallate, a polyphenol found in high concentrations in green tea, has been shown to have the ability to inhibit cancer at any stage. Encapsulating epigallocatechin-3-gallate with polylactic acid (PLA) and polyethylene glycol nanoparticles has demonstrated promising results against prostate cancer, due to the nanoparticles target specificity and increased bioavailability. *In vivo* studies using mice with tumor sizes of 854 and 707 mm^3 showed a decrease in tumor size of about 10-fold when using the nanoparticle-encapsulated form of epigallocatechin-3-gallate compared to the free form. These results highlight the potential of using nanoparticle-encapsulated epigallocatechin-3-gallate for the treatment of prostate cancer [38]. Curcumin is a hydrophobic polyphenol from *curcuma longa* L have numerous bioactive properties including anticarcinogenic, anti-inflammatory and antioxidant. Poly (lactic-coglycolic acid) (PLGA) encapsulated curcumin represent 55% and 21-folds increased activity through intravenous and oral administration respectively. In curcumin loaded microsphere increases bioavailability, induce apoptosis and induced tumour regression in Balb/c mice [36]. Likewise, nanoemulsion of camptothecin derivative is prepared by han et al (2009) using oil emulsion technique. Its delivery increased about 28.6 and 23.6-folds against A549 lung cancer cells and S180 sarcoma cell lines respectively. Antitumor activity also increased rapidly [39]. *In vivo* studies have shown that resveratrol encapsulated in polylactic acid nanoparticles demonstrates a significant reduction in tumor growth due to enhanced permeation and retention effects. Another study showed that entrapping resveratrol in a nanocarrier increases its cytotoxicity against prostate cancer cells in a dose-dependent manner. These results propose that the use of resveratrol-loaded nanoparticles has potential as a treatment for prostate cancer due to its ability to increase drug uptake and efficacy [40, 41].

Hyaluronic acid based nanocarrier coupled with siRNA attacking Bcl2 and survivin in NSCLC tumor model in both *invitro* and *invivo* studies provides nice payload delivery and antitumor activity. A monomethoxy-poly (ethylene glycol)-block-poly (D, L-lactide) (mPEG-PDLLA) formulation of paclitaxel represents efficient response against ovarian cancer and small cell lung cancer. Non-small cell lung cancer (NSCLC) overexpresses epidermal growth factor receptor (EGFR). So chitosan based nanoparticles surface modified to target EGFR and will delivers siRNA. NSCLC cells selectively uptake the surface modified chitosan nanocarrier which results in massive apoptosis in cancer cells. Because siRNA knocked out Mad2 induces the apoptosis [42]. Combination therapy, as the name suggests, involves the use of multiple drugs to treat a condition or disease. This approach is often used to treat cancer, as it can be more effective than using a single drug alone. One example of combination therapy for cancer is the use of liposomes, which are tiny bubbles made of a special type of fat, to deliver multiple drugs to tumor cells. Liposomes can be used to deliver a combination of drugs, such as rapamycin and doxorubicin, which have been shown to have a synergistic anti-cancer effect. Additionally, the use of a photo-absorbent agent in combination with the liposomal drug delivery system can further enhance the anti-cancer effects of the drugs. This is because the photo-absorbent agent absorbs energy from light, which can be used to trigger the discharge of the drugs from the liposomes and target the tumor cells more effectively. Overall, combination therapy with liposomes and photo-absorbent agents can be an effective approach for treating cancer, especially when tumors become resistant to single-drug treatments [43]. Resveratrol is a natural compound found in certain foods and plants, including grapes, berries, and red wine. It has been proven to have potential anti-cancer properties, and has been studied for its potential use in the treatment of various types of cancer.

One way to deliver resveratrol to cancer cells is by using nanoparticles, which are very small particles that can be used to carry drugs or other agents to specific areas of the body. In the case of resveratrol, nanoparticles made of poly lactic-co-glycolic acid (PLGA) have been used to deliver the compound to tumor cells. In addition to the PLGA nanoparticles, the resveratrol can also be attached to other agents, such as folic acid and indocyanine green, which can help to extend the drug circulation time in the bloodstream. This can help to increase the amount of time the drug is available to target cancer cells, potentially increasing its effectiveness. Studies have shown that using resveratrol-loaded PLGA nanoparticles can result in significant cell death in glioblastoma cells, compared to using free resveratrol alone. Additionally, co-encapsulating the nanoparticles with a combination of two drugs, betulinic acid and gemcitabine, can induce apoptosis in cancer cells and it has been shown to have antitumor efficacy in mice models. the use of resveratrol-loaded PLGA nanoparticles with surface attachments has the potential to be an effective approach for the treatment of cancer, by prolonging the circulation time of the drug and increasing its ability to target and kill tumor cells [8].

Irinotecan is a camptothecin derivative having an additional bis-piperidine group. When the PEG-irinotecan conjugate drug administrated in the site it produces SN38. It is also a camptothecin analogue derivative. Both derivatives are good anticancer agents. When it is conjugated with poly ethylene glycol nanoparticles (PEG), PEG-irinotecan conjugate shows improved half-life in circulation about 15 days and the PEG-SN38 also demonstrated anticancer activity towards pancreatic, breast and colorectal cancer with 245 fold more effective than PEG-irinotecan [48, 49]. Hyaluronic acid (HA) is a natural compound found in the body that plays a vital role in maintaining the structural integrity of tissues, including the joints and skin. It is also a potential drug delivery vehicle, as its gel-like consistency allows it to easily incorporate and transport various drugs and other agents. One way to use HA for drug delivery is to create nanoparticles made of the compound. These nanoparticles can be used to deliver drugs, such as doxorubicin, which is a commonly used chemotherapy drug for the treatment of cancer. One advantage of using HA nanoparticles for drug delivery is that they can be designed to release the drug in response to specific conditions, such as a low pH environment. This allows for more controlled and targeted delivery of the drug to tumor cells, potentially increasing its effectiveness. Reports have shown that HA nanoparticles can be effective in delivering doxorubicin to cancer cells, and have demonstrated efficient activity against tumors in animal models. Additionally, HA nanoparticles modified with glycyrrhetinic acid, which is a derivative of HA, may have potential as a targeting material for tumors in the liver, due to the presence of a glycyrrhetinic acid receptor in hepatocyte membranes [44]. Methotrexate is a commonly used chemotherapy drug for the treatment of cancer, including breast and laryngeal epidermoid carcinoma. It works by inhibiting the activity of an enzyme called dihydrofolate reductase, which is essential for the synthesis of nucleic acids, the building blocks of DNA.

One way to improve the effectiveness of methotrexate as a cancer treatment is to deliver the drug using a nanoparticle delivery system. One such system is glycosylated one-step PAMAM dendrimers, which are highly branched, tree-like molecules that can be used to carry and deliver drugs to specific areas of the body. It is reported that using PAMAM dendrimers to deliver methotrexate can increase its anticancer efficacy compared to administering the drug alone. This is likely due to the ability of the dendrimers to target the drug to cancer cells and improve its uptake, leading to a greater concentration of the drug at the site of the tumor. Another approach for delivering methotrexate to cancer cells is to use nanoparticles made of poly lactic-co-glycolic acid (PLGA) coupled with transferrin (Tf), a protein that binds to and transports iron in the body. These Tf-PLGA nanoparticles can be used to deliver the drug, and have been shown to have a good antiproliferative activity against laryngeal epidermoid carcinoma and breast cancer cells [45]. Carboxymethyl dextran nanoparticles coupled with doxorubicin and disulfide bond which in drug release during the presence of GSH displayed higher cytotoxicity towards SCC7 cells. It delivered doxorubicin to the nuclei of SCC7 cells with the enhanced permeation and retention effects. Curcumin-dextran nanoparticles along with Methotrexate fabricated by Methotrexate. With great sustained manner of Methotrexate and curcumin moieties release increased the toxicity against breast cancer MCF-7 cells [46].

Cationic bovine serum albumin (CBSA) nanoparticle is a stable one, capable to deliver siRNA and protect it from degradation. It also promotes accumulation in lungs inducing apoptosis and reducing tumor growth in mice model. Cow milk based lactoferrin with doxorubicin for oral delivery administration to target hepatocellular carcinoma. Hepatocellular carcinoma cells have lactoferrin receptors attracts lactoferrin in nanoparticles and it is reported to exhibit bioavailability and efficacy compared to free doxorubicin [47]. Probable combination therapy for multidrug-resistant breast cancer is the use of a nanocarrier to deliver a combination of two drugs, metformin and doxorubicin. Metformin is a drug frequently used to treat type 2 diabetes, and has been shown to have potential anticancer properties. Doxorubicin is a chemotherapy drug that is commonly used to treat a variety of cancers, including breast cancer. The nanocarrier used to deliver the drugs is made of poly (lactic-co-glycolic acid)-d-α-tocopheryl polyethylene glycol 1000 succinate (PLGA-TPGS), which has been shown to be an effective delivery system for both drugs. The combination of metformin and doxorubicin in the nanocarrier can inhibit the P-glycoprotein efflux system, which is a mechanism by which cancer cells can pump drugs out of the cell and prevent them from being effective. The use of a PLGA-TPGS nanocarrier to deliver a combination of metformin and doxorubicin has the potential to be an effective treatment for multidrug-resistant breast cancer, by overcoming the P-glycoprotein efflux system and increasing the effectiveness of the drugs has been reported by Shafiei–Irannejad [50]. Alginate is a natural polysaccharide that is derived from seaweed and is commonly used in

Table 13.1 Polymeric anticancer therapeutic agent, experiment model and effect of conjugation.

Nanocarrier system	Therapeutic agents	*In vivo/In vitro*	Effects of conjugation	Reference
Polylactic acid (PLA)–polyethylene glycol	Epigallocatechin-3-gallate	*In vitro*	Increased anti-proliferative properties	[38]
		In vivo	Selective accumulation in the tumor	[38]
Poly (lactic-coglycolic acid)	Curcumin	*In vivo*	Anticarcinogenic, anti-inflammatory and antioxidant	[36]
Nanoemulsion	Camptothecin	*In vitro*	Increased drug delivery and anti-tumor activity	[39]
Polylactic acid	Resveratrol	*In vivo*	Induction of cytotoxicity	[40]
Hyaluronic acid	siRNA	*In vitro*	Increased antitumor activity	[42]
Chitosan	siRNA	*In vitro*	Induction of apoptosis	[42]
Polyethylene glycol-poly lactic-co-glycolic acid	Betulinic acid, gemcitabine	*In vitro*	Induction of apoptosis	[8]
		In vivo	Antitumor properties	[8]
Hyaluronic acid	doxorubicin	*In vivo*	Great Antitumor activity	[44]
Transferrin poly lactic-co-glycolic acid	Quercetin	*In vitro*	Antiproliferative property against larynx epidermoid carcinoma and breast	[45]
Curcumin-dextran	Methotrexate	*In vitro*	Increased cytotoxicity	[46]
Lactoferrin	doxorubicin	*In vivo*	bioavailability and efficacy	[47]

a variety of applications, including in the food industry as a thickening and stabilizing agent. For example, alginate nanoparticles can be used to deliver a combination of two drugs, doxorubicin and pheophorbide A, to cancer cells. Pheophorbide A is a photosensitizer that rapidly produces cytotoxic singlet oxygen when directly exposed to light. It is reported that alginate nanoparticles to deliver doxorubicin and pheophorbide A has been shown to be effective in murine melanoma cells (B16 cells), with a dose-dependent release of the drugs. The combination of doxorubicin and pheophorbide A, has a strong anti-tumor effect, and the nanoparticles can accumulate in the tumor in mice without any off-target effects [51]. In polymeric nano based cancer therapy, These are some of the major applications which is also listed in Table 13.1.

13.5 Conclusion

In this chapter, we have discussed about the various structures and their applications as nanocarrier for anticancer therapy. Polymeric nanocarriers attracted lot of attention due to their advantages like biodegradability, less toxic, stability and control using stimuli. Polymeric nanocarriers have established to shelter and safeguard bioactive compound, protein and even nucleosides from different factors. With its biodegradable properties there is no chance for the settling which may lead to toxicity. Size of the nanocarriers helps it to pass through many biological barriers such as mucous, blood brain barrier, skin, etc. Multifunctional nanocarriers also reported can deliver drugs and also have fluorescence tag which help during imaging. Drug release at the site of interest controlled over many external stimuli are multi stimuli responsive nanocarriers. It also reported that combination drug fabricated nanoparticles show higher activity and helpful in administration of poorly water-soluble drugs. It reduces the symptomatic side effect of drug and the chance of development of drug resistance. The therapeutic effects of the medicine are being promoted by these various benefits. There are certain restrictions. Polymeric nanocarriers opens many ways in the field of drug delivery. In future, we believe that nanocarrier with natural product is economic and safer treatment for cancer therapy.

Acknowledgement

We would Like to thank RUSA2.0 (MHRD) Government of India, New Delhi (Ref. No. 02BDU/RUSA2.0/TRP/BS/Date: 22/11/2021) for financially supported this study. Special thanks to Bharathidasan University, Tiruchirappalli, India for providing facility.

References

1. Surendiran, A., Sandhiya, S., Pradhan, S.C., Adithan, C., Novel applications of nanotechnology in medicine. *Indian J. Med. Res.*, 130, 6, 689–701, 2009.
2. Hofmann-Amtenbrink, M., Hofmann, H., Hool, A., Roubert, F., Nanotechnology in medicine: European research and its implications. *Swiss Med. Wkly.*, 144, 1, 2014.
3. Colone, M., Calcabrini, A., Stringaro, A., Drug Delivery Systems of Natural Products in Oncology. *Molecules*, 25, 4560, 2020.
4. Avramović, N., Mandić, B., Savić-Radojević, A., Simić, T., Polymeric Nanocarriers of Drug Delivery Systems in Cancer Therapy. *Pharmaceutics*, 12, 298, 2020.
5. Lu, H., Zhang, S., Wang, J., Chen, Q., A Review on Polymer and Lipid-Based Nanocarriers and Its Application to Nano-Pharmaceutical and Food-Based Systems. *Front. Nutr.*, 8, 1, 783831, 2021.
6. Patra, J.K. *et al.*, Nano based drug delivery systems: recent developments and future prospects. *J. Nanobiotechnology*, 16, 71, 2018.
7. Ahmadu, T. and Ahmad, K., An Introduction to Bioactive Natural Products and General Applications BT, in: *Bioactive Natural Products for Pharmaceutical Applications*, D. Pal and A.K. Nayak (Eds.), pp. 41–91, Springer International Publishing, United States, Cham, 2021.
8. Ahuja, R., Panwar, N., Meena, J., Singh, M., Sarkar, D.P., Panda, A.K., Natural products and polymeric nanocarriers for cancer treatment: a review. *Environ. Chem. Lett.*, 18, 2021, 2020.
9. Devadasu, V.R., Bhardwaj, V., Kumar, M.N.V.R., Can Controversial Nanotechnology Promise Drug Delivery? *Chem. Rev.*, 113, 1686, 2013.
10. De, R., Mahata, M.K., Kim, K.T., Structure-Based Varieties of Polymeric Nanocarriers and Influences of Their Physicochemical Properties on Drug Delivery Profiles. *Adv. Sci.*, 9, 1, 2105373, 2022.
11. Yan, G., Li, A., Zhang, A., Sun, Y., Liu, J., Polymer-Based Nanocarriers for Co-Delivery and Combination of Diverse Therapies against Cancers. *Nanomaterials*, 8, 85, 2018.
12. Tang, Z., He, C., Tian, H., Ding, J., Hsiao, B.S., Chu, B., Chen, X., Polymeric nanostructured materials for biomedical applications. *Libk.*, 60, 86–128, 2016.
13. Letchford, K. and Burt, H., A review of the formation and classification of amphiphilic block copolymer nanoparticulate structures: micelles, nanospheres, nanocapsules and polymersomes. *Eur. J. Pharm. Biopharm.*, 65, 259, 2007.

14. Li, M., Song, W., Tang, Z., Lv, S., Lin, L., Sun, H., Li, Q., Yang, Y., Hong, H., Chen, X., Nanoscaled poly(l-glutamic acid)/doxorubicin-amphiphile complex as pH-responsive drug delivery system for effective treatment of nonsmall cell lung cancer. *ACS Appl. Mater. Interfaces*, 5, 1781, 2013.
15. Lee, H., Zeng, F., Dunne, M., Allen, C., Copolymer Micelles for Formulation of Hydrophobic Drugs. *Society*, 3119, 2005.
16. La, S.B., Okano, T., Kataoka, K., Preparation and Characterization of the Micelle-Forming Polymeric Drug Indomethacin-Incorporated Poly(ethylene oxide)–Poly(-benzyl. *J. Pharm. Sci.*, 85, 85, 1996.
17. Chen, T., Guo, X., Zhao, A., Wang, J., Shi, C., Zhou, S., Morphological transition of self-assembled architectures from PEG-based ether-anhydride terpolymers. *Soft Matter*, 9, 3021, 2013.
18. An, J.Y., Yang, H.S., Park, N.R., Koo, T., Shin, B., Lee, E.H., Cho, S.H., Development of Polymeric Micelles of Oleanolic Acid and Evaluation of Their Clinical Efficacy. *Nanoscale Res. Lett.*, 15, 133, 2020.
19. Tiarks, F., Landfester, K., Antonietti, M., Preparation of polymeric nanocapsules by miniemulsion polymerization. *Langmuir*, 17, 908, 2001.
20. Lee, J.S., Hwang, Y., Oh, H., Kim, S., Kim, J.H., Lee, J.H., Shin, Y.C., Tae, G., Choi, W., II, A novel chitosan nanocapsule for enhanced skin penetration of cyclosporin A and effective hair growth *in vivo*. *Nano Res.*, 12, 3024, 2019.
21. Deng, S., Gigliobianco, M.R., Censi, R., Di Martino, P., Polymeric Nanocapsules as Nanotechnological Alternative for Drug Delivery System: Current Status, Challenges and Opportunities. *Nanomater. (Basel, Switzerland)*, 10, 847, 2020.
22. Chen, G., Abdeen, A.A., Wang, Y., Shahi, P.K., Robertson, S., Xie, R., Suzuki, M., Pattnaik, B.R., Saha, K., S. Gong, A biodegradable nanocapsule delivers a Cas9 ribonucleoprotein complex for *in vivo* genome editing. *Nat. Nanotechnol.*, 14, 974, 2019.
23. Liu, G.Y., Chen, C.J., Ji, J., Biocompatible and biodegradable polymer- somes as delivery vehicles in biomedical applications. *Soft Matter*, 8, 8811, 2012.
24. Wang, X., Liu, G., Hu, J., Zhang, G., Liu, S., Concurrent block copolymer polymersome stabilization and bilayer permeabilization by stimuli-regulated «traceless» crosslinking. *Angew. Chemie - Int. Ed.*, 53, 3138, 2014.
25. Pawar, P., Gohil, S., Jain, J.P., Kumar, N., Functionalized polymersomes for biomedical applications. *Polym. Chem.*, 4, 3160, 2013.
26. D., N., R., S., Dhand, C., Prabhakaran, M.P., Beuerman, R.W., Lakshminarayanan, R., Role of Size of Drug Delivery Carriers in Pulmonary and Intravenous Administration with Emphasis on Cancer Therapeutics and Lung- Targeted Drug Delivery. *RSC Adv.*, 4, 32673–32689, 2014.
27. Noriega-Luna, B., Godínez, L.A., Rodríguez, F.J., Rodríguez, A., Zaldívar-Lelo De Larrea, G., Sosa-Ferreyra, C.F., Mercado-Curiel, R.F., Manríquez, J., Bustos, E., Applications of dendrimers in drug delivery agents, diagnosis, therapy, and detection. *J. Nanomater.*, 2014, 507273, 2014.

28. Nikzamir, M., Hanifehpour, Y., Akbarzadeh, A., Panahi, Y., Applications of Dendrimers in Nanomedicine and Drug Delivery: A Review. *J. Inorg. Organomet. Polym. Mater.*, 31, 2246, 2021.
29. Ihre, H.R., Padilla de Jesús, O.L., Szoka Jr., F.C., Fréchet, J.M.J., Polyester dendritic systems for drug delivery applications: Design, synthesis, and characterization. *Bioconjugate Chem.*, 13, 443, 2002.
30. Preman, N.K., Barki, R.R., Vijayan, A., Sanjeeva, S.G., Johnson, R.P., Recent developments in stimuli-responsive polymer nanogels for drug delivery and diagnostics: A review. *Eur. J. Pharm. Biopharm.*, 157, 121, 2020.
31. Ribovski, L., de Jong, E., Mergel, O., Zu, G., Keskin, D., van Rijn, P., Zuhorn, I.S., Low nanogel stiffness favors nanogel transcytosis across an *in vitro* blood–brain barrier. *Nanomed. Nanotechnol. Biol. Med.*, 34, 102377, 2021.
32. Costa, D., Valente, A.J.M., Queiroz, J., Plasmid DNA nanogels as photoresponsive materials for multifunctional bio-applications. *J. Biotechnol.*, 202, 98, 2015.
33. Vinogradov, S.V., Nanogels in the race for drug delivery. *Nanomedicine*, 5, 165, 2010.
34. Li, K., Zhan, R., Feng, S.S., Liu, B., Conjugated polymer loaded nanospheres with surface functionalization for simultaneous discrimination of different live cancer cells under single wavelength excitation. *Anal. Chem.*, 83, 2125, 2011.
35. Ojijo, V. and Ray, S.S., Processing strategies in bionanocomposites. *Prog. Polym. Sci.*, 38, 1543, 2013.
36. Chronakis, I.S., Novel nanocomposites and nanoceramics based on polymer nanofibers using electrospinning process - A review. *J. Mater. Process. Technol.*, 167, 283, 2005.
37. Sahoo, B.P. and Tripathy, D.K., *Properties and applications of polymer nanocomposites: Clay and carbon based polymer nanocomposites*, Springer, Berlin, Heidelberg, p. 1, 2017.
38. Siddiqui, I.A., Adhami, V.M., Bharali, D.J., Hafeez, B.B., Asim, M., Khwaja, S., II, Ahmad, N., Cui, H., Mousa, S.A., Mukhtar, H., Introducing nanochemoprevention as a novel approach for cancer control: Proof of principle with green tea polyphenol epigallocatechin-3-gallate. *Cancer Res.*, 69, 1712, 2009.
39. Han, M. *et al.*, A novel camptothecin derivative incorporated in nano-carrier induced distinguished improvement in solubility, stability and anti-tumor activity both *in vitro* and *in vivo*. *Pharm. Res.*, 26, 926, 2009.
40. Arora, D. and Jaglan, S., Therapeutic applications of resveratrol nanoformulations, Environ. *Chem. Lett.*, 16, 35, 2018.
41. Sanna, V., Siddiqui, I.A., Sechi, M., Mukhtar, H., Resveratrol-loaded nanoparticles based on poly(epsiloncaprolactone) and poly(D,L-lactic-co-glycolic acid)-poly(ethylene glycol) blend for prostate cancer treatment. *Mol. Pharm.*, 10, 3871, 2013.

42. Nascimento, A.V., Singh, A., Bousbaa, H., Ferreira, D., Sarmento, B., Amiji, M.M., Mad2 checkpoint gene silencing using epidermal growth factor receptor-targeted chitosan nanoparticles in non-small cell lung cancer model. *Libk.*, 11, 3515–3527, 2014.
43. Thapa, R.K., Byeon, J.H., Choi, H.-G., Yong, C.S., Kim, J.O., PEGylated lipid bilayer-wrapped nano-graphene oxides for synergistic co-delivery of doxorubicin and rapamycin to prevent drug resistance in cancers. *Nanotechnology*, 28, 295101, 2017.
44. Tian, G., Sun, X., Bai, J., Dong, J., Zhang, B., Gao, Z., Wu, J., Doxorubicin-loaded dual-functional hyaluronic acid nanoparticles: Preparation, characterization and antitumor efficacy *in vitro* and *in vivo*. *Mol. Med. Rep.*, 19, 133, 2019.
45. Halder, A., Mukherjee, P., Ghosh, S., Mandal, S., Chatterji, U., Mukherjee, A., Smart PLGA nanoparticles loaded with Quercetin: Cellular uptake and *in-vitro* anticancer study. *Mater. Today Proc.*, 5, 9698, 2018.
46. Curcio, M., Cirillo, G., Tucci, P., Farfalla, A., Bevacqua, E., Vittorio, O., Iemma, F., Nicoletta, F.P., Dextran-curcumin nanoparticles as a methotrexate delivery vehicle: A step forward in breast cancer combination therapy. *Pharmaceuticals*, 13, 2, 2020.
47. Golla, K., Bhaskar, C., Ahmed, F., Kondapi, A.K., A target-specific oral formulation of doxorubicin-protein nanoparticles: efficacy and safety in hepatocellular cancer. *J. Cancer*, 4, 644, 2013.
48. Nowotnik, D.P. and Cvitkovic, E., ProLindacTM(AP5346): a review of the development of an HPMA DACH platinum polymer therapeutic. *Adv. Drug Deliv. Rev.*, 61, 1214, 2009.
49. Sapra, P., Zhao, H., Mehlig, M., Malaby, J., Kraft, P., Longley, C., Greenberger, L.M., Horak, I.D., Novel delivery of SN38 markedly inhibits tumor growth in xenografts, including a camptothecin-11–refractory model. *Clin. Cancer Res.*, 14, 1888, 2008.
50. Shafiei-Irannejad, V., Samadi, N., Salehi, R., Yousefi, B., Rahimi, M., Akbarzadeh, A., Zarghami, N., Reversion of Multidrug Resistance by Co-Encapsulation of Doxorubicin and Metformin in Poly(lactide-co-glycolide)-d-α-tocopheryl Polyethylene Glycol 1000 Succinate Nanoparticles. *Pharm. Res.*, 35, 119, 2018.
51. Zhang, C., Shi, G., Zhang, J., Niu, J., Huang, P., Wang, Z., Wang, Y., Wang, W., Li, C., Kong, D., Redox- and light-responsive alginate nanoparticles as effective drug carriers for combinational anticancer therapy. *Nanoscale*, 93304, 3304–3314, 2017.

14

Current Trends in Dissolvable Microneedles: Transdermal Applications

Mohamed Sheik Tharik Abdul Azeeze[1]*, Sai Eswar Bondalakunta[2], Krati Shukla[2] and Abhay Raizaday[2]

[1]Department of Pharmaceutical Analysis, Faculty of Pharmacy, Karpagam Academy of Higher Education, Coimbatore, Tamil Nadu, India
[2]College of Pharmacy-JSS Academy of Technical Education, Noida, Uttar Pradesh, India

Abstract

Introduction: The use of microneedles for transdermal drug delivery is rapidly advancing, offering benefits like extended drug release and bypassing the first-pass effect. However, the stratum corneum remains a barrier for drug distribution. Microneedles provide a painless and minimally invasive solution, increasing drug bioavailability. Coated and hollow microneedles have shown effectiveness, but dissolving microneedles are particularly promising due to their easy manufacturing, disposability, and ability to be made from the therapeutic agent itself, eliminating the need for additional excipients.

Coverage: This book chapter discusses microneedle drug formulations, design, dosage control, efficacy, and delivery rate. It also covers the different types of microneedles, their designs, fabrication materials, and production processes. Clinical trials and recent applications in various therapeutic areas are highlighted. These applications include inflammatory skin conditions, scalp issues, cancer, inflammation and pain management, diabetes, cosmetics, vaccine distribution, diagnostics, and monitoring. The chapter also presents the latest trends in the microneedle landscape, emphasizing evidence-based literature, and recent clinical studies.

Conclusion: To promote commercialization, it is essential to establish regulations related to the production of microneedle products promptly. The promising potential of dissolving microneedles in drug delivery makes it crucial to continue research and development in this field.

**Corresponding author*: mtharik@jssaten.ac.in

Sathish-Kumar Kamaraj, Arun Thirumurugan, Muthuchamy Maruthupandy, Mercedes Guadalupe López Pérez and Shanmuga Sundar Dhanabalan (eds.) Nanoscience and Nanotechnology for Smart Prevention, Diagnostics and Therapeutics: Fundamentals to Applications, (339–356) © 2024 Scrivener Publishing LLC

***Keywords*:** Transdermal, microneedles, drug delivery, design, clinical trials, applications, IVIVC, treatment

14.1 Introduction

The parenteral and oral routes are commonly used to deliver the majority of small molecule drugs, with approximately 80% of such compounds facing solubility and bioavailability issues. Transdermal delivery (TDD) and percutaneous absorption are well-established methods, and recent advancements include technologies like microneedles (MNs), iontophoresis, microdermabrasion, ultrasound cavitation, and chemical permeability enhancers. TDD has gained popularity due to its ability to circumvent problems associated with oral and systemic drug delivery, such as pH fluctuations, liver metabolism, and enzymatic activity [1].

The skin, along with glands, hair follicles, and nails, forms the integumentary system and acts as a static barrier protecting internal organs from external environmental hazards. Traditionally seen as a barrier for drug administration, the skin has become an intriguing area of study enabling various transdermal and intradermal technologies. MNs have emerged as a promising solution to deliver medications intradermally, overcoming the limitations of traditional approaches [2]. These small, conical, or pyramidal protrusions, typically 1000 μm in length, offer several advantages for intradermal drug administration. They can deliver medicinal compounds that would be difficult to distribute via the transdermal route, yet are painless as they only penetrate the stratum corneum and cannot reach deeper nerve terminals, making them suitable for patients with needle phobia [3].

Human skin's extensive surface area makes it an accessible, targeted, and non-invasive method for drug delivery, addressing problems associated with oral drug administration, such as gastrointestinal discomfort, and providing a sustained release over time [4]. However, the stratum corneum (SC) poses the main challenge to effective drug absorption in transdermal distribution. MNs have emerged as a potential solution, bypassing the superficial skin barrier and offering a less invasive alternative to traditional injections [5]. MNs can be categorized based on their function, including solid, coated, hollow, dissolving, and hydrogel-forming MNs [6].

The success of MNs depends on the mechanical properties and biocompatibility of the materials used in their design. Polymers are a preferred choice due to their customizability, mechanical strength, and low production costs, while also avoiding immune reactions [7]. In the medical field, MNs play a crucial role in delivering pharmaceuticals ranging from small to macromolecules, with biotechnology companies dedicating efforts to develop MNs-loaded protein therapeutics [8]. However, challenges remain in effectively delivering large-sized biotherapeutics due to poor absorption [9].

Advancements in research and production technology for MNs hold the promise of enabling the future delivery of clinically significant pharmaceuticals [10]. This book chapter discusses evidence-based literature and clinical trials, highlighting various therapeutic areas where dissolving microneedles could be used for drug delivery. Applications will be explored for conditions such as inflammatory skin disorders, cancer, inflammation, pain management, hair issues, diabetes, cosmetics, vaccine distribution, diagnostics, and monitoring.

14.2 Types of MNs

With rapid advancements in science and technology, microneedles (MNs) for transdermal drug delivery have emerged as an innovative solution to overcome the poor patient compliance associated with conventional syringe injections. MNs work by penetrating the stratum corneum with minimal discomfort to the patient [11]. Figure 14.1 showcases the different types of MNs.

MNs come in a variety of structural configurations, morphologies, and materials, with heights ranging from 25 to 2000 μm [12]. External factors, such as physiochemical characteristics, skin physiology, and ambient conditions, can influence the drug delivery route of microneedles. Dry environments (low humidity) may slow down drug penetration into the skin, while excessive moisture (e.g., sweat) can affect drug release kinetics, impacting the osmotic gradient for transdermal drug delivery [13]. However, increased skin temperature can enhance drug permeation through the skin by increasing diffusivity and dilating skin vessels [14].

MNs offer precise dosage loading and metering accuracy for delicate medications like insulin and chemotherapeutics. Compared to oral consumption, microneedle patches typically require smaller dosages to achieve equivalent therapeutic efficacy, as they bypass digestion and first-pass metabolism [15]. The specific drug type, preferred treatment strategy, and patient profile influence the amount of medication to be loaded into MNs, which are manufactured using strict control procedures to ensure precise drug delivery. Skin physiology, environmental conditions, and the method of application to the skin surface also influence the speed of drug dissolution into skin interfaces [16, 17].

There are different types of MNs, including solid, coated, hollow, and dissolving MNs, each with distinct characteristics, benefits, drawbacks, uses, and materials [18]. Solid MNs are designed to pierce the stratum corneum, promoting drug delivery to the dermis and enhancing kinetic transport and bioavailability across the skin. They are particularly effective for delivering vaccines, providing longer-lasting and stronger immune responses compared to intramuscular injections. Solid MNs can be made from various materials, including silicon, metals, and polymers [19, 20].

Hollow MNs have an empty core or chamber used to inject or hold medication fluid, enabling them to store larger doses or quantities of medicine compared to solid MNs. They act as conduits for drug diffusion into the dermis, and their production settings and material composition can be tailored to provide tunable release kinetics. Hollow MNs have been successfully used for various vaccines and inoculations, but they may experience technical issues during injection, such as leakage and blockage [21, 22].

Coated MNs are solid-type MNs coated with a medication solution. The thickness of the coating determines the amount of medicine delivered. Coated MNs enable minimally invasive delivery of proteins and DNA, and they offer quick drug administration to the skin. However, there is a risk of drug spreading to other patients at the needle's tip [23, 24]. Dissolving MNs, which emerged in 2005, offer promising features such as one-step medication application and quick release of macromolecules. After insertion into the epidermis, the needle tip of dissolving MNs dissolves and releases the drug, facilitating diffusion. The best materials for dissolvable MNs are water-soluble ones. The micro-mould method is the preferred fabrication process for dissolving MNs, which may require expertise and face challenges with total insertion and delayed breakdown [16, 25–27].

14.3 Biodegradable MNs

Biodegradable MNs utilize various biodegradable polymers, including chitosan, polylactic acid, and polyglycolic acid, to disintegrate in the skin. Research has shown the potential of these biodegradable MNs as a patient-friendly alternative to conventional sustained-delivery strategies.

Arsalan Abu-Much *et al.* highlighted the frequent use of biodegradable MNs in controlled-release medication delivery. To overcome the drawbacks of intravenous diuresis, the authors incorporated the model drug furosemide (FUR). These MNs were developed using a casting-mold technique and consisted of flexible backing layers made of sodium alginate and

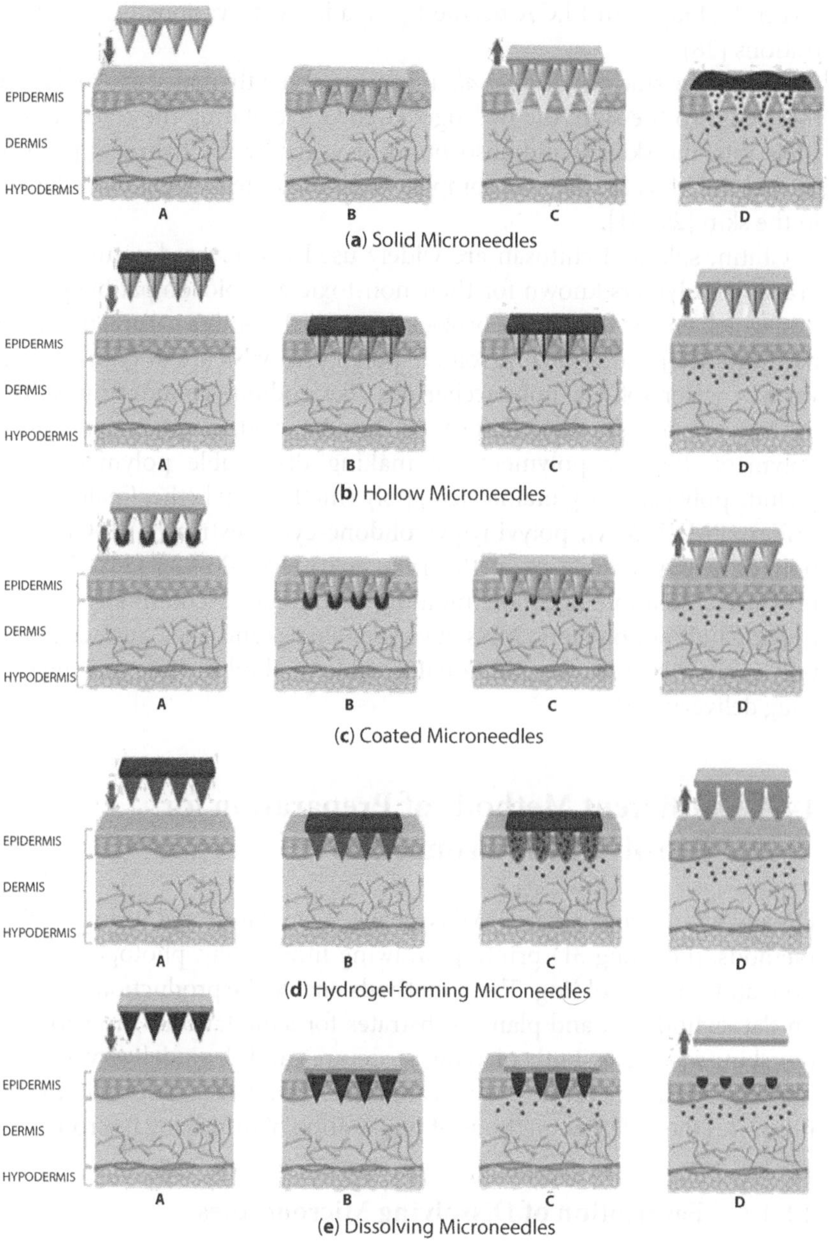

Figure 14.1 Various types of microneedles [reprinted from [1] with permission from Elsevier].

glycerol, along with PLGA needle tips loaded with various FUR concentrations [28].

In another study by Kim *et al.*, it was observed that once the MNs were inserted into the skin, the hydrogel swelled, facilitating the separation of MNs into the skin. Rapid absorption of water by the hydrogel particles caused the MNs to split and completely disintegrate, leaving the MNs tips in the skin [29–31].

Chitin, silk, and chitosan are widely used and favored natural biodegradable polymers known for their non-toxic and biocompatible properties, which enable them to be broken down by the body's natural metabolic pathways [32]. Additionally, soluble polymeric MNs are developed using various water-soluble polysaccharides like sodium chondroitin sulfate, dextran, carboxymethylcellulose, hyaluronic acid, and amylopectin as polymers. Popular polymers for making dissolvable polymeric MNs include poly gamma-glutamic acid, poly (methyl vinyl ether/maleic anhydride) (PMVE/MA), polyvinylpyrrolidone-cyclodextrin (PVP-CD), and poly (vinylpyrrolidone-co-methacrylic acid) (PVP-MAA) [33]. The biopolymeric solution injection molding technique is commonly used to develop hollow polymeric MNs in various shapes and sizes. Surface-coated polymeric MNs offer an additional feasible method for safe and effective drug delivery [34].

14.4 Different Methods of Preparation for Dissolvable Microneedles

Dissolving microneedles (DMNs) have been fabricated using various methods, including 3D printing, drawing lithography, photopolymerization, and micro-molding. These methods enable the production of DMNs on flat, cylindrical, and planar substrates for simultaneous insertion into the skin, serving as both piercing structures and drug delivery systems. The appropriate medication doses are delivered into the tissue. Figure 14.2 illustrates the different methods of preparation of dissolving microneedles.

14.4.1 Fabrication of Dissolving Microneedles

Lin *et al.* utilized a modified two-step molding process with bovine serum albumin (BSA) to create DMNs. The TP5-DMNA produced with BSA

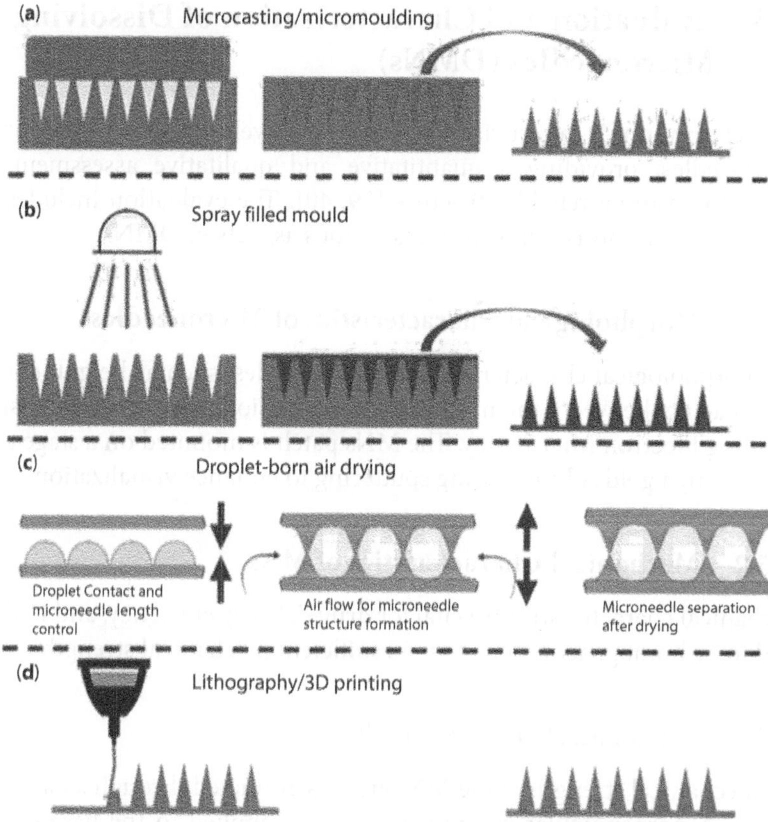

Figure 14.2 Various methods used in the preparation of dissolvable microneedles [reprinted from figure 2 with permission from Elsevier].

showed improved skin penetration and greater drug loading effectiveness [35]. Kim *et al.* proposed the droplet-born air blowing method, which provides rapid (10 min) and moderate (4-25 °C) manufacturing conditions for DMNs without drug loss. This method molds droplets into microneedles through air blowing, providing enough force for skin penetration. The pressure and timing of the droplet dispenser can be adjusted to modify the drug amount in the microneedles [36, 37]. Antonara *et al.* discussed the use of FDM 3D printing technology to streamline the manufacture of DMNs by directly 3D printing negative molds. Optimized method using statistical design and 3D printed molds to create PVA/PVP MNs with galantamine hydrobromide [38].

14.5 Evaluation and Characterization of Dissolving Microneedles (DMNs)

Characterization is a crucial step in the development of optimized microneedles, providing a quantitative and qualitative assessment of DMNs to ensure a reliable structure [39, 40]. The evaluation includes *in vitro* and *in vivo* procedures to assess various aspects of DMNs.

14.5.1 Morphological Characteristics of Microneedles

The morphological characteristics of microneedles, such as height, diameter, spacing between them, and base dimensions, are assessed using Scanning Electron Microscopy. The MNs patch is mounted on a stage and coated with a gold solution using sputtering to enhance visualization.

14.5.2 Mechanical Characteristics of MNs

Mechanical characteristics are vital for drug delivery efficiency, and DMNs need to be homogeneous and possess sufficient mechanical strength.

14.5.2.1 Axial Mechanical Strength Test

The mechanical strength of the MNs array is evaluated through axial force testing. A perpendicular compressive force is applied to the base of the MNs, and the force and displacement difference are measured when the MNs are pulled in one direction. The force-displacement (FD) curve illustrates the force changes during MNs breakdown.

14.5.2.2 Transverse Mechanical Strength Test

A metal probe applies pressure to a specific point on the MNs until a fracture is observed. The force-displacement curve is analyzed to determine the mechanical strength of the MNs.

14.5.3 Insertion Test

The penetration of MNs is assessed through staining of MN-created pores and monitoring of Trans-epidermal Water Loss (TEWL), both of which help estimate the *in vitro* and *in vivo* activity of DMNs. Techniques like confocal microscopy and histological cryosectioning are used to overcome limitations in quantitative evaluation of insertion depth.

14.5.3.1 *Staining of MNs*

This technique visually evaluates MN penetration in the skin using dyes like trypan blue or methylene blue. The staining agent colors only healthy epidermis, allowing accurate identification of microchannels. The staining method offers quick quantitative analysis.

14.5.3.2 *Histological Tissue Staining/Cryosectioning*

Determining the depth of MN penetration in the dermal layers is challenging. Skin treated with MNs is taken out and placed in a specific medium. Cryosectioning using a cryostat produces thin slices (6–12 μm), which are stained with hematoxylin and eosin (H&E) to identify MNs and microchannels.

14.5.3.3 *Confocal Microscopy*

Confocal laser-scanning microscopy (CLSM) is an advanced optical microscopic technique used to magnify sample images with high resolution and clarity. It provides information on the sizes of pores created by MNs. A solution containing fluorescent particles is applied to the treated skin, passing through the microchannels created by MNs. Confocal microscopy measures the fluorescence produced by these particles to identify penetration distances up to 250 micrometers from the skin's surface. This technique is suitable for transparent MNs.

14.5.4 *In Situ* Dissolution of DMNs

In *in-situ* dissolving studies, human cadaver skin is used, and DMNs are examined at specific intervals (e.g., 1, 2, 3, 4, and 5 minutes) using Scanning Electron Microscope (SEM) to observe MN dissolution and changes in microneedle geometry.

14.6 Applications of Dissolvable Microneedles

Dissolvable microneedles (DMNs) find extensive applications in the medical and cosmetic industries due to their painless, easy-to-use, and

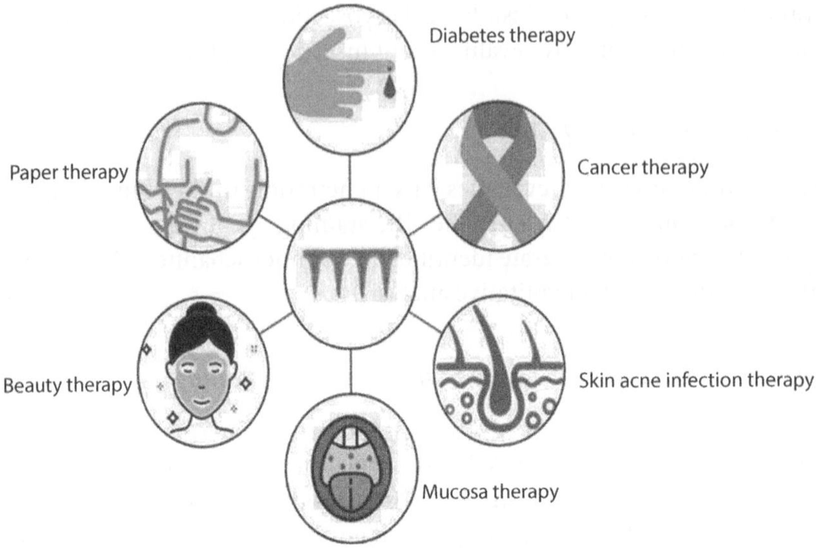

Figure 14.3 Diverse applications of microneedles (MNs). These applications encompass their usage in the cosmetic industry for reducing wrinkles and aiding weight loss, in treating skin acne infections, for cancer therapies involving drug delivery and vaccination, as pain relievers utilizing local anesthetics, in diabetes management for insulin delivery, and for mucosa therapy to target specific regions in the oral cavity for drug administration.

safe nature, as well as their ability to bypass gastrointestinal degradation. Figure 14.3 displays the various applications of MNs.

14.6.1 Cosmetic Industry

In the cosmetic industry, DMNs are utilized for wrinkle reduction and weight loss. Clinical trials assessing adenosine-loaded dissolving microneedle patches have shown promising results in reducing skin wrinkles and improving skin elasticity [41, 42].

14.6.2 Skin Acne Infection

Studies have demonstrated the therapeutic efficacy of synthetic salicylic acid (SA)-loaded polymeric microneedles for the treatment of *P. acnes* infection, suggesting their potential use for various skin ailments [43].

14.6.3 Cancer Applications

DMNs are utilized to inject nanomaterials, enzymes, genes, antibodies, vaccinations, and other medicinal molecules. They enable drug delivery to deeper skin layers and can trigger the necessary immune response. Microneedles offer a non-painful self-administration method for delivering anti-cancer vaccines and drugs, providing an alternative to traditional systemic administration using hypodermic needles. Their use in the treatment of superficial tumors shows promise and improved response rates in certain cancer mouse models [44, 45].

14.6.4 Pain Reliever

Microneedles have been used to release lidocaine, a local anesthetic, for pain relief during surgical procedures, providing acute and long-term pain relief [46, 47].

14.6.5 Diabetes Applications

For diabetes therapy, DMNs offer a simple, non-invasive, and repeatable approach to deliver insulin transdermally, potentially reducing the need for multiple daily injections. Additionally, hydrogel-forming MNs patches have been investigated for the transdermal delivery of metformin hydrochloride, which plays a crucial role in diabetes treatment [48].

14.6.6 Mucosa Therapy

Highly drug-loaded MN patches have been designed for the buccal region to transfer macromolecules more effectively than through the skin. Studies have demonstrated successful delivery of human growth hormone and insulin to the buccal cavity using DMNs [49, 50].

14.7 Dissolvable MNs Undergoing Clinical Trials: Therapeutic Applications

Dissolving microneedles (MNs) have attracted considerable attention in therapeutic research [51]. Ongoing academic investigations and clinical trials highlight their significance in various therapeutic applications. Currently, there are 13 clinical trials registered on https://clinicaltrials.gov/ that involve the use of dissolving MNs. Table 14.1 presents a selection of

Table 14.1 Selected results from clinical trials focusing on the treatment of diabetes, skin, and scalp diseases.

NCT Number	Conditions	Phases	Study Type	URL
NCT05377905	Skin cancers - squamous cell carcinoma cutaneous squamous cell carcinoma	1 and 2	Interventional	https://ClinicalTrials.gov/show/NCT05377905
NCT04249115	Benign skin, tumor lesion skin, skin lesion seborrheic keratosis	Not applicable	Interventional	https://ClinicalTrials.gov/show/NCT04249115
NCT03380845	Acne scars	Not applicable	Interventional	https://ClinicalTrials.gov/show/NCT03380845
NCT05071274	Acne vulgaris	Not applicable	Interventional	https://ClinicalTrials.gov/show/NCT05071274
NCT02596750	Pain	Not applicable	Interventional	https://ClinicalTrials.gov/show/NCT02596750
NCT01145326	Phlebotomy | pain perception	2 and 3	Interventional	https://ClinicalTrials.gov/show/NCT01145326
NCT04066023	Cluster headache	2 and 3	Interventional	https://ClinicalTrials.gov/show/NCT04066023
NCT01257763	Healthy	1	Interventional	https://ClinicalTrials.gov/show/NCT01257763
NCT00837512	Type 1 diabetes mellitus	2 and 3	Interventional	https://ClinicalTrials.gov/show/NCT00837512
NCT02682056	Diabetes	Not applicable	Interventional	https://ClinicalTrials.gov/show/NCT02682056

(Continued)

Table 14.1 Selected results from clinical trials focusing on the treatment of diabetes, skin, and scalp diseases. (*Continued*)

NCT Number	Conditions	Phases	Study Type	URL
NCT01684956	Type 1 diabetes	2	Interventional	https://ClinicalTrials.gov/show/NCT01684956
NCT02949024	Diabetic macular edema	1 and 2	Interventional	https://ClinicalTrials.gov/show/NCT02949024
NCT03126786	Diabetic macular edema	2	Interventional	https://ClinicalTrials.gov/show/NCT03126786

extracted results from these trials, focusing on the treatment of diabetes, skin, and scalp diseases.

14.8 Conclusion and Future Direction

Recently, there has been a notable increase in both clinical trials and academic studies investigating the possibilities of dissolving MNs. Indeed, MNs demonstrate significant potential in a wide range of medical fields, including cosmetics, vaccine distribution, diagnostics, cancer therapy, pain relief, scalp and hair treatments, and diabetes management. The increasing focus on the commercial feasibility of these applications is propelling the progress of microneedles technology. With enormous market potential, the field of MNs is expected to expand continuously, making dissolving MNs a subject of continued interest and research in the foreseeable future.

Conflict of Interest

The authors declare no conflict of interest, financial or otherwise.

Funding

None

Acknowledgment

The authors thank to College of Pharmacy-JSS Academy of Technical Education, Noida, Uttar Pradesh, India providing the facilities.

References

1. Chatterjee, B., Reddy, A., Santra, M., Khamanga, S., Amorphization of Drugs for Transdermal Delivery-a Recent Update. *Pharmaceutics*, 14, 5, 983, 2022 May.
2. Lane, M.E. and Parisi, N., Dermal Formulations for Local Treatment, in: *Specialised Pharmaceutical Formulation*, vol. 2, p. 75, 2022 May 4.
3. Yong, L., Guangzhi, G., Yanni, W., Fengsen, M., Drug delivery with dissolving microneedles: Skin puncture, its influencing factors and improvement strategies. *J. Drug Deliv. Sci. Technol.*, 7, 103653, 2022 Aug.
4. Benson, H.A., Grice, J.E., Mohammed, Y., Namjoshi, S., Roberts, M.S., Topical and transdermal drug delivery: from simple potions to smart technologies. *Curr. Drug Deliv.*, 16, 5, 444–60, 2019 Jun 1.
5. Hu, X. and He, H., A review of cosmetic skin delivery. *J. Cosmet. Dermatol.*, 20, 7, 2020–30, 2021 Jul.
6. Parhi, R., Recent advances in microneedle designs and their applications in drug and cosmeceutical delivery. *J. Drug Deliv. Sci. Technol.*, 5, 103639, 2022 Aug.
7. Gorantla, S., Dabholkar, N., Sharma, S., Rapalli, V.K., Alexander, A., Singhvi, G., Chitosan-based microneedles as a potential platform for drug delivery through the skin: Trends and regulatory aspects. *Int. J. Biol. Macromol.*, 184, 438–53, 2021 Aug 1.
8. Lombardo, D., Kiselev, M.A., Caccamo, M.T., Smart nanoparticles for drug delivery application: development of versatile nanocarrier platforms in biotechnology and nanomedicine. *J. Nanomater.*, 2019, 1–26, 2019 Feb 27.
9. Al-Japairai, K.A., Mahmood, S., Almurisi, S.H., Venugopal, J.R., Hilles, A.R., Azmana, M., Raman, S., Current trends in polymer microneedle for transdermal drug delivery. *Int. J. Pharm.*, 587, 119673, 2020 Sep 25.
10. Jiao, Y., Stevic, M., Buanz, A., Uddin, M.J., Tamburic, S., Current and Prospective Applications of 3D Printing in Cosmetics: A Literature Review. *Cosmetics*, 9, 6, 115, 2022 Dec.
11. Xiu, X., Gao, G., Liu, Y., Ma, F., Drug delivery with dissolving MNs: skin puncture, its influencing factors and improvement strategies. *J. Drug Deliv. Sci. Technol.*, 76, 103653, 2022.
12. Karim, Z., Karwa, P., Hiremath, S.R.R., Polymeric microneedles for transdermal drug delivery- a review of recent studies. *J. Drug Deliv. Sci. Technol.*, 77, 103760, 2022.

13. Priya, S. and Singhvi, G., Microneedles-based drug delivery strategies: A breakthrough approach for the management of pain. *Biomed. Pharmacother.*, 155, 113717, 2022.
14. Nguyen, T.T., Nguyen, T.T.D., Tran, N.-M.-A., Vo, G.V., Advances of microneedles in hormone delivery. *Biomed. Pharmacother.*, 145, 112393, 2022.
15. Ali, M., Namjoshi, S., Benson, H.A.E., Mohammed, Y., Kumeria, T., Dissolvable polymer microneedles for drug delivery and diagnostics. *J. Control. Release*, 347, 561–589, 2022.
16. Sartawi, Z., Blackshields, C., Faisal, W., Dissolving microneedles: Applications and growing therapeutic potential. *J. Control. Release*, 348, 186–205, 2022.
17. Zhang, L., Du, W., Li, X., Ling, G., Zhang, P., Dissolving microneedles based on polysaccharide for dermatological diseases therapy. *J. Drug Deliv. Sci. Technol.*, 78, 103913, 2022.
18. Tucak A, Sirbubalo M, Hindija L, Rahić O, Hadžiabdić J, Muhamedagić K, Čekić A, Vranić E. Microneedles: Characteristics, materials, production methods and commercial development. Micromachines. 2020;11(11):961. doi:10.3390/mi11110961.
19. Ananda, P.W.R., Elim, D., Zaman, H.S., Muslimin, W., Tunggeng, M.G.R., Permana, A.D., Combination of transdermal patches and solid microneedles for improved transdermal delivery of primaquine. *Int. J. Pharm.*, 609, 121204, 2021.
20. Abiandu, I. and Ita, K., Transdermal delivery of potassium chloride with solid microneedles. *J. Drug Deliv. Sci. Technol.*, 53, 101216, 2019.
21. Jin, Z., Zhao, B., Wan, W., Li, Y., Bai, X., Hu, C., Wang, J., Li, Y., Xin, W., Kang, L., Yang, H., Wang, J., Gao, S., Application of microneedle-based vaccines in biosecurity. *J. Biosaf. Biosecur.*, 4, 1, 75–83, 2022.
22. Ita, K., Ceramic microneedles and hollow microneedles for transdermal drug delivery: Two decades of research. *J. Drug Deliv. Sci. Technol.*, 44, 314–322, 2018.
23. Jakka, D., Matadh, A.V., Shivakumar, H.N., Maibach, H., Murthy, S.N., Polymer Coated Polymeric (PCP) microneedles for sampling of drugs and biomarkers from tissues. *Eur. J. Pharm. Sci.*, 175, 106203, 2022.
24. Halder, J., Rath, G., Rai, V.K., Cyclosporine coated microneedle for transcutaneous delivery: Characterization, *in vitro* evaluation, and *in vivo* anti-psoriatic efficacy against IMQ-induced psoriasis. *J. Drug Deliv. Sci. Technol.*, 73, 103450, 2022.
25. Ouyang, M., Wang, X., Fu, Y., Xie, G., Du, S., Li, Y., Zhang, L., Tao, J., Zhu, J. Skin optical clearing enabled by dissolving hyaluronic acid microneedle patches. *Int. J. Biol. Macromol.*, 220, 1188–1196, 2022.
26. Huang, Y., Lai, H., Jiang, J., Xu, X., Zeng, Z., Ren, L., Liu, Q., Chen, M., Zhang, T., Ding, X., Zhao, C., Cui, S., pH-activatable oxidative stress amplifying dissolving microneedles for combined chemo-photodynamic therapy of melanoma. *Asian Asian J. Pharm. Sci.*, 17, 5, 679–696, 2022.

27. Li, X.-J., Li, Y., Meng, Y., Pu, X.-Q., Qin, J.-W., Xie, R., Wang, W., Liu, Z., Jiang, L., Ju, X.-J., Chu, L.-Y., Composite dissolvable microneedle patch for therapy of oral mucosal diseases. *Biomater. Adv.*, 139, 213001, 2022.
28. Abu-Much, A., Darawshi, R., Dawud, H., Kasem, H., Ammar, A.A., Preparation and characterization of flexible furosemide-loaded biodegradable microneedles for intradermal drug delivery. *Biomater. Sci.*, 10, 22, 6486–99, 2022.
29. Kim, M., Jung, B., Park, J.H., Hydrogel swelling as a trigger to release biodegradable polymer microneedles in skin. *Biomaterials.*, 33, 2, 668–78, 2012 Jan 1.
30. Cheung, K. and Das, D.B., Microneedles for drug delivery: Trends and progress. *Drug Deliv.*, 23, 7, 2338–54, 2016 Sep 1.
31. Battisti, M., Vecchione, R., Casale, C., Pennacchio, F.A., Lettera, V., Jamaledin, R., Profeta, M., Di Natale, C., Imparato, G., Urciuolo, F., Netti, P.A., Non-invasive production of multi-compartmental biodegradable polymer microneedles for controlled intradermal drug release of labile molecules. *Front. Bioeng. Biotechnol.*, 7, 1–14, 2019 November.
32. Wang, M., Hu, L., Xu, C., Recent advances in the design of polymeric microneedles for transdermal drug delivery and biosensing. *Lab. Chip*, 17, 8, 1373–1387, 2017.
33. Tuan-Mahmood, T.-M., McCrudden, M.T.C., Torrisi, B.M., McAlister, E., Garland, M.J., Singh, T.R., Donnelly, R.F., Microneedles for intradermal and transdermal drug delivery. *Eur. J. Pharm. Sci.*, 50, 5, 623–637, 2013.
34. Lhernould, M.S., Deleers, M., Delchambre, A., Hollow polymer microneedles array resistance and insertion tests. *Int. J. Pharm.*, 480, 1-2, 152–7, 2015 Mar 1.
35. Lin, S., Cai, B., Quan, G., Peng, T., Yao, G., Zhu, C., Wu, Q., Ran, H., Pan, X., Wu, C., Novel strategy for immunomodulation: Dissolving microneedle array encapsulating thymopentin fabricated by modified two-step molding technology. *Eur. J. Pharm. Biopharm.*, 122, 104–112, 2018.
36. Kim, J.D., Kim, M., Yang, H., Lee, K., Jung, H., Droplet-born air blowing: Novel dissolving microneedle fabrication. *J. Control. Release*, 170, 3, 430–436, 2013.
37. Ye, Z., Xiang, Y., Monroe, T., Yu, S., Dong, P., Xian, S., Webber, M.J., Polymeric Microneedle Arrays with Glucose-Sensing Dynamic-Covalent Bonding for Insulin Delivery. *Biomacromolecules*, 23, 10, 4401–4411, 2022.
38. Antonara, L., Dallas, P.P., Rekkas, D.M., A novel 3D printing enabled method for fast and reliable construction of polymeric microneedles using experimental design. *J. Drug Deliv. Sci. Technol.*, 68, 102888, 2022.
39. Narayana Reddy, P.V., Coumar, M.S., Sudhakar, Y., Dissolving Microneedles for Transdermal Drug Delivery Systems: A Brief Review. *Adv. Mater. Lett.*, 8, 4, 323–332, 2017.
40. Nguyen, H.X., Bozorg, B.D., Kim, Y., Wieber, A., Birk, G., Lubda, D., Banga, A.K., Poly (vinyl alcohol) microneedles: Fabrication, characterization, and

application for transdermal drug delivery of doxorubicin. *Eur. J. Pharm. Biopharm.*, 129, 88–103, 2018 Aug 1.
41. Zhi, D., Yang, T., Zhang, T., Yang, M., Zhang, S., Donnelly, R.F., Microneedles for gene and drug delivery in skin cancer therapy. *J. Control. Release*, 335, 158–77, 2021.
42. Kang, G., Tu, T., Kim, S., Yang, H., Jang, M., Jo, D. *et al.*, Adenosine-loaded dissolving microneedle patches to improve skin wrinkles, dermal density, elasticity and hydration. *Int. J. Cosmet. Sci.*, 40, 2, 199–206, 2018.
43. Zhang, T., Sun, B., Guo, J., Wang, M., Cui, H., Mao, H. *et al.*, Active pharmaceutical ingredient poly (ionic liquid)-based microneedles for the treatment of skin acne infection. *Acta Biomater.*, 115, 136–47, 2020.
44. Vora, L.K., Moffatt, K., Tekko, I.A., Paredes, A.J., Volpe-Zanutto, F., Mishra, D. *et al.*, Microneedle array systems for long-acting drug delivery. *Eur. J. Pharm. Biopharm.*, 159, 44–76, 2021.
45. Lan, X., Zhu, W., Huang, X., Yu, Y., Xiao, H., Jin, L. *et al.*, Microneedles loaded with anti-PD-1–cisplatin nanoparticles for synergistic cancer immunochemotherapy. *Nanoscale*, 12, 36, 18885–98, 2020.
46. Liu, H., A clinical mini-review: Clinical use of Local anesthetics in cancer surgeries. *Gaz. Med. Sci.*, 1, 3, 30–034, 2020.
47. Kathuria, H., Li, H., Pan, J., Lim, S.H., Kochhar, J.S., Wu, C. *et al.*, Large size microneedle patch to deliver lidocaine through skin. *Pharm. Res.*, 33, 11, 2653–67, 2016.
48. Migdadi, E.M., Courtenay, A.J., Tekko, I.A., McCrudden, M.T., Kearney, M.-C., McAlister, E. *et al.*, Hydrogel-forming microneedles enhance transdermal delivery of metformin hydrochloride. *J. Control. Release*, 285, 142–51, 2018.
49. Ma, Y., Boese, S.E., Luo, Z., Nitin, N., Gill, H.S., Drug coated microneedles for minimally-invasive treatment of oral carcinomas: development and *in vitro* evaluation. *Biomed. Microdevices*, 17, 2, 1–14, 2015.
50. Caffarel-Salvador, E., Kim, S., Soares, V., Tian, R.Y., Stern, S.R., Minahan, D. *et al.*, A microneedle platform for buccal macromolecule delivery. *Sci. Adv.*, 7, 4, eabe2620, 2021.
51. W. Faisal and A. Crean, inventors; University College Cork, assignee. Microneedles, and methods for the manufacture thereof. United States patent application US 17/617, 382, 2022 Aug 11.

15

Advantages of Synthesized Nanocarriers from Saltmarsh Phytochemicals and Their Biological Applications

Balamurugan Palanisamy, Mohankumar Narayanan, Ramachandran Govindhan, Chackaravarthi Gnanasekaran, Rajivgandhi Govindhan* and Manoharan Natesan†

Department of Marine Science, Bharathidasan University, Tiruchirappalli, Tamil Nadu, India

Abstract

Nanotechnology provides the tools and technology platform for the investigation and transformation of biological systems, and biology offers inspiration models and bio-assembled components to nanotechnology. Nanotechnology is described as the combination of important materials from chemical, physical, and biological approaches. Nanomedicine is one of the attractive and challenging ones focused on the employment of nanomaterials in medical applications such as drug delivery. Nanoparticles are used as transport modules for other substances, they are called nanocarriers. A nanocarrier is a nanomaterial being used as a transport module for another substance, such as a drug. Commonly used nanocarriers include micelles, polymers, carbon-based materials, liposomes, and other inorganic substances like gold nanoparticles, mesoporous silica nanoparticles (MSNs), porous silicon nanoparticles, and metal-organic frameworks (MOFs). Nanocarriers are colloidal drug carrier systems having submicron particle sizes typically <500 nm. Nanocarriers have been extensively investigated in the past few decades as they showed great promise in the area of drug delivery. Metal nanocarriers are a matter of significant interest because of their unique characteristics, such as customizable size, large surface-to-volume ratio, easy synthesis, noble optical properties, and easy surface functionalization. In recent years, inorganic materials as nanocarriers

**Corresponding author*: rajivagndhimicro@yahoo.com
†*Corresponding author*: biomano21@yahoo.com

Sathish-Kumar Kamaraj, Arun Thirumurugan, Muthuchamy Maruthupandy, Mercedes Guadalupe López Pérez and Shanmuga Sundar Dhanabalan (eds.) Nanoscience and Nanotechnology for Smart Prevention, Diagnostics and Therapeutics: Fundamentals to Applications, (357–378) © 2024 Scrivener Publishing LLC

have shown great potential in drug delivery systems. Salt marshes are salt-tolerant vegetation and are found in the upper coastal intertidal zone, which is regularly or occasionally flooded by tides. Recent studies have shown that salt marsh plants possess high potential bioactive compounds due to their environmental factors which have various rich niches and it also shows that they were used in traditional medicines for wound healing, antispasmodic, and fever. But still salt marshes and their associated microorganisms are understudied in India compared to other countries. Phytochemical or phytoconstituents are secondary metabolites produced by plants that consist of a variety of bioactive compounds such as alkaloids, vitamins, flavonoids, phenols, glycosides, taxol, saponins, etc. Various studies on phytochemicals have revealed their biopotential abilities in the treatment of cardiovascular disease, diabetes mellitus, microbial infections, fever, etc. The aim of this review is the discussion about the potential of phytoconstituents in the Saltmarsh plant and their biological applications like anticancer activity, antimicrobial properties.

Keywords: Nanocarriers, bioactive sources, metal nanoparticles, drug delivery, biomedical application

15.1 Introduction

The term nanotechnology concept was introduced by American Physicist Richard Feynman in 1959 which leads to the development of the field of nanoparticle studies. The term nanotechnology is used to describe the manipulation of matter on an atomic, molecular, and supramolecular scale where a unique quantum mechanical effect takes place. Nanoparticles possess various potential properties compared to their bulk structures. The use of nanotechnology in the field of health and medicine had led to an establishment of a new area in the field of nanoscience as nanomedicine [1]. Nanotechnology and nanobiotechnology are future aspects in this rapidly developing field of nanoparticle synthesis, due to their increasingly new beneficial properties. Nanotechnology is used as a crossover technology due to its small size, and high-volume ratio than those of bulk materials. Metallic nanoparticles like silver (Ag) and gold (Au) nanoparticles (NPs) have played a major role in the transformation of the medicinal field due to their widespread applications, such as targeted drug delivery, diagnosis, and therapeutics due to their extremely small sizes, noncytotoxicity, and easy alternative optical, physical, and chemical properties. Green synthesized nanoparticles have shown great potent anticancer activity against several cancer cells, antioxidant activity, etc. They have also proven to be used as biomarkers for diseases due to their strong affinity toward biomolecules like peptides, antibodies, and oligonucleotides [2]. Metal nanoparticles have various applications in photonics, catalysis, data storage and management, and drug delivery. In recent years, development in the field of nanoparticles has led to various nanoparticle-based diagnostics and treatments for cancer, allergies, diabetes, pathological infections, and other diseases [3].

15.2 Saltmarsh

Salt marshes are common habitats of estuaries that occur worldwide, mainly in middle to high latitudes. These are the coastal wetlands, where upper coastal intertidal marshy zones exist between the open sea and land. Salt marsh components are tidal flats, tidal creeks, low marsh, and high marsh formed due to variations in elevation, salinity, and patterns of halophyte vegetation zones [4]. The definition of salt marshes stresses that the flooding waters are saline but need not be seawater, i.e., more saline than seawater. The salt marsh ecosystem is highly regarded as the most modified terrestrial ecosystem. Salt marsh resides between land and sea, and it has various biota of both land and sea, both marine and terrestrial elements, and both flora and fauna [5]. Salt marshes are coastal wetlands in the upper tidal zones and are subject to inundation of tides at regular or seasonal intervals. They are one of the most productive coastal ecosystems with diverse ecosystem functions like prevention of coastal soil erosion, intrusion of seawater, storm barrier, carbon sequestration, nutrient cycling, and providing habitat for various biota [6]. The global extent of salt marsh varies from 2.2 to 40 Mha, colonized by salt-tolerant vegetation or halophytic macrophytes, such as herbs, shrubs, grasses, and sedges which differ from their associates by growing only in salt conditions. Salt marsh vegetation is differentiated into various types based on biogeographical characteristics such as arctic, boreal, West Atlantic, dry coast, temperate, and tropical types [5]. Saltmarshes are widespread in temperate regions, whereas they are mostly substituted by patchy distributions nearby mangroves in tropical and subtropical regions. Studies on tropical salt marshes are fewer compared to the studies on temperate salt marshes, particularly in Asia [7]. Salt marsh coast has several specific features, and they display high tolerance to harsh environments. As an adaptive mechanism, they produce high-potential secondary metabolites which possess various biopotential properties [8].

15.3 Phytochemicals and Their Role

Phytochemical or phytoconstituents are secondary metabolites produced by plants that consist of a variety of bioactive compounds such as alkaloids, vitamins, flavonoids, phenols, glycosides, taxol, saponins, etc. These phytochemicals produced by plants are not essential for their growth but they are well known for their major role as immunostimulants in human disease treatment as a single drug or combination with other drugs [9]. Phytochemicals either in their indigenous state or their constituents show prominent biopotential efficacy in human health by involving in the mechanism of metal chelation, free radical scavenging, and protease inhibitory activities. The phytochemicals involve in the speeding up of protective enzyme production [10]. Phytochemicals are proactively involved in the inhibition of certain types of proteins, enzymes, and other metabolites causing diseases or disorders [11]. The advantage of phytochemicals is their non-toxicity on normal cells and their mode of action through multiple pathways resulting in effective inhibitory activities. Various studies on phytochemicals have revealed their biopotential abilities in the treatment of cardiovascular disease, diabetes mellitus, microbial infections, fever, etc. The major disadvantage of phytochemicals is their compound's huge size and polar nature which makes them cross the endothelial lining of blood vessels, gastrointestinal tract, brain blood barriers (BBB), and mucosa more difficult. Another drawback is their enzymatic digestion in the gastrointestinal tract which results in the high usage of plant secondary metabolites in the treatment leading to the cause peripheral organ toxicity in some cases [12].

15.4 Nanocarriers

In simple terms, nanoparticles are referred to as particles with at least one dimension of the order between 1 and 100 nm, and they are currently defined with a volume-specific surface area equal to or greater than 60 m^2/cm^3. Nanocarriers are termed as the nanoparticles used as the module for the transportation of desired substances to a specific area [13]. Nanocarriers are also referred to as colloidal nano-scale devices that are used for transiting medicine and contain the small molecular weight of macromolecules as genetics or enzymes.

Nanocarriers prevent the drug from degradation, decrease the renal clearance and its half-life in the bloodstream boost the usefulness of cytotoxic drugs, regulate the discharge kinetics of drugs, and enhance the solubility of insoluble drugs. The size of nanocarriers ranges between 10 and 400 nm, which makes them suitable medication carriers because of their ability to carry a large number of drugs, and extended flow time. Nanocarriers can also be used for the resolution of conventional drug limitations having weak aqueous solubility, poor bioavailability, and undesirable pharmacokinetic characteristics [14]. Drug delivery using nanotechnology possess immense potential in improving the efficacy of drug delivery through the development of suitable carriers and also enhances a piece of knowledge about drug-related issues like side effects of the drugs and their mode of action. Drug delivery systems possess the ability to entrapment of drugs and their potential in retaining and transporting the drugs which makes them the first choice in pharmaceutical applications. Encapsulated drugs can also be successfully administered for drug delivery systems due to their ability to protect drugs from external factors [15].

15.5 Synthesis of Nanocarriers

Nanocarriers are synthesized by two main methods as top-down approach and the bottom-up approach due to their differences in size, shape, and stability (Figure 15.1). During the synthesis of nanoparticles their size, shape, and conductivity can be easily modified by making alterations in the reaction. The top-down approach is a destruction method where the bigger size of nanoparticles is reduced to the desired size, which involves many processes, such as thermal decomposition and mechanical milling. The bottom-up approach is a constructive method where the nanoparticles are

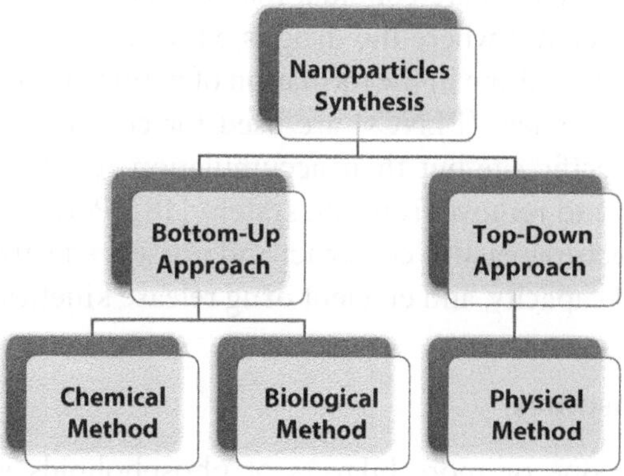

Figure 15.1 Synthesis of nanoparticles.

built from the atoms to a desired nanoparticle by the most commonly used method known as chemical vapor deposition spinning laser pyrolysis [15].

The synthesis of nanoparticles is basically classified into three types—physical, chemical, and biological methods. The physical method follows the top-down approach, such as laser ablation where the formed nanoparticles are monodispersed nanoparticles. One of the drawbacks of this method is the wastage of resources, which limits its use in some biomedical applications [16]. The chemical method follows the bottom-up approach where the metal ions are reduced by reducing agents, and surfactants or stabilizing agents which results in the production of well-defined nanoparticles with desired sizes and shapes. One of the drawbacks of this method is the toxic nature of the chemicals implemented which restricts their use in some biomedical applications. Finally, the biological method, which is also known as green synthesis, follows the bottom-up approach by implementing the unicellular and extracellular organisms, such as bacteria, algae, and plants. These biologically synthesized nanoparticles are cost-effective, easy to produce, eco-friendly, and less toxic in nature which makes them more suitable for various suitable biomedical applications [17].

15.6 Types of Nanocarriers

Nanocarriers boost the physiological characteristics of widely used compounds in the diagnosis. Various types of nanocarriers are schematically represented in Figure 15.2.

15.6.1 Polymeric Nanocarriers

Polymeric nanoparticles are colloidal macromolecules which has specific structural architecture where the drug is either entrapped or attached to the outer surface resulting in the formation of nanocapsules or nanosphere. Polymeric nanocarriers (PLNCs) are used for controlled drug release at the targeted specific site but their accumulation results in toxicity making them harder to remove from the system [18]. Polymeric nanocarriers are flexible structures easy to construct and possess effective drug delivery, large drug load capacity, and efficient drug release kinetics [19].

15.6.2 Liposomes

Liposomes are one or more bilayers of phospholipids vesicles that are spherical in shape. They can be generated from natural or synthetic

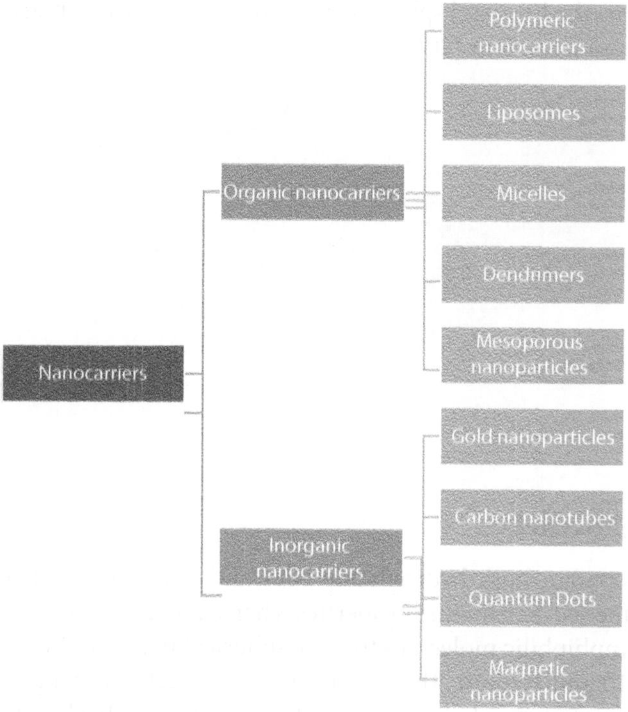

Figure 15.2 Types of nanocarriers.

phospholipids and cholesterol. Liposomes are phospholipid-based amphipathic nanocarriers possessing advantages such as biodegradability, prolonged release of bioactive compounds, and biocompatible. Liposomes possess high drug delivery potential due to their size of 50 to 450 nm, which makes them suitable for easy delivery of different kinds of drug molecules. The phospholipids are major components of the cell membrane consisting of a fatty acid-based hydrophobic tail and a phospholipid-based hydrophilic head [9]. The liposome is classified into two types based on its bilayers and size such as Multilamellar vesicles (MLV), and Unilamellar vesicles. Uni-lamellar vesicles are of two types, large unilamellar vesicles (LUV) and small unilamellar vesicles (SUV) [13]. Liposomes coated with polyethylene glycol show increased circulation time in the systematic circulation than the conventional liposomes [14]. Liposomes produced by using saturated fatty acids are more stable but show less permeability and the liposomes produced by using unsaturated fatty acids are more permeable but have less stability. Liposomes possess a unique character of gathering in the liver and spleen, which gives them an advantage in pathogenic disease treatment [20].

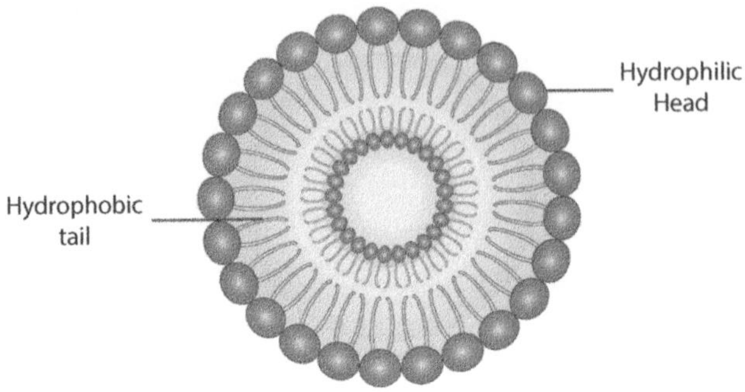

Liposome

15.6.3 Micelles

Micelles contain both hydrophobic core and hydrophilic shell elements with unique self-assembling properties when exposed to a solvent. Micelles contain amphiphilic molecules that are spherical in shape, with size ranges between 10 and 100 nm. The structure of the micelles is controlled by the interaction of hydrophobic/hydrophilic molecules. Micelles are used as drug carriers for water-insoluble drugs due to their properties. Polymeric micelles have high stability and efficient biocompatibility, which is used for solubilizing many poor soluble compounds for drug delivery [21]. Polymeric micelles, lipid-core micelles, and immuno-micelles are showing great potential in targeted drug delivery systems with high target specificity and binding capability. Various advantages of micelles include prolonged self-life, unaffected drug character during drug delivery, etc [22].

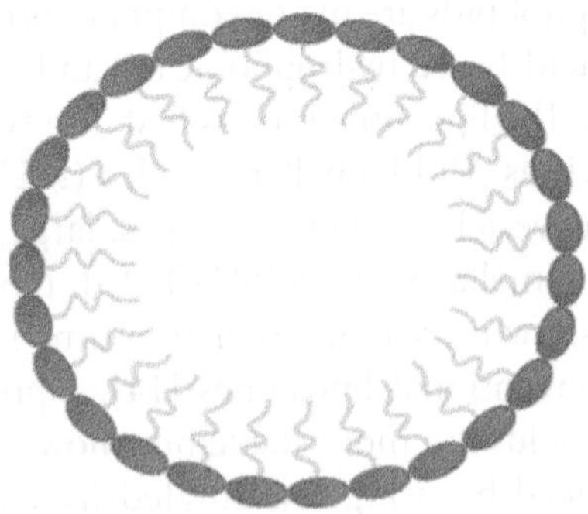

Micelles

15.6.4 Dendrimers

The word dendrimer originally originated from the Greek word dendron, which means "tree or branch" and meros which means "part." Dendrimers are nano-sized symmetric molecules with distinct monodispersing structures and well-defined branches which makes them suitable for drug delivery. Their size ranges between 2 and 10 nm in diameter. The main components of dendrimers are the central nucleus, branches, and different terminal classes. The most commonly used molecule for dendrimers in polyamidoamine (PAMAM) is ethylenediamine and other molecules used are diaminohexane or diaminobutane, and diaminododecane [14].

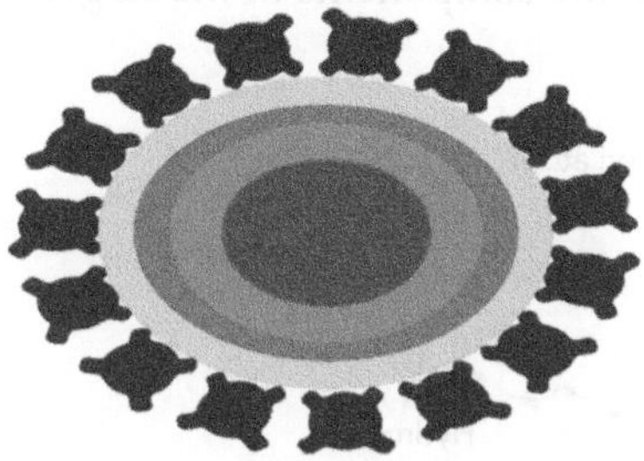

Dendrimers

15.6.5 Mesoporous Silica Nanoparticles

Mesoporous silica nanoparticles (MSNPs) are special mesoporous, which are widely used for their controlled release and supply of various drug molecules without affecting the chemical properties of the drugs. MSNs are well known for their characteristics like large pore diameter, increased surface, the small size of pores, and adjustable pore volume. The diameter of mesoporous ranges between 2 and 50 nm. The mesoporous nanoparticle has a strong base with a porous framework and a big surface area that enables various functional groups to get attached to the targeted site for drug delivery [23].

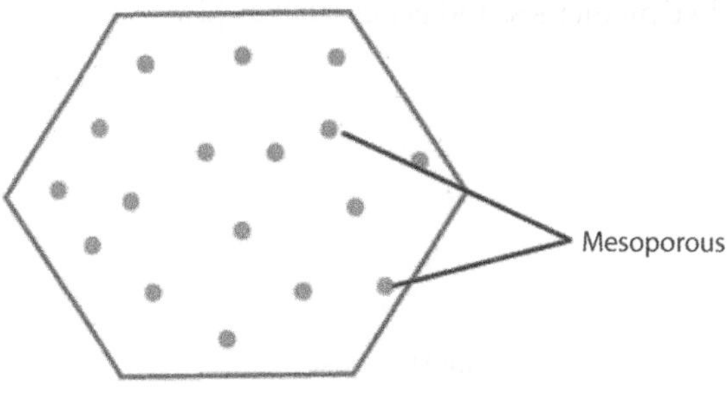

Mesoporous Silica Nanoparticles

15.6.6 Carbon Nanotubes

Carbon nanotubes (CNTs) belong to the fullerene structural family with one-atom-thick sheet walls made up of carbon, known as graphene [24]. These are one of the building blocks of nanotechnology due to their powerful optical absorption characteristics. It is classified into two types, single-walled carbon nanotubes (SWCNT) and multiwalled carbon nanotubes (MWCNT). Carbon nanotube size ranges between 50 and 100 nm in diameter which makes it easier to absorb nanoparticles. Carbon nanotubes are widely used due to their flexibility, strength, and distinctive needle-like structure which gives an advantage in the attachment and absorption of a wide range of therapeutic components in the target cell [25].

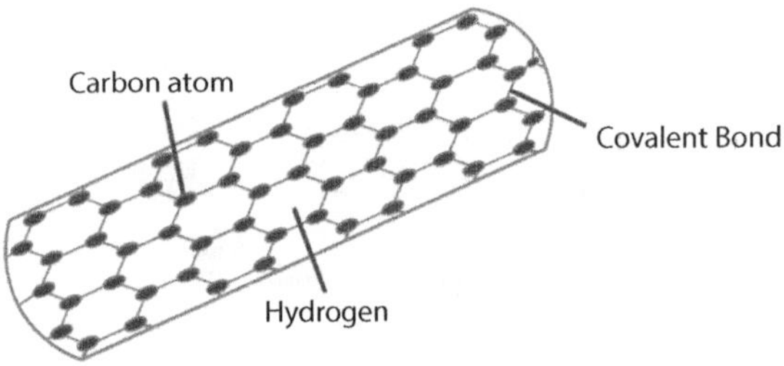

Carbon Nanotubes

15.6.7 Gold Nanoparticles (AuNPs)

Inorganic nanoparticles exhibit more tuneable surface adjustment and size-adjustable strategies. Inorganic nanoparticles play a major role in various fields due to their characteristics like sensing, bioimaging, and drug processing. Gold nanoparticles are one of the most important inorganic nanoparticles used in the drug industry due to their surface plasmon resonance (SPR), tuneable and optical characteristics. Their size ranges between 1 and 150 nm, which make their dispersion very simple to regulate. Gold nanoparticle (AuNP) nontoxicity and biocompatibility make them a better choice as drug carriers for delivering molecules like medicines, and genes such as [2].

Gold Nanoparticles

15.6.8 Superparamagnetic Iron Oxide Nanoparticles

Superparamagnetic iron oxide nanoparticles (SPIONs) are magnetite (Fe_3O_4) and maghemite (Fe_2O_3) excellent nanoparticles due to their superparamagnetic behavior, and biocompatibility. Superparamagnetic is a process that occurs when the volume is smaller than the normal magnetic material which does not have any magnetization property after the removal of an external magnetic field. Nanoparticles become superparamagnetic at room temperature where their core particle diameter is 20 nm or less [26].

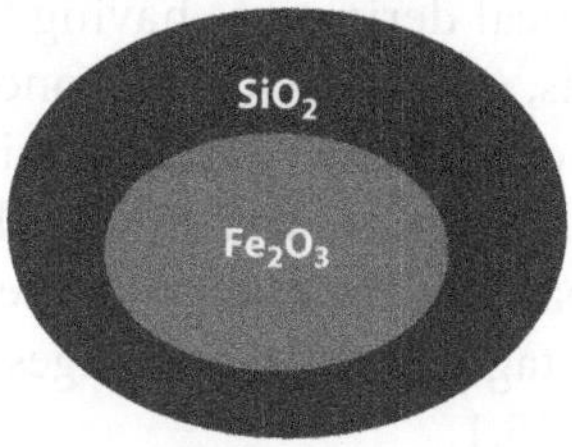

SPIONs

15.6.9 Quantum Dots

Quantum dots (QDs) are nothing but semi-conductive nanoparticles fabricated by the elements in II-VI or III-V groups of the periodic table (i.e., CdS, ZnS, CdTe, etc). Their size ranges between 2 and 10 nm in diameter and their inner and outer shells are made from different semiconductors. Quantum dots have distinctive electronic and optical characteristics due to their quantum containment. Quantum dots (QDs) has many biological applications such as monitoring various molecular objectives at the same time, large coefficients of molar extinction, and regulation of peak emission wavelengths by particle composition. Quantum dot nanocarriers can improve drug efficacy and decrease their adverse effects [27]. Liu *et al.* (2018) [28] reveal the ability of silicon quantum dot nanocarriers to show a sustained cellular uptake of Nrf2 into the nucleus conjugated with allyl isothiocyanate a phytochemical in the treatment of cancer.

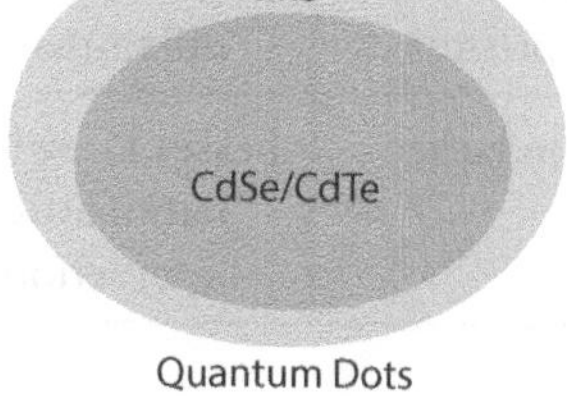

Quantum Dots

15.7 Application of Phytochemical Nanocarriers

Nanoparticles provide various biological applications due to their surface properties uniqueness and size and shape. But one of the drawbacks is the toxicity produced during their synthesis by metals or chemical agents. This can be rectified by loading them along with phytochemicals which can efficiently decrease the toxicity and also improve therapeutic properties. The non-nutritive plant chemicals that have a protective or disease-preventing property are known as phytochemicals [12]. Plants possess numerous chemical derivatives having high potential medicinal properties like alkaloids, amino acids, flavonoids, amines, carboxylic acid, anthranoids, carbohydrates, vitamins, minerals, glycosides, peptidoglycan, polyphenol, saponins, and so on. These compounds can be found in the different parts of the plant like the stem, roots, leaf, fruits, seed, etc [29]. The advantages and disadvantages of the nanocarriers have been discussed in Table 15.1.

Table 15.1 Advantages and disadvantages of nanocarriers.

Advantages	Disadvantages
Controlled drug delivery	High cost for synthesis and storage
Targeted drug delivery and multidrug delivery	May turn toxic with some solvents during preparation
Reduced drug dosage and intake periods	Self-aggregation
Increased drug efficiency	Poor long-term drug release
Effective drug activity and less toxicity	Some nanocarriers like nanospheres need to be removed surgically
Decreased drug resistance and improvised healing process	Might induce an immunological or inflammatory response
Increased production in large quantity	Lesser knowledge about nanoparticles and human-body interaction

15.7.1 Antimicrobial Activity

Microbial resistance or Antimicrobial resistance is a major threat that increases the search for newer antibiotics. Even though phytochemicals possess various bioactive compounds which have antimicrobial activities their low solubility in the aqueous phase makes their usage difficult in various diverse applications. Most of the studies involve the synthesis of metal nanoparticles utilizing plant extract and their biological application. In one study the green synthesized Ag NPs using polyvinyl alcohol extracts from the tissue-cultured callus and the leaf of *Sesuvium portulacastrum L.* showed distinct antibacterial activity against *Staphylococcus aureus* and antifungal activity against *Penicillium* sp. Silver nanoparticles are well-known for their antimicrobial activity due to their silver ion and their compounds which are more toxic to microorganisms [30]. Another study on the green synthesized zinc oxide (ZnO) nanoparticles from the *Seidlitzia rosamarinus* plant extract showed high irradiation power of 270w on the gram-positive and gram-negative bacteria which was higher than other reported cases due to the intruding effects of nanoparticles on bacterial cell walls. The green synthesized CuO nanoparticles showed great activity against microbial pathogens *Bacillus subtilis* with the highest zone of inhibition at 17.1 mm at 100 μg/ml and the highest zone of inhibition against *Staphylococcus aureus* at 9.8 mm at 10 μg/ml [31]. The methanolic extract of *Suaeda maritima* showed a maximum zone of inhibition against the three human pathogens when compared with other plants in the Selvam *et al.* [32] study about the biopotential abilities of methanol solvent extract of salt marsh plants viz. *Sesuvium portulacastrum, Suaeda maritima,* and *Salicornia brachiata* against the three human pathogens *Pseudomonas aeruginosa, Bacillus cereus,* and *Escherichia coli.* Green synthesized gold nanoparticles (AuNPs) using the aqueous extract of *Salicornia brachiata* had shown effective antibacterial activity against *Pseudomonas aeruginosa, Salmonella typhi, Staphylococcus aureus,* and *Escherichia coli* bacterial pathogens. Al-Azzawi *et al.* [33] had done work on preliminary phytochemical screening analysis of the *Sesuvium portulacastrum* salt marsh plant in the United Arab Emirates with different solvents like aqueous, ethanolic, and dichloromethane and studied their antibacterial activity against *Staphylococcus aureus, Escherichia coli.* This study reveals the presence of phytoconstituents like tannins, flavonoids, terpenoids, proteins, phenol, and saponins in the three different extracts. The study also revealed the presence of 22, 23-dihydrostigmasterol, benzoic acid 3, 4, 5-trihydroxy-, epicatechin, and capsaicin in *S. portulacastrum* salt marsh plant and their potential antibacterial activity against *E. coli*, and *S. aureus* major factors for nosocomial infections and gastrointestinal disorders. Dinani *et al.* [34] revealed the presence of phytochemicals like alkaloids, terpenoids, steroids, tannins, saponins, flavonoids, and phenols in *Seidlitzia rosamarinus*. They identified two cinnamic acid derivatives N-cis-feruloyl tyramine, and N-cis-caffeoyl tyramine along with *p-hydroxy acetophenone* (Piceol) from the aerial parts of the *Seidlitzia rosamrinus*. These reported derivatives had been reported in the other plant species with effective hepatoprotective, cytotoxicity, antimicrobial, and antioxidant activities. Several techniques can be utilized for the use of nanoparticles in antimicrobial activity which improves drug efficiency.

15.7.2 Anticancer Activity

The wide usage of nanoparticles as cancer agents for their targeted drug delivery of chemotherapeutic drugs has grabbed global attention. During their synthesis addition of phytochemicals boost the nanoparticle preparation. Saleem *et al.* [35] revealed the higher-level presence of phenol and flavonoid contents in the *Suaeda fruticosa* plant which corresponded to the higher radical scavenging potential and showed moderate cytotoxic against MCF-7 and MDA-MB-231 breast cancer cell lines. In another study [36], the green synthesized zinc oxide (ZnO) nanoparticles from the *Limonium pruinosum* have shown moderate anticancer activity against A-431 cell lines and also shown no cytotoxicity against normal cells even in higher concentrations of 250 *μg/ml. Another study* [37] reveals the importance of the silver nanoparticles' antiangiogenic properties which block abnormal protein signaling. The synthesized Ag NPs from the *Suaeda monoica* plants with a nanoparticle size of 31nm had shown cytotoxic effects on the Hep-2 tumor cells which decreased to 50% (IC_{50}) at the concentration level of 500 nM under *in vitro* conditions. Thus, the nanoparticles synthesized with phytochemicals along with other combinational drugs can be used in cancer therapy.

15.7.3 Photocatalytic Activity

Belen Perez Adassus *et al.* [38] had done work on the green synthesis of zinc oxide nanoparticles from *Sarcocornia ambigua* collected from the Bahía Blanca estuary region, Argentina, and studied their photocatalytic activity using methylene blue as an organic pollutant. This study revealed the ability of *Sarcocornia ambigua* in zinc oxide nanoparticles by the green synthesis method. The synthesized zinc oxide (ZnO) nanoparticles showed great potential photocatalytic activity.

Peddi *et al.* [31] had done work on the phytochemical screening and green synthesis of copper oxide (CuO) nanoparticles using the aqueous extract of *Suaeda maritima* salt marsh plant. This study revealed the potentiality of *Suaeda maritima* to synthesize copper oxide nanoparticles and showed good antioxidant activity and was also found to be potent against microbial pathogens. They also exhibited excellent photocatalytic activity by degrading methylene blue dye with the highest degradation % rate than reported. Ayaz Ahmad *et al.* [39] had done work on the green synthesis of gold nanoparticles using the aqueous extract of the *Salicornia brachiata* salt marsh plant and studied the catalytic activity. This study revealed the ability of *Salicornia brachiata* to synthesize gold nanoparticle and their stability. The synthesized gold nanoparticles exhibited strong catalytic activity in the reduction of methylene blue. Thus, it is significant that nanoparticles can also be used in the bioremediation process.

15.7.4 Antioxidant

Arockiya Arthi Rajathi *et al.* [40] had done work on the phytofabrication of gold (Au) nanoparticles using the leaves of *Suaeda monoica* collected from the Vellar region of the Cuddalore district and studied its free radical scavenging property. This study revealed the gold nanoparticle synthesis ability of the *Suaeda monoica* plant, even at room temperature. The synthesized gold nanoparticles displayed an effective antioxidant property that can be used for various biological applications.

Saleem *et al.* [35] had done work on the phytochemical screening of the *Suaeda fruticosa* salt marsh plant collected from the Bahawalpur region of Pakistan using methanol and dichloromethane solvents and studied their biological and toxicological activities. This study revealed the higher-level presence of phenol and flavonoid contents in the *Suaeda fruticosa* plant which corresponded to the higher radical scavenging potential. The extracts showed strong to moderate inhibitory activities against acetylcholinesterase, butyrylcholinesterase, tyrosinase, α-amylase, and α-glucosidase enzymes, and both extracts showed moderate cytotoxic against MCF-7 and MDA-MB-231 breast cancer cell lines. Docking studies also confirmed the inhibition ability of selected compounds against acetylcholinesterase, butyrylcholinesterase, tyrosinase, α-amylase, and α-glucosidase enzymes, which can be used for further investigation studies. Eswaraih *et al.* [41] had done the work on the phytochemical analysis of selected species of mangrove along with *Suaeda nudiflora* collected from the Nizampatnam region of Andhra Pradesh using methanol extract and studied their antimicrobial and antioxidant activities. This study revealed the presence of phytoconstituents like tannins, and phenols present in the *Suaeda nudiflora* species along with free radical scavenging activity. The antioxidant activity of *Suaeda nudiflora* and *Ipomea tuba* had shown great potential against *Micrococcus luteus, Pseudomonas aeruginosa, Staphylococcus aureus, Klebsiella pneumonia,* and *Bacillus subtilis* bacterial pathogens. Ouselati *et al.* [42] had done work on the evaluation of phenolic compounds and antioxidant properties of *Suaeda* species *Suaeda fruticosa, Suaeda pruinosa, Suaeda monoica,* and *Suaeda mollis* using acetonic extracts. This study revealed the presence of polyphenol compounds rutin hydrate, and 2,5-dihydroxybenzoic acid in the *Suaeda* species. *Suaeda fruticosa* showed the highest activity in the inhibition of β-carotene bleaching assays. These results show the potential of polyphenol compounds in the *Suaeda* species' ability in the role of antioxidant activity which can be used in the prevention of various diseases such as cancer, diabetes, arteriosclerosis, and neurodegenerative diseases. Sellem *et al.* [8] had done work on phytochemical screening analysis in *Inula graveolens* salt marsh species collected from the Chebba region in Tunisia in different solvent extracts and their biopotential activities like antimicrobial activity against eight different bacteria and *candida* species *Staphylococcus aureus, Micrococcus luteus, Listeria monocytogenes, Salmonella typhimurium, Pseudomonas aeruginosa, Enterobacter aerogenes, Candida tropicalis, Candida albicans*, antioxidant activity, and enzymatic inhibition activities like Acetylcholinesterase inhibitory potential and Anti-tyrosinase activity. This study reveals that *Inula graveolens* acetone extracts had shown the highest total polyphenol, total flavonoids, and total proanthocyanidins contents and also had the strongest antioxidant, and acetylcholinesterase inhibitory activity. The ethyl acetate extract of *I. graveolens* demonstrated tyrosinase inhibition activity. Less polar solvent extracts have shown more antimicrobial activity than polar solvents.

15.8 Conclusion

This review mainly focuses on the overview of the various methods developed over a period in the utilization of phytochemicals through nanocarriers for their stability and biocompatibility. Recent studies reveal the potential importance of nanotechnology in the field of drug delivery due to its ability in improving phytochemical stability and control drug release. Although it has many advantages, some of the drawbacks are the difficulty in converting the laboratory into clinical trials and cytotoxicity effects *in vivo* in the case of some inorganic nanoparticles. Another drawback is their high cost of production which reduces their wide range of applications. Thus, the customized formulation of nanoparticles matches the physical and chemical properties of the drug of interest and is very much required for efficient drug activity. Although salt marsh plants possess excellent phytoconstituents like antimicrobial, anthelmintic, antitumor, and antioxidant properties, they are understudied in India. In this present review, the role of phytoconstituents in plants and their biological activities along with the nanocarriers were discussed. In recent years nanocarriers are one of the most advanced nanotechnology techniques used in the field of research and development. The phytochemicals along with nanocarriers have shown increased drug efficiency and decreased drug toxicity, which can be used in diverse applications like nanomedicine, and imaging agents in disorders. In conclusion with the rapid increase in knowledge in the field of nanotechnology, a new generation of nanocarriers brings the potential application of phytochemicals in the clinical field.

Acknowledgement

We would Like to thank RUSA2.0 Biological sciences (Ref. No. 02BDU/RUSA2.0/TRP/BS/Date: 22/11/2021) for providing fellowship and Bharathidasan University, Tiruchirappalli, India for providing facility.

References

1. Simos, Y., Spyrou, K., Patila, M., Karouta, N., Stamatis, H., Gournis, D., Dounousi, E., Peschos, D., Trends of nanotechnology in type 2 diabetes mellitus treatment. *Asian. J. Pharm. Sci.*, 16, 1, 62–76, 2021.
2. Noah, N., Green synthesis: Characterization and application of silver and gold nanoparticles, in: *Green Synthesis, Characterization and Applications of Nanoparticles*, pp. 111–135, Elsevier, Amsterdam, The Netherlands, 2019.
3. Nazaripour, E., Mousazadeh, F., Moghadam, M.D., Najafi, K., Borhani, F., Sarani, M., Ghasemi, M., Rahdar, A., Iravani, S., Khatami, M., Biosynthesis of lead oxide and cerium oxide nanoparticles and their cyto-toxic activities against colon cancer cell line. *Inorg. Chem. Commun.*, 131, 108800, 2021.

4. Janardan, P., Microbiology of Salt Marshes in the UK: Insights into Life in Gradient-Rich Environments, Edge Hill University, UK, 2021.
5. Adam, P., General features of saltmarshes and their environment, in: *Saltmarsh Ecology*, pp. 1–71, Cambridge University Press, Cambridge, England, 2011.
6. Viswanathan, C., Purvaja, R., Jeevamani, J.J.J., Samuel, V.D., Sankar, R., Abhilash, K.R., Geevarghese, G.A., Muruganandam, R., Gopi, M., Raja, S., Das, R.R., Patro, S., Krishnan, P., Ramesh, R., Salt marsh vegetation in India: Species composition, distribution, zonation pattern and conservation implications. *Estuarine Coastal Shelf Sci.*, 242, 106792, 2020.
7. Reis, A., Krull, M., Carvalho, L.R.S., Barros, F., Effects of Tropical Saltmarsh Patches on the Structure of Benthic Macrofaunal Assemblages. *Wetlands*, 39, 5, 945–954, 2019.
8. Sellem, I., Chakchouk-Mtibaa, A., Smaoui, S., Mellouli, L., Total Polyphenol, Flavonoid, and Proanthocyanidin Contents and Biological Activities of Inula Graveolens Collected from Chebba (Tunisia) Salt Marsh. *J. Herbs Spices Med. Plants*, 27, 4, 426–444, 2021.
9. Ahmad, R., Srivastava, S., Ghosh, S., Khare, S.K., Phytochemical delivery through nanocarriers: a review. *Colloids Surf., B*, 197, 111389, 2021.
10. More, M.P., Pardeshi, S.R., Pardeshi, C.V., Sonawane, G.A., Shinde, M.N., Deshmukh, P.K., Naik, J.B., Kulkarni, A.D., Recent advances in phytochemical-based Nano-formulation for drug-resistant Cancer. *Med. Drug Discov.*, 10, 100082, 2021.
11. Xie, J., Yang, Z., Zhou, C., Zhu, J., Lee, R.J., Teng, L., Nano- technology for the delivery of phytochemicals in cancer therapy. *Biotechnol. Adv.*, 34, 4, 343–353, 2016.
12. Xiao, J., Phytochemicals in Food and Nutrition. *Crit. Rev. Food Sci. Nutr.*, 56, sup1, S1–S3, 2016.
13. Hossen, S., Hossain, M.K., Basher, M.K., Mia, M.N.H., Rahman, M.T., Uddin, M.J., Smart nanocarrier-based drug delivery systems for cancer therapy and toxicity studies: A review. *J. Adv. Res.*, 15, 1–18, 2019.
14. Huda, S., Alam, M.A., Sharma, P.K., Smart nanocarriers-based drug delivery for cancer therapy: An innovative and developing strategy. *J. Drug Deliv. Sci. Technol.*, 60, 102018, 2020.
15. Alqosaibi, A.I., Nanocarriers for anticancer drugs: Challenges and perspectives. *Saudi J. Biol. Sci.*, 29, 6, 2022.
16. Parboosing, R., Govender, T., Maguire, G.E.M., Kruger, H.G., Synthesis, characterization and biocompatibility of a multifunctional gold nanoparticle system for the delivery of single-stranded RNA to lymphocytes. *S. Afr. J. Chem.*, 71, 1–14, 2018.
17. Ayipo, Y.O., Bakare, A.A., Badeggi, U.M., Jimoh, A.A., Lawal, A., Mordi, M.N., Recent advances on therapeutic potentials of gold and silver nanobiomaterials for human viral diseases. *Curr. Res. Chem. Biol.*, 2, 100021, 2022.

18. Yadav, H., Mahalvar, A., Pradhan, M., Yadav, K., Kumar Sahu, K., Yadav, R., Exploring the potential of phytochemicals and nanomaterial: a boon to antimicrobial treatment. *Med. Drug Discov.*, 17, 100151, 2023.
19. Gagliardi, M., Bardi, G., Bifone, A., Polymeric nanocarriers for controlled and enhanced delivery of therapeutic agents to the CNS. *Ther. Deliv.*, 3, 7, 875–887, 2012.
20. Pandya, P., Giram, P., Bhole, R.P., Chang, H.I., Raut, S.Y., Nanocarriers based oral lymphatic drug targeting: Strategic bioavailability enhancement approaches. *J. Drug Deliv. Sci. Technol.*, 64, 102585, 2021.
21. Torchilin, V.P., Micellar Nanocarriers: Pharmaceutical Perspectives. *Pharm. Res.*, 24, 1, 1–16, 2006.
22. Cao, Y., Rewatkar, P., Wang, R., Hasnain, S.Z., Popat, A., Kumeria, T., Nanocarriers for oral delivery of biologics: small carriers for big payloads. *Trends Pharmacol. Sci.*, 42, 11, 957–972, 2021.
23. Gupta, S. and Gupta, M.K., Possible role of nanocarriers in drug delivery against cervical cancer. *Nano Rev. Exp.*, 8, 1, 1335567, 2017.
24. Shi, Z., Zhou, Y., Fan, T., Lin, Y., Zhang, H., Mei, L., Inorganic nano-carriers based smart drug delivery systems for tumor therapy. *Smart Mater. Med.*, 1, 32–47, 2020.
25. Zhang, Q., Kuang, G., Zhang, L., Zhu, Y., Nanocarriers for platinum drug delivery. *Biomed. Technol.*, 2, 77–89, 2023.
26. Yang, X., Hong, H., Grailer, J.J., Rowland, I.J., Javadi, A., Hurley, S.A., Xiao, Y., Yang, Y., Zhang, Y., Nickles, R.J., Cai, W., Steeber, D.A., Gong, S., cRGD-functionalized, DOX-conjugated, and 64Cu-labeled superparamagnetic iron oxide nanoparticles for targeted anticancer drug delivery and PET/MR imaging. *Biomaterials*, 32, 17, 4151–4160, 2011.
27. Rahman, M.M., Islam, M.R., Akash, S., Harun-Or-Rashid, M., Ray, T.K., Rahaman, M.S., Islam, M., Anika, F., Hosain, M.K., Aovi, F.I., Hemeg, H.A., Rauf, A., Wilairatana, P., Recent advancements of nanoparticles application in cancer and neurodegenerative disorders: At a glance. *Biomed. Pharmacother.*, 153, 113305, 2022.
28. Liu, P., Behray, M., Wang, Q., Wang, W., Zhou, Z., Chao, Y., Bao, Y., Anticancer activities of allyl isothiocyanate and its conjugated silicon quantum dots. *Sci. Rep.*, 8, 1084, 2018.
29. Bharti, S.K., Krishnan, S., Kumar, A., Kumar, A., Antidiabetic phytoconstituents and their mode of action on metabolic pathways. *Ther. Adv. Endocrinol. Metab.*, 9, 3, 81–100, 2018.
30. Nabikhan, A., Kandasamy, K., Raj, A., Alikunhi, N.M., Synthesis of antimicrobial silver nanoparticles by callus and leaf extracts from salt-marsh plant, Sesuvium portulacastrum L. *Colloids Surf., B*, 79, 2, 488–493, 2010.

31. Peddi, P., Ptsrk, P.R., Rani, N.U., Tulasi, S.L., Green synthesis, characterization, antioxidant, antibacterial, and photocatalytic activity of Suaeda maritima (L.) Dumort aqueous extract-mediated copper oxide nanoparticles. *J. Genet. Eng. Biotechnol.*, 19, 1, 2021.
32. Panneer Selvam, S., Gurudhathan, K., Gopinath, J., Kothandaraman, G., Dinakarkumar, Y., Phytochemical investigation and antibacterial activity of salt marsh plant extracts. *J. Chem. Pharm. Sci.*, 9, 292–294, 2016.
33. Al-Azzawi, A., Alguboori, A., Hachim, M.Y., Najat, M., al Shaimaa, A., Sad, M., Preliminary phytochemical and antibacterial screening of Sesuvium portulacastrum in the United Arab Emirates. *Pharmacogn. Res.*, 4, 4, 219–224, 2012.
34. Zolfaghari, B., Mazaheri, M., Dinani, S., Dinani, M.S., Isolation of cinnamic acid derivatives from Seidlitzia Isolation and Identification of Cinnamic Acid Derivatives from the Aerial Parts of Seidlitzia Rosmarinus Ehrenb. Ex Boiss. *Med. Sci. J. Rep. Pharm. Sci.*, 6, 1, 77–83, 2017.
35. Saleem, H., Khurshid, U., Sarfraz, M., Tousif, M.I., Alamri, A., Anwar, S., Alamri, A., Ahmad, I., Abdallah, H.H., Mahomoodally, F.M., Ahemad, N., A comprehensive phytochemical, biological, toxicological and molecular docking evaluation of Suaeda fruticosa (L.) Forssk.: An edible halophyte medicinal plant. *Food Chem. Toxicol.*, 154, 112348, 2021.
36. Naiel, B., Fawzy, M., Halmy, M.W.A., Mahmoud, A.E.D., Green synthesis of zinc oxide nanoparticles using Sea Lavender (Limonium pruinosum L. Chaz.) extract: characterization, evaluation of anti-skin cancer, antimicrobial and antioxidant potentials. *Sci. Rep.*, 12, 1, 2022.
37. Satyavani, K., Gurudeeban, S., Ramanathan, T., Balasubramanian, T., Toxicity Study of Silver Nanoparticles Synthesized from Suaeda monoica on Hep-2 Cell Line. *Avicenna J. Med. Biotechnol.*, 4, 1, 35–39, 2012.
38. Adassus, M.B.P., Spetter, C.V., Lassalle, V.L., Biofabrication of ZnO nanoparticles from Sarcocornia ambigua as novel natural source: A comparative analysis regarding traditional chemical preparation and insights on their photocatalytic activity. *J. Mol. Struct.*, 1256, 132460, 2022.
39. Ayaz Ahmed, K.B., Subramanian, S., Sivasubramanian, A., Veerappan, G., Veerappan, A., Preparation of gold nanoparticles using Salicornia brachiata plant extract and evaluation of catalytic and antibacterial activity. *Spectrochim. Acta, Part A*, 130, 54–58, 2014.
40. Arockiya Aarthi Rajathi, F., Arumugam, R., Saravanan, S., Anantharaman, P., Phytofabrication of gold nanoparticles assisted by leaves of Suaeda monoica and its free radical scavenging property. *J. Photochem. Photobiol., B*, 135, 75–80, 2014.

41. Eswaraiah, G., Peele, K.A., Krupanidhi, S., Kumar, R.B., Venkateswarulu, T.C., Studies on phytochemical, antioxidant, antimicrobial analysis. *J. King Saud. Univ. Sci.*, 32, 1, 842–847, 2020.
42. Oueslati, S., Trabelsi, N., Boulaaba, M., Legault, J., Abdelly, C., Ksouri, R., Evaluation of antioxidant activities of the edible and medicinal Suaeda species and related phenolic compounds. *Ind. Crops Prod.*, 36, 1, 513–518, 2012.

Index

Printed and bound by CPI Group (UK) Ltd, Croydon, CR0 4YY

17/09/2026

14964279-0002